Subwavelength Optics

Theory and Technology

Yongqi Fu

School of Physical Electronics
University of Electronic Science and Technology of China

Publisher: Bentham Science, UAE

Table of Contents

Foreword

There is an astonishing explosion of interest in recent years in near-field optics, nano-optics, nano-photonics and plasmonics, which appear as emerging optical science branches dealing with evanescent waves confined in a nanometer scale volume. Such waves can be manipulated with a nano-probe located in the optical near field of structures. The optical near field imaging can reach a nano-scale resolution beyond the conventional diffraction limit (Airy disk), so does as well the high-density data storage.

The book 'Subwavelength Optics' is structured with both basic theory and practical applications elucidating from elements/devices design/numerical simulation, micro/nanofabrication to geometrical and optical characterization so that it can be of value to educators teaching undergraduate and graduate courses in multiple disciplines. For them, it will serve as a textbook that illustrates basic principles and multidisciplinary approaches. Most chapters are essentially independent of each other, providing flexibility in choice of topics to be covered. Thus, the book can also readily be adopted for training and tutorial short courses. Both fresh students and engineers/researchers can be benefited from reading the book. The book, written by a leading practitioner of near-field optics, arrives at an opportune time because there is a substantial body of results available in the field that need to be gathered together in a systematic fashion, although the book can cover only a part of this emerging area, where the new knowledge and new research discoveries come out in explosion.

Professor Yunlong Sheng
Docteur ès Science Physique (France)
Department of Physics, Physical
Engineering and Optics
Laval University,
Québec,
Canada

Preface

In recent years, two concepts: metamaterials/left-handed materials/negative refraction index, and plasmonics, appeared as emerging sciences and technology in optics. On the basis of exponential increase in the number of published papers every year, it is clear that we are at the eve of a new revolution that will impact many fields of science and technology, including photonics, computation, communication, biology, medicine, materials science, physics, chemistry, and photovoltaics. The book truly reflects the present status of this rapidly developing area of science and technology and highlights some of the important historic developments. Most of the chapters discussing ongoing scientific research, and promising future directions were identified.

Subwavelength optics, defined by the fusion of microfabrication and nanotechnology as well as near-field optics and nanophotonics, is an emerging frontier providing challenges for fundamental research and opportunities for new technologies. Subwavelength optics has already made its impact in the marketplace. It is a multidisciplinary field. It creates opportunities in physics, chemistry, applied sciences, engineering, and biology, as well as biomedical technology. However, a few nano-optics relevant books involve micro-/nanofabrication technologies and characterizations for manufacturing near-filed optical components, nanophotonic devices, and plasmonic structures together so far. Combining my tens years hands-on experiences on micro-/nanofabrication technologies which were applied in the fields of micro- and nano-optics, plasmonics, and nanophotonics, I collected and edited the fabrication and application issues systematically with respect to subwavelength optics from theory to technology.

Subwavelength optics has meant different things to different people, in each case being defined with a narrow focus. Several books and reviews exist that cover selective aspects of subwavelength optics. However, there is a need for an up-to-date monograph that provides a unified synthesis and systematization of this subject. This book fills this need by providing a unifying, multifaceted description of subwavelength optics to benefit a multidisciplinary readership. Relevant fundamental theory, structures design, micro/nanofabrication, and characterization as well as applications surrounding subwavelength optics were systematically described in this book. The objective is to provide a basic knowledge of a broad range of topics so that individuals in all disciplines can rapidly acquire the minimal necessary background for research and development in subwavelength optics. The author prefers that this book can serve as both a textbook for education and training and a reference book that aids research and development in those areas integrating light, photonics, nanotechnology, semiconductor, chemistry, and biology. Another aim of the book is to stimulate the interest of researchers, industries, and businesses to foster collaboration through multidisciplinary programs in this frontier science, leading to development and transition of the resulting technology.

This book encompasses the fundamentals and various applications involving the integration of microfabrication, nanotechnology, photonics, semiconductor, materials, chemistry, and biology. Each chapter begins with an abstract and introduction section describing literature review of the contents in the chapter and what a reader will find in that chapter, and ends with a brief summary and may serve as a review of the materials presented.

In writing this book, which covers a very broad range of topics, I received help from a large number of individuals at Institute of Optics and Electronics, Chinese Academy of Sciences, University of Electronic Science and Technology of China, and Sichuan University. This help has consisted of furnishing technical information, creating illustrations, providing critiques, and preparing the manuscript. A separate acknowledgement recognizes these individuals.

Here I would like to acknowledge the individuals whose broad-based support has been of paramount value in completing the book. I owe a great deal of sincere gratitude to my wife, Di Zhang. She has been a constant source of inspiration, providing support and encouragement for this writing. As a perfect wife, she has bravely undertaken heavy burden of our family in overseas and looked after my son. There will be no my success without her sacrifice and great contribution to the family. I am also indebted to our son, Wenzhe Fu, for showing his love and understanding by sacrificing his quality time with me.

I express my sincere appreciation to Dr. Haofei Shi and Dr. Xuefeng Yang for reviewing Section 3-2 of Chapter 3. Finally, I thank Mr. Yu Liu, whose clerical help in manuscript preparation was invaluable.

Yongqi Fu
In Vancouver
Feb. 2009

About the Author

Yongqi Fu was born in Lanzhou, P.R. China. He received his B. Eng, M.S(Eng), and Ph.D from Jilin University, Changchun University of Science and Technology, and the Changchun Institute of Optics, Fine Mechanics and Physics, Chinese Academy of Sciences in 1988, 1994, and 1996, respectively. Then he worked at State Key Laboratory of Applied Optics, Chinese Academy of Sciences from 1996 to 1998 as a Postdoctoral Fellow. Next, he worked at Precision Engineering and Nanotechnology (PEN) Center, School of Mechanical and Aerospace Engineering, Nanyang Technological University, Singapore, and Singapore-MIT Alliance as a Research Fellow and Senior Research Fellow respectively from Aug. 1998 to Jun. 2007. He is a leader of FIB group at PEN Center. After that, he worked at State Key Laboratory of Optical Technology for Microfabrication, Institute of Optics and Electronics, Chinese Academy of Sciences as professor under "BAIREN Program". Currently, he works at School of Physical Electronics, University of Electronic Science and Technology of China as professor in nanophotonics and subwavelength optics. His current research interests include micro/nanofabrication, applications of FIB in microelectronics/ semiconductors, nano-photonics, biosensing, SPR/LSPR-based immunoassay, MOEMS/NEMS, micro/nano-optics, photonic crystals, optical measurement/nanometrology, and optical system instrumentation (*e.g.* nano-inspection/detection system). He authored and co-authored more than 130 peer reviewed journal papers with *h*-index of 13 generated from SCI citation database and 26 invited technical papers for symposiums and conference proceedings. In addition, he authored and coauthored three books and book chapters with titles of "*Subwavelength Optics: Theory and Technology*", "*Encyclopedia of Nanoscience and Nanotechnology*", and "*Ion beams in Nanoscience and Technology*", respectively. Most recently, he was invited to be candidate of Associate Editor of *Optics Express*. He is referee of 12 journals such as *Applied Physics Letters*, *Journal of Optical Society of America A/B*, *IEEE Photonic Technology Letters*, and *Journal of Vacuum Science and Technology B*.

List of Acronyms

ABS	=	Air bearing surface
APTMS	=	(3-Aminopropyl) trimethoxysilane
ALD	=	Atomic layer deposition
ASR	=	Asymmetric split-ring
AFM	=	Atomic force microscope
a-NSOM	=	Aperture-less near field scanning optical microscope
BPM	=	Beam propagation method
CCD	=	Charge coupled device
CE	=	Chemical enhancement
CRLH	=	Composite right/left-handed metamaterial
CT	=	Computed tomography
CV	=	Cylindrical vector
CVD	=	Chemical vapor deposition
DASD	=	Direct access storage devices
DDA	=	Discrete dipole approximation
DLSPP	=	Dielectric-loaded surface plasmon polariton
DNG	=	Double metamaterials
DOF	=	Depth of focus
DVD	=	Digital versatile disc
EBG	=	Electromagnetic band gap
EBSD	=	Electron back-scattering diffraction
EDC	=	1-Ethyl-3-[3-dimethylaminopropyl] carbodiimidehydrochloride
EM	=	Electromagnetic
ERI	=	Effective refractive index
EME	=	Electromagnetic enhancement
FDTD	=	Finite difference and time domain
FIB	=	Focused ion beam
FIBM	=	Focused ion beam milling
FMM	=	Fourior model method
FRET	=	Fluorescence resonance energy transfer
FSL	=	Far-field optical superlens
FSO	=	Fluorocarbon surfactant
FWHM	=	Full width at half maximum
GAE	=	Gas-assistant etching
GMT	=	Generalized multipole technique
HDD	=	Hard disk drives
ITO	=	Indium-tin-oxide
FDNWS	=	Funnel-shaped nanocylinder waveguide structures
LHM	=	Left-handed metamaterial
LIL	=	Laser interference lithography
LNPs	=	Localized nanoparticle plasmons
LPSR	=	Localized surface Plasmon resonance
MAPMT	=	Multi-anode photomultiplier tubes
MBE	=	Molecular beam epitaxy
MCP	=	Multi-channel plate
MEMS	=	Micro electrical and mechanical system
MIR	=	Mid-infrared range
MIMO	=	Metamaterial multiple-input-multiple-output
MMM	=	Multifocal multiphoton microscopy
MMP	=	Multiple multipole program
MRI	=	Magnetic resonance imaging
NEMS	=	Nano-mechanical-electronic system
NIM	=	Negative index materials
NIR	=	Near-infrared
NHS	=	Amidohydroxysuccinimide group
NRI	=	Negative-Refractive-Index
NSL	=	Nanosphere lithography
NSOM	=	Near field scanning optical microscope
NR-SG	=	Silica sol–gel matrix
OCT	=	Optical coherence tomography

OI	=	Optical imaging
PAMM	=	Patch antenna and microstrip method
PBS	=	Phosphate-buffered saline
PCB	=	Printing circuit board
PCF	=	Photonic Crystal Fibers
PCs	=	Photonic crystals
PDMS	=	Polydimethylsiloxane
PEC	=	Perfect electronic conductor
PET	=	Positron emission tomography
PMMA	=	Poly methyl methacrylate
PMT	=	Photomultiplier tube;
PM	=	Photo mask
PS	=	Polystyrene nanospheres
QDs	=	Quantum dots
QCLs	=	Quantum cascade lasers
RCWA	=	Rigorous coupling wave analysis
RF	=	Radio frequency
RH	=	Right-handed
RIE	=	Reactive ion etching
SAM	=	Self-assembly monolayer
SE	=	Surface enhancement
SEM	=	Scanning electron microscopy
SERS	=	Surface enhanced Raman spectroscopy
SHG	=	Second harmonic generation
SIMS	=	Secondary ion mass spectroscopy
SNOM	=	Scanning near field optical microscope
s-NSOM	=	Scattering near field scanning optical microscope
SP	=	Surface plasmon
SPM	=	Probe scanning microscope
SPR	=	Surface Plasmon resonance
SPPs	=	Surface Plasmon polaritons
SOPs	=	States of polarization
SRR	=	Surface ring resonator
STM	=	Scanning tunneling microscope
TIR	=	Total internal reflection
TFSF	=	Total-field scattered-field
TMCTS	=	1, 3, 5, 7- Tetramethylcyclotetrasiloxane
TPL	=	Two-photon luminescence
UHV	=	Ultra high vacuum
USI	=	Ultrasound imaging
VCSEL	=	Vertical cavity surface emitting laser
XRD	=	X-ray diffraction

Summary

From beginning of this century, there has been a dramatic increase in interest in the study of surface plasmon polaritons-based metallic subwavelength structures and learning. This is a refreshing concise book on issues and considerations in designing, nanofabrication and characterization of subwavelength plasmonic structures as well as their applications in imaging, superfocusing, semiconductor lasers, data storage, optical communications, biosensing, and immunoassay. This book can serve as a textbook for education and training as well as a reference book that aids research and development in those areas such as integrating light, photonics, nanotechnology, semiconductor, chemistry, and biology. Another aim of the book is to stimulate the interest of researchers, engineers, and businessmen to foster collaboration through multidisciplinary programs in this frontier science, leading to development and transition of the resulting technology. Both basic and applied aspects were presented in this book. Many illustrative worked-out examples and instructive exercises were given.

1 INTRODUCTION

Abstract: A brief historic evolution and literature review of subwavelength optics was given. Then some descriptions were presented for multidisciplinary education, training, and research. Finally, scope of this book and brief contents for each chapter were introduced in this chapter.

1.1 Subwavelength optics — a brief history

Subwavelength optics is a new optical science dealing with evanescent waves which are confined in a nanometer scale volume smaller than the spot size determined by classical diffraction theory. Such waves can be manipulated with a nanoprobe which is located in the optical near field of structures. The optical systems work in the near field can reach to a nanolevel resolution because it breakthroughs limitation of the conventional diffraction limitation (*i.e.* Airy disk). High-density data storage can be performed with subwavelength optics for state-of-art techniques such as microscope imaging with nanometric resolution, high-resolution photo-fabrications/photolithography, single-molecule detections, and local spectral analyses. Subwavelength optics also called near-field optics, or nanophotonics because the essential issue is that photons interacts with sample surface or subsurface in the near field region.

Basic foundation of nanophotonics involves the interaction between light and matters. [1] This interaction occurs for most opto-electronic systems, that is, it involves changes in the properties of the electrons presented in the systems. Hence, an important starting point is a parallel discussion of the natural issues between photons and electrons.

Near-field optics and related techniques appear and behave on the basis of electromagnetic interactions of matters in the quasi-static regime, where the electromagnetic fields in a mode coupled with the matters play the fundamental role. Here, the quasi-static regime implies that the dominant electromagnetic interactions among the matters take place across a distance which is much smaller than the optical wavelength. Recent developments in micro/nanofabrication provide a diversity of probes/tips with the potential to pick up the tiny local field out of the sea of macroscopic electromagnetic interactions. [1] In the near-field regime, the optical properties of the matters and associated near-fields depend strongly on sizes and shapes of the matters being involved. In fact, optical response of the matters is determined by internal electronic processes including the interactions with optical fields which are consistent with both electronic and electromagnetic boundary conditions. The resulting scattered fields reflect the properties of these internal processes in the illuminated objects, especially when they are observed in an optical near-field region. The scattered fields exhibitting asymptotic behavior in the far-field region are limited as propagating of the optical waves in a retarded nature that carry electromagnetic energy out of the objects. [2] In this case, the optical response of the matters can be represented in macroscopic quantities such as dielectric functions, which enable us to reproduce the macroscopic electromagnetic boundary conditions correctly. [3] In contrast, when an observation is carried out in a subwavelengh vicinity of objects, the scattered fields involve strong near-fields showing steep decay, which reveals the details of optical interactions taking place inside the objects. Theoretical descriptions of optical near-field processes should therefore be presented on the basis of detailed studies of the electromagnetic interactions of matters at the microscopic level.

It is possible to separate theoretically the near-filed relevant components from those of a propagating nature within ordinary scattering problems in the approximation that the observation process exerts a small or negligible disturbance in the optical response of the objects. The basic optical near-field properties can be found in evanescent waves that arise at illuminated material boundaries and decay exponentially in the direction perpendicular to the boundary surface. [3] The evanescent waves mediate optical interactions only in a short range in the direction normal to the surface and have a wavevector along the surface with the magnitudes larger than those of optical waves in free space. This makes it possible to drive an optical interaction localized in a spatial extent which is narrower than the optical wavelength in vacuum which corresponds to the origin of ultrahigh spatial resolution realized in optical near-field techniques. Optical fields exhibit such a localized behavior only as a result of interaction with matters, so that optical near-fields can be considered modes of optical fields coupled with matter. Here, we have three different characteristic scales with respect

Yongqi Fu (Ed.)

to the observation processes of scattered optical fields: the size of the scatterer, the distance between the observation point and the object, and the optical wavelength under consideration. We can observe several different characteristics involved in optical interactions of matters depending on the relationships among these parameters. The definition of the observation points also depends on our techniques of observation and involves a number of important issues discussed with respect to the way of far-field techniques. Therefore, the definition was put forth based on nanofabricating the probes/tips that enable us to localize the observation point according to the requirement of observation.

In the opto-electronic systems on macroscopic scales, the size and shape of objects are reflected significantly in the optical response since a macroscopic electronic excitation due to collective motion comes into resonance with a cavity mode in the object. Examples are plasmons, excitons, and their polaritons excited in metallic and semiconductor small objects or thin films. The resonant behavior sometimes strongly enhances the optical near-field interaction, which can be used as a probe/tip with high resolution and sensitivity to the specific modes of excitation. In the microscopic limit, the optical response depends strongly on the shape of the objects whose sizes are close to electronic de Broglie wavelengths, due to the non-local responses of quantum mechanical opto-electronic systems in optical fields. The optical near-field probes in this microscopic region provide us the possibility of observing and controlling the local excitations and observations in nonlocal modes. [4, 5]

Besides the local behaviors, optical near-field techniques also involve important issues of the ways by which local modes and local material response can be excited and observed using macroscopic apparatus. As a matter of fact, for optical near-field observation using an optical system, we should firstly prepare a setup that can illuminate near-field objects or probes/tips by use of a macroscopic light source in the far-field. And we should then have a system which is appropriate for selectively collecting the scattered light from the small objects or probes/tips by aid of a macroscopic photodetector. It is very difficult to consider such a complicated mixture of microscopic and macroscopic phenomena as a whole from both the experimental and theoretical pint of view. The problems can, however, be very much simplified if the systems under consideration can be separated into several characteristic subsystems, each of them has a different characteristic scale described as an optical mode. The near-field studies, therefore, involve considerations of the reparability of the system into characteristic subsystems with clear meaning and also of the relationship between the properties of electromagnetic interactions and the scales of matter and observation. Note that such separation into meaningful subsystems is possible because the optical wavelength is usually quite large compared with the sizes of the microscopic objects and probes/tips available in our observations.

Recently, a new term, metamaterials, appeared in optics and microwaves. It was come up with firstly by a British scientist, J.B. Pendry in 2000.[6] He was the first one to imagine a practical way to make a novel artificial material, named "left-handed metamaterial (LHM)". "Left-handed" in this context means a material in which the "right-hand rule" is not obeyed, allowing an electromagnetic wave to convey energy in the opposite direction by wave propagation. Pendry's initial idea was that metallic wires aligned along the propagation direction could provide a metamaterial with negative permittivity ($\varepsilon < 0$). Note however that natural materials (such as ferroelectrics) were already known to exist with negative permittivity. The challenge was to construct a material that also showed negative permeability ($\mu < 0$). In 1999, Pendry demonstrated that an open ring ('C' shape) with axis along the propagation direction could provide a negative permeability. In the same paper, he showed that a periodic array of wires and rings could give rise to a negative refractive index.[7] After that, superlens was put forth on the basis of the negative refractive index for the purpose of imaging by X. Zhang group.[8] The new challenge is how to collect the objective signals from far-field to field of view of the superlens. Further exploration is still necessary in this emerging area. It is reasonable to believe that this is a promising area with high potential market in the future, especially for the LHM imaging.

1.2 Multidisciplinary education, training, and research

We live in a complex world. Interdisciplinary crossing is unavoidable. The naissance of each new subject must be accompanied with one or more breakthroughs in other areas or technologies. Similarly, subwavelength optics appeared under the circumstances of breakthrough and development of nanotechnology especially in nanofabrication technology such as e-beam lithography and focused ion beam milling, and nanometrology such as scanning probe microscopy. Pushed under this breakthrough, not only optics area, many other areas like material, mechanical, electronics, life science, and chemistry, all generate their own new branches, *e.g.*, nano-materials, nano-electronics, and nano-mechanics. Therefore, contents in this book naturally cover some other areas besides optics itself. Classified by applications, biology, mechanics, electronics, communication, nanofabrication, and chemistry will be involved in this book. Thus it provides a reference for the people in multidisciplinary education, training and research. For instance, the people in life science will benefit from the chapter describing biosensing and immunoassay on the basis of localized surface plasmon resonance and metallic nanoparticles array. The people in digital science will benefit from the chapter regarding application of plasmonic structures-based data recording techniques. And those people in micro- and nanofabrication area will benefit from the chapters illustrating relevant technologies of fabrication of plasmonic structures and nanophotonic devices. Some chapters were presented independently such as Chapters 6, 7, and 11-14. They can be used individually for the readers in different areas. Of course, for the purpose of teaching, it may increases difficulty of fully understanding for students because of their limited knowledge background. But on the other hand, it might be suitable for those students study in multidisciplinary courses.

A major challenge for researchers working in a multidisciplinary area is the need to learn relevant concepts outside of their expertise. This may require searching through a vast amount of literature, often leading to frustrations of not being able to extract pertinent information quickly. Apparently, researchers will benefit from this book for the synthesized knowledge and application examples from the different areas involved in the relevant chapters.

The book is structured so that it can also be of value to those educators teaching both undergraduate and graduate courses in multiple departments. For them, it will serve as a textbook that elucidates basic principles and multidisciplinary approaches. Most chapters are essentially independent on each other, providing flexibility in choice of topics to be covered. Thus, the book can also readily be adopted for training and tutorial short courses at universities as well as various professional society symposiums and conferences.

1.3 Scope of this book

Each chapter begins with abstract and introduction describing its contents and ends with a relevant summary. This introduction also provides a guide to what may be omitted by a reader familiar with the specific contents or by someone who is less inclined to go through in detail. Our target is that the new researchers in subwavelength optics can obtain a clear outline. In addition, a systematic frame from theoretical background, devices design, fabrication, and characterization to applications can be formed after reading. Based on this consideration, we focused our sight on theory and technology issues in this book. On the basis of electromagnetic field analyses and Maxwell equations, design, fabrication and characterization of plasmonic structures were described in detail. Then some typical emerging applications of the plasmonic structures were illustrated in the following chapters.

In Chapter 1, we gave an introduction of subwavelength optics from its brief history to current relevant emerging application technologies firstly, and then showed an outline of this book.

In Chapter 2, basic theory about electromagnetic field analyses and Maxwell equations were illustrated. Then we gave another concept, surface plasmon (SP), and SP wave propagation in dielectric materials and metals.

In Chapter 3, wave propagation through subwavelength metallic structures and numerical algorithms were presented. Approaches of both stationary and dynamic analyses were illustrated. On the basis of Mie theory, some analyzing methods such as multiple multipole program (MMP), and beam propagation method (BPM) as well as Fourior model method (FMM) were introduced birefly. In addition, some computational numerical algorithms were described such as discrete dipole approximation (DDA), and finite difference and time domain (FDTD).

After introduction of the relevant fundamental theory, near-field microscopy and its applications were presented in Chapter 4. Working principle and system setup were highlighted. Then some applications in areas of nanophotonics, life science, biochemistry, and metrology were introduced.

Chapter 5 presented a frontier topic for microwave and optics— metamaterials. Concept and transmission line theory were presented, including a new concept of negative refraction. Then some typical applications were listed.

Chapter 6 and 7 illustrated subwavelength optics relevant effective fabrication technologies. We classified them into two parts: one is top-down technology, and the other is bottom-up technology. The former includes focused ion beam technology, and laser interference photolithography. The latter involves self-assembly monolayer and electrochemistry techniques. Fabrication principles, examples and existing problems were illustrated.

Chapter 8 discussed characterization techniques regarding plasmonic and nanophotonic devices, including two parts: geometrical characterization and optical characterization. The main method for geometrical characterization: atomic force microscope, is described. Optical characterization involves near-field scanning optical microscope (NSOM), Raman spectroscope, confocal microscopy, and multiphoton microscopy. Amongst, NSOM is highlighted in the chapter because it plays an important role in optical characterization.

As a typical nanostructure, nanoholes array was highlighted in Chapter 9. Transmission properties and polarization effect originating from the size matter and shape issue were addressed. It is a key component in the biochip-based optofluidic systems. Representative works were reported by R. Gordon group in University of Victoria, Canada. Application for cancer cell detection was discussed.

In Chapter 10, one of typical applications of the plasmonic structures, imaging and superfocusing is illustrated. Negative refraction-based superlens which was frequently reported by X. Zhang group in Berkeley, California University in recent years. Plasmonic structures for superfocusing/sueperlensing which originated from research works of author of this book were mainly described in detail. In addition, for systematism, some other structures such as plasmonic tapered waveguide and metal nanocylinders funnel-shaped array were introduced also.

Chapter 11 discussed metallic nanoparticles array for biosensing, localized surface plasmon resonance (LSPR)-based immunoassay, photothermal therapy, and bio-imaging. Design and fabrication issues of the nanboparticles were addressed by the way.

Chapter 12 discussed plasmonic structures being used in lasers especially semiconductor lasers for amplification, emission, and collimation. The surface plasmon-based femtosecond laser for ablation fabrication was illustrated by the way.

In Chapter 13, metamaterials-based microwave antennas were presented. Currently, the metamaterials-based antenna becomes one of hot spots in metamaterials study. Relevant theoretical background of the antenna was introduced first. Then design and testing issues of the antennas were given. Electromagnetic band-gap structures-based antennas and waveguide slit array antenna were highlighted in this chapter.

Chapter 14 discussed plasmonic structures applied in data storage. Three types recording techniques: thermal assisted magnetic recording, atomic force microscope (AFM)-based data storage, and plasmonic lens array-based recording, were presented in detail.

Finally, future prospects and challenges in the field of subwavelength optics were described in Chapter 15.

For more information regarding basic theory of subwavelength optics and relevant subjects, the suggested materials for reference were listed as follows:

- Paras N. Prasad, eds.: *Nanophotonic*, Wiley Interscience, 2004.
- S.Kawata, Ohtsu, and Irie, eds.: *Nano-optics*, Springer, 2002.
- Lukas Novotny, eds.: *Principle of Nano-Optics*, 7th edition, Cambridge University Press, 2006.
- *Nanophotonics for surface plasmon polaritons,*
- G. Eleftheriades and K. Balmain, eds.: *Neagtive-Refraction Metamaterials*, Wiley Interscience, 2005.

Reference

[1]   M.Ohtsu and H.Hori, *Near-Field Subwavelength optics*, Kluwer Academic / Plenum, New York, 1999.

[2]   J.D. Jackson, *Classical Electromagnetics*, 2nd edition, John Wiley & Sons, New York, 1975.

[3]   M. Born and E. Wolf, *Principle of Optics*, 3rd edition, Pergamon Press, Oxford, 1965.

[4] S.Kawata, Ohtsu, and Irie, *Nano-optics*, Springer, 2002.

[5] Paras N. Prasad: *Nanophotonic*, Wiley Interscience, 2004.

[6] J. B. Pendry, "Negative refraction makes a perfect lens", Phys. Rev. Lett. 85, 3966-3969 (2000).

[7] J. A. Porto, F. J. García-Vidal, and J. B. Pendry, "Transmission Resonances on Metallic Gratings with Very Narrow Slits," Phys. Rev. Lett. 83, 2845 – 2848 (1999).

[8] N. Fang, H. Lee, C. Sun, X. Zhang, "Sub–Diffraction-Limited Optical Imaging with a Silver Superlens", Science 308, 34-537 (2005).

2 MAXWELL EQUATIONS AND ELECTROMAGNETIC WAVE PROPAGATION

Abstract: Basic theory, electromagnetic field, and surface plasmons, were described briefly in this chapter. Maxwell equations and electromagnetic wave propagation were illustrated. Finally, metal optics and surface plasmon poloritons excitation were presented. They are essential issues for the theory of plasmonics and Nanophotonics.

2.1 Introduction [1]

Early nineteenth century physical thought was dominated by the action at a distance concept, formulated by Newton more than 100 years earlier in his immensely successful theory of gravitation. In this view, the influence of individual bodies extends across space, instantaneously affects other bodies, and remains completely unaffected by the presence of an intervening medium. Such an idea was revolutionary; until then action by contact, in which objects are thought to affect each other through physical contact or by contact with the intervening medium, seemed the obvious and only means for mechanical interaction. Priestly's experiments in 1766 and Coulomb's torsion-bar experiments in 1785 seemed to indicate that the force between two electrically charged objects behaves in strict analogy with gravitation: both forces obey inverse square laws and act along a line joining the objects. Oersted, Ampere, Biot, and Savart soon showed that the magnetic force on segments of current-carrying wires also obeys an inverse square law. The experiments of Faraday in the 1830s placed doubt on whether action at a distance really describes electric and magnetic phenomena. When a material (such as a dielectric) is placed between two charged objects, the force of interaction decreases; thus, the intervening medium does play a role in conveying the force from one object to the other. To explain this, Faraday visualized "lines of force" extending from one charged object to another. The manner in which these lines were thought to interact with materials they intercepted along their path was crucial in understanding the forces on the objects. This also held for magnetic effects. Of particular importance was the number of lines passing through a certain area (the flux), which was thought to determine the amplitude of the effect observed in Faraday's experiments on electromagnetic induction.

Faraday's ideas presented a new world view: electromagnetic phenomena occur in the region surrounding charged bodies, and can be described in terms of the laws governing the "field" of his lines of force. Analogies were made to the stresses and strains in material objects, and it appeared that Faraday's force lines created equivalent electromagnetic stresses and strains in media surrounding charged objects. His law of induction was formulated not in terms of positions of bodies themselves, but in terms of lines of magnetic force. Inspired by Faraday's ideas, Gauss restated Coulomb's law in terms of flux lines, and Maxwell extended the idea to time changing fields through his concept of displacement current.

In 1860s Maxwell created what Einstein called "the most important invention since Newton's time"— a set of equations describing an entirely field-based theory of electromagnetism. These equations do not model the forces acting between bodies, as do Newton's law of gravitation and Coulomb's law, but rather describe only the dynamic, time-evolving structure of the electromagnetic fields. Thus bodies are not seen to interact with each other, but rather with the electromagnetic fields they create, an interaction described by a supplementary equation (the Lorentz force law). To better understand the interactions in terms of mechanical concepts, Maxwell also assigned properties of stress and energy to the field. Using constructions that we now call the electric and magnetic fields and potentials, Maxwell synthesized all known electromagnetic laws and presented them as a system of differential and algebraic equations. By the end of the nineteenth century, Hertz had devised equations involving only the electric and magnetic fields, and had derived the laws of circuit theory (Ohm's law and Kirchoff's laws) from the field expressions. His experiments with high-frequency fields verified Maxwell's predictions of the existence of electromagnetic waves propagating at finite velocity, and helped solidify the link between electromagnetism and optics. But one problem remained: if the electromagnetic fields propagated by stresses and strains on a medium, how could they propagate through a vacuum? A substance called the "luminiferous aether", long thought to support the transverse waves of light, was put forth to the task of carrying the vibrations of the electromagnetic fields as well. However, the pivotal experiments of Michelson and

Morely showed that the aether was fictitious, and the physical existence of the field was firmly established. The essence of the field concept can be conveyed through a simple thought experiment. Consideing two stationary charged particles in free space, since the charges are stationary, we know that (1) another force is presented to balance the Coulomb force between the charges, and (2) the momentum and kinetic energy of the system are zero. Now supposing that one charge is quickly moved and returned to rest at its original position, action at a distance will require the second charge to react immediately (Newton's third law), but it does not by Hertz's experiments. There it appears to be no change in energy of the system: both particles are again at rest in their original positions. However, after a time (given by the distance between the two charges divided by the speed of light) we find that the second charge does experience a change in electrical force and begins to move away from its state of equilibrium. But by doing so it has gained net kinetic energy and momentum, and the energy and momentum of the system seem to be larger than that of the charge at the start. This can only be reconciled through field theory. If we regard the field as a physical entity, then the nonzero work required to initiate the motion of the first charge and return it to its initial state can be seen as increasing the energy of the field. A disturbance propagates at finite speed and, upon reaching the second charge, transfers energy into kinetic energy of the charge. Upon its acceleration, this charge also sends out a wave of field disturbance, carrying energy with it, eventually reaching the first charge and creating a second reaction. At any given time, the net energy and momentum of the system, composed of both the bodies and the field, remain constant. We thus come to regard the electromagnetic field as a true physical entity: an entity that is capable of carrying energy and momentum.

There are numerous books regarding electromagnetics already. It is not necessary to present some lengthy pages to describe electromagnetic again here. Considering scope of this book, we give a brief introduction about electromagnetic field propagation and Maxwell equations only in this chapter. We assume that the readers have the background knowledge of electromagnetics already. Then we will directly lead readers stepping into our highlighted issues of this book: surface plasmon (SP) wave and left-handed materials, and illustrate the SP wave propagation and corresponding applications.

2.2 Maxwell equations [2]

The state of excitation which is established in space by the presence of electric charges is said to constitute an electromagnetic field. It is represented by two vectors, **E** and **B**, called the electric vector and the magnetic induction respectively. To describe the effect of the field on material objects, it is necessary to introduce a second set of vectors, the electric current density **J**, the electric displacement **D**, and the magnetic vector **H**. The space and time derivatives of the five vectors are related by Maxwell equations, which hold at every point in whose neighborhood the physical properties of the medium are continuous.

$$\mathbf{curl}\ H - \frac{1}{c}\dot{D} = \frac{4\pi}{c}J, \tag{2.1}$$

$$\mathbf{curl}\ E + \frac{1}{c}\dot{B} = 0, \tag{2.2}$$

The dot denotes differentiation with respect to time. They are supplemented by two scalar relations:

$$\text{div } \mathbf{D} = 4\pi\rho, \tag{2.3}$$

$$\text{div } \mathbf{B} = 0, \tag{2.4}$$

Eq. (2.3) may be regarded as a defining equation for the electric charge density ρ and Eq. (2.4) may be said to imply that no free magnetic poles exist.

From Eq. (2.1) it follows (since div curl≡0) that

$$\mathbf{div}\ J = -\frac{1}{4}\mathbf{div}\ \dot{D},$$

or, using Eq. (2.3),

$$\frac{\partial \rho}{\partial t} + \mathbf{div}\, \boldsymbol{J} = 0, \qquad (2.5)$$

By analogy with a similar relation encountered in hydrodynamics, Eq. (2.5) is called equation of continuity. It expresses the fact that the charge is conserved in the neighborhood of any point. For if one integrates Eq. (2.5) over any region of space, one obtains with the help of Gauss theorem as

$$\frac{d}{dt} \int \rho dV + \int \boldsymbol{J} \cdot \boldsymbol{n} dS = 0, \qquad (2.6)$$

The second integral being taken over the surface bounding the region and the first integral throughout the volume, and $\boldsymbol{n}$ denoting the unit outwards normal. This equation implies that the total charge

$$e = \int \rho dV, \qquad (2.7)$$

contained within the domain can only increases on account of the flow of electric current

$$\Im = \int \boldsymbol{J} \cdot \boldsymbol{n} dS \qquad (2.8)$$

If all the field quantities are independent of time, and if, moreover, there are no current ($\boldsymbol{J}=0$), the field is said to be static. If all the field quantities are time independent, but current is presented ($\boldsymbol{J}\neq 0$), one speaks of a stationary field. In optical fields, the field vectors are very rapidly varying functions of time; but the sources of the field are usually such that, when averages over any macroscopic time interval are considered rather than the instantaneous values, the properties of the field are found to be independent of the instant of time at which the average is taken. The word stationary is often used in a wider sense to describe a field of this type. An example is a field constituted by the steady flux of radiation (say from a distant star) through an optical system.

Table 1 Differential form with/without magnetic and/or polarizable media

Items	Differential form in the absence of magnetic or polarizable media	Differential form with magnetic and/or polarizable media
Gauss' law for electricity	$\nabla \cdot E = \dfrac{\rho}{\varepsilon_0} = 4\pi k \rho$	$\nabla \cdot \mathbf{D} = \rho, \quad \mathbf{D} = \varepsilon_0 \mathbf{E} + \mathbf{P}$
Gauss' law for magnetism	$\nabla \cdot B = 0$	$\nabla \cdot B = 0$
Faraday's law of induction	$\nabla x\, E = -\dfrac{\partial B}{\partial t}$	$\nabla x\, E = -\dfrac{\partial B}{\partial t}$
Ampere's law	$\nabla x\, B = \dfrac{4\pi k}{c^2} J + \dfrac{1}{c^2}\dfrac{\partial E}{\partial t}$ $= \dfrac{J}{\varepsilon_0 c^2} + \dfrac{1}{c^2}\dfrac{\partial E}{\partial t}$	$\nabla x\, H = J + \dfrac{\partial D}{\partial t}$ $\mathbf{B} = \mu_0(\mathbf{H} + \mathbf{M})$

Table 2 Integral form of Maxwell equations in the absence of magnetic or polarizable media

Items	Integral form in the absence of magnetic or polarizable media
Gauss' law for electricity	$\oint \vec{E} \cdot d\vec{A} = \dfrac{q}{\varepsilon_0}$
Gauss' law for magnetism	$\oint \vec{B} \cdot d\vec{A} = 0$
Faraday's law of induction	$\oint \vec{E} \cdot d\vec{s} = -\dfrac{d\Phi_B}{dt}$
Ampere's law	$\oint \vec{B} \cdot d\vec{s} = \mu_0 i + \dfrac{1}{c^2}\dfrac{\partial}{\partial t}\int \vec{E} \cdot d\vec{A}$

In summary, Maxwell equations can be concluded by Table 1 and 2. Table 1 is a form in which involves the Maxwell equations for the two cases of differential form in the absence of magnetic or polarizable media, and differential form with magnetic and/or polarizable media, respectively.

2.3 Electromagnetic wave propagation

2.3.1 Light interaction with matters

Changing **E**-field results in variation of **H**-field, and similarlly, the changed **H**-field will lead to variation of **E**-field as well. This process can be described by the curl equations in the form of Eq. (2.1) and (2.2). A wave equation for **E** and **H** can be expressed as

$$\nabla^2 U(r,t) = \frac{1}{v^2}\frac{\partial^2 U(r,t)}{\partial t^2} \tag{2.9}$$

Solution of Eq. (2.9) is waves propagating with a (phase) velocity v as

$$U(r,t) = \mathrm{Re}\{U_0(r)e^{i\omega t}\} \tag{2.10}$$

We start analysis from the wave equation for the **E**-field which can be written as

$$\nabla^2 E(r,t) = \frac{1}{v^2}\frac{\partial^2 E(r,t)}{\partial t^2} \tag{2.11}$$

The curl equations for **E** and **H** (materials with **M**=0 only) can be written as the following expressions respectively

$$\nabla \times \boldsymbol{E} = -\frac{\partial B}{\partial t} = -\mu_0 \frac{\partial \boldsymbol{H}}{\partial t} \tag{2.12}$$

$$\nabla \times \boldsymbol{H} = \frac{\partial D}{\partial t} + J \tag{2.13}$$

Apply curl on both sides of Eq. (2.12), we obtain partial differential equation that just depends on **E**

$$\nabla \times \nabla \times \boldsymbol{E} = \nabla \times (-\mu_0 \frac{\partial \boldsymbol{H}}{\partial t}) \tag{2.14}$$

Substitute Eq. (2.12) into Eq. (2.13) and applied **D**=ε₀**E**+**P**

$$\nabla \times \nabla \times \boldsymbol{E} = -\mu_0 \frac{\partial^2 D}{\partial t^2} - \mu_0 \frac{\partial J}{\partial t} = -\mu_0 \varepsilon_0 \frac{\partial^2 \boldsymbol{E}}{\partial t^2} - \mu_0 \frac{\partial^2 \boldsymbol{P}}{\partial t^2} - \mu_0 \frac{\partial J}{\partial t} \tag{2.15}$$

Compare Eq. (2.11) with (2.15), use vector identity of ∇×∇×**E**=∇(∇•**E**)-∇²**E**, and verify that ∇•**E**=0 at conditions of (a) ρ_f=0; and (b) ε(r) does not vary significantly within a λ distance, Eq. (2.15) can be changed to

$$\nabla^2 \boldsymbol{E} = \mu_0 \varepsilon_0 \frac{\partial^2 \boldsymbol{E}}{\partial t^2} + \mu_0 \frac{\partial^2 \boldsymbol{P}}{\partial t^2} + \mu_0 \frac{\partial J}{\partial t} \tag{2.16}$$

To solve this equation, we need to find **P(E)** and **J(E)**. First of all, we analyze dielectric linear, homogeneous and isotropic media. **P** is linearly proportional to **E**, **P**=ε₀χ**E**, where χ is a scalar constant called the "electric susceptibility". Eq. (2.16) can be rewritten as

$$\nabla^2 \boldsymbol{E} = \mu_0 \varepsilon_0 \frac{\partial^2 \boldsymbol{E}}{\partial t^2} + \mu_0 \frac{\partial^2 \boldsymbol{P}}{\partial t^2} = \mu_0 \varepsilon_0 \frac{\partial^2 \boldsymbol{E}}{\partial t^2} + \mu_0 \varepsilon_0 \chi \frac{\partial^2 \boldsymbol{E}}{\partial t^2} = \mu_0 \varepsilon_0 (1+\chi) \frac{\partial^2 \boldsymbol{E}}{\partial t^2} \tag{2.17}$$

Defining relative dielectric constant as ε_r=1+χ, Eq. (2.16) can be changed to

$$\nabla^2 \boldsymbol{E} = \mu_0 \varepsilon_0 \varepsilon_r \frac{\partial^2 \boldsymbol{E}}{\partial t^2} \tag{2.18}$$

where ε_r results from **P** and represents all the materials properties. It is noted that in anisotropic media **P** and **E** are not necessarily parallel, $P_i = \sum_j \varepsilon_0 \chi_{ij} E_j$; and in non-linear media, **P**=ε₀χ**E**+ε₀χ$^{(2)}$**E**$^{(2)}$+ε₀χ$^{(3)}$**E**$^{(3)}$+….

Compare Eq. (2.11) and (2.18), we have

$$v^2 = \frac{1}{\varepsilon_0 \mu_0}\frac{1}{\varepsilon_r} = \frac{c_0^2}{\varepsilon_r} \tag{2.19}$$

where $c_0^2 = 1/(\varepsilon_0\mu_0) = 1/((8.85 \times 10^{-12} \text{ C}^2/\text{m}^3\text{kg}) (4\pi \times 10^{-7}\text{m kg/C}^2)) = (3.0 \times 10^8\text{m/s})^2$.

Optical refractive index is defined by $n = c/v = \sqrt{\varepsilon_r} = \sqrt{1+\chi}$, whereas includes polarization which results in the same wave equation with a different ε_r, then c becomes v accordingly, and thus Eq. (2.11) becomes

$$\nabla^2 E(r,t) = \frac{n^2}{c^2}\frac{\partial^2 E(r,t)}{\partial t^2} \tag{2.20}$$

We get solution of Eq. (2.20) for monochromatic wave to be

$$E(r,t) = \text{Re}\{E(z,\omega)e^{-ikr+i\omega t}\} \tag{2.21a}$$

$$H(r,t) = \text{Re}\{H(z,\omega)e^{-ikr+i\omega t}\} \tag{2.21b}$$

Group velocity can be changed to be

$$v_g \equiv \frac{d\omega}{dk} = \frac{c}{\sqrt{\varepsilon_r}} = \frac{c}{\sqrt{1+\chi}} \tag{2.22}$$

and phase velocity is $v_{ph} \equiv \omega/k$.

Relation between $\mathbf{P}$ and $\mathbf{E}$ is dynamic, $\mathbf{P}(r, t) = \varepsilon_0\chi\mathbf{E}(r, t)$, assumes an instantaneous response, we get $\mathbf{P}$ in real life as

$$P(r.t) = \varepsilon_0 \int_{-\infty}^{+\infty} dt' x(t-t')E(r,t') \tag{2.23}$$

where $\mathbf{P}$ results from response to $\mathbf{E}$ over some characteristic time τ, and Function $x(t)$ is a scalar function lasting a characteristic time τ.

In frequency domain, relation between $\mathbf{P}$ and $\mathbf{E}$ can be changed to be, $\mathbf{P}(k, \omega) = \varepsilon_0\chi\mathbf{E}(k, \omega)$ with dielectric constant $\varepsilon(\omega) = \varepsilon_0[1+\chi(\omega)]$. Transparent materials can be described by a purely real refractive index n. From dispersion relation $\omega^2 = c^2k^2/n^2$, we get $k = \pm n\omega/c$. Absorbing materials can be described by a complex $n = n'+in''$. It follows that

$$k = \pm\frac{\omega}{c}(n'+in'') = \pm(\frac{\omega}{c}n'+i\frac{\omega}{c}n'') \equiv \pm(\beta - i\frac{\alpha}{2}) \tag{2.24}$$

where $\beta = n'\omega/c = k_0n'$, $\alpha = -2 n''\omega/c = -2k_0n''$, n' acts as a regular refractive index, and α is the absorption coefficient. Investigate sign "+" of Eq. (2.24), (2.21a) can be changed to

$$E(z,t) = \text{Re}\{E(z,\omega)e^{-i\beta z-\alpha z/2+i\omega t}\} \tag{2.25}$$

n is derived quantity from χ (next subsection we determine χ for different materials). Complex n results from a complex $\chi = \chi'+i\chi''$, then it has the following expression

$$n = n'-i\frac{\alpha}{2k_0} = \sqrt{1+\chi'+\chi''} \tag{2.26}$$

For weakly absorbing media, when $\chi' \ll 1$ and $\chi'' \ll 1$

$$\sqrt{1+\chi'+\chi''} \approx 1+\frac{1}{2}(\chi'+\chi'') \tag{2.27}$$

Then we have refractive index $n' = 1+\chi'/2$, and absorption coefficient $\alpha = -2k_0n'' = -k_0\chi''$.

2.3.2 Dispersion in materials

Currently, there are several models describing the material dispersion such as perfect electrical conductor (PEC), conductive model, Debye, Drude (for free electron gas), Lorentz (for bound electron gas), and Sellmeier (The resulting material is not dispersive. It can be used for single wavelength only, typically used to calculate fiber dispersion). For PEC, it is simple, $\mathbf{E} = 0$, which is equivalent to a conductor with $\sigma \to \infty$. For conductive model, it has the form $\varepsilon_r(\omega) = \varepsilon_\infty + i\sigma/(\omega\varepsilon_0)$; Debye model: $\varepsilon_r(\omega) = \varepsilon_\infty + \varepsilon_{debye}\gamma/(\gamma-i\omega)$; Drude model: $\varepsilon_r(\omega) = \varepsilon_\infty - \omega_p/\omega(i\gamma+\omega)$; Lorentz model: $\varepsilon_r(\omega) = \varepsilon_\infty + \varepsilon_{lorentz}\omega_0^2/(\omega_0^2 - 2i\omega\delta_0 - \omega^2)$, where $\omega = 2\pi f$, f is the frequency, and γ is the damping factor. For Sellmeier model, it has the form

$$\varepsilon = n^2 = A_1 + \frac{B_1\lambda_s^2}{\lambda_s^2 - C_1} + \frac{B_2\lambda_s^2}{\lambda_s^2 - C_2} + \frac{B_3\lambda_s^2}{\lambda_s^2 - C_3} \tag{2.28}$$

where $\lambda_s = c/f_s$, f_s is center frequency of the sources.

So far we have χ, ε, and n. All pairs (n' and n'', χ' and χ'', ε' and ε'') describe the same optical properties of materials. For some problems one set is preferable, and for others may be another set is considered. n' and n'' are used when discuss wave propagation, see Eq. (2.25). χ' and χ'', ε' and ε'' are used when discussing microscopic origin of optical effects. Internal relation between n and ε can be expressed from $\varepsilon'_r = (n')^2 - (n'')^2$, and $\varepsilon''_r = 2n'n''$, we have the following relations

$$n' = \sqrt{\frac{\sqrt{\varepsilon_r'^2 + \varepsilon_r''^2} + \varepsilon_r'}{2}} \tag{2.29a}$$

$$n'' = \sqrt{\frac{\sqrt{\varepsilon_r'^2 + \varepsilon_r''^2} - \varepsilon_r'}{2}} \tag{2.29b}$$

Now we discuss the behavior of bound electrons in an electromagnetic field. Optical properties of insulators are determined by bound electrons using Lorentz model, as shown in Fig. 2-1. Charges in a material are treated as harmonic

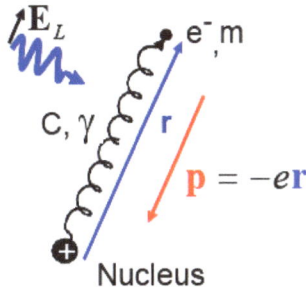

Fig. 2-1 Physical model of bound electrons in an electromagnetic field — Lorentz model.

oscillators and can be expressed in mechanical point of view as (one oscillator) $ma_{el} = F_{E,Local} + F_{damping} + F_{spring}$, which can be further modeled as

$$m\frac{d^2r}{dt^2} + m\gamma\frac{dr}{dt} + Cr = -eE_L e^{-i\omega t} \tag{2.30}$$

The electric dipole moment of this system is $\mathbf{p} = -e\mathbf{r}$

$$m\frac{d^2p}{dt^2} + m\gamma\frac{dp}{dt} + Cp = e^2 E_L e^{-i\omega t} \tag{2.31}$$

Let's guess the solution of above equation as follows

$$p = p_0 e^{-i\omega t} \tag{2.32a}$$

$$\frac{dp}{dt} = -i\omega p_0 e^{-i\omega t} \tag{2.32b}$$

$$\frac{d^2p}{dt^2} = -\omega^2 p_0 e^{-i\omega t} \tag{2.32c}$$

Then Eq. (2.31) can be rewritten

$$-m\omega^2 p_0 - im\gamma\omega p_0 + Cp_0 = e^2 E_L \tag{2.33}$$

and $p_0(\mathbf{E}_L)$ therefore can be solve from Eq. (2.33) easily.

p_0 can be derived by defining as $\omega_0^2 = C/m$ (turns out to be the resonance ω) as

$$p_0 = \frac{e^2}{m} \frac{1}{\omega_0^2 - \omega^2 - i\gamma\omega} E_L \qquad (2.34)$$

Atomic polarizability can be defined as

$$\alpha(\omega) \equiv \frac{p_0}{\varepsilon_0 E_L} = \frac{e^2}{\varepsilon_0 m} \frac{1}{\omega_0^2 - \omega^2 - i\gamma\omega} = A e^{i\theta(\omega)} \qquad (2.35)$$

Response of matter (**P**) is not instantaneous due to ω-dependent response. Amplitude and phase lag of α with **E** can be written as

$$A = \frac{e^2}{\varepsilon_0 m} \frac{1}{\sqrt{(\omega_0^2 - \omega^2)^2 + \gamma^2 \omega^2}} \qquad (2.36)$$

$$\theta = \tan^{-1} \frac{\gamma\omega}{\omega_0^2 - \omega^2} \qquad (2.37)$$

Firstly, we consider rarified media such as gasses. For dipole moment of one atom, j, we have $p_j = \varepsilon_0 \alpha_j E_L$, where E_L is **E**-field photon. Then for all atoms the polarization vector is

$$P = \frac{1}{V} \sum_j p_j = \frac{\varepsilon_0}{V} \sum_j \alpha_j E_L = \varepsilon_0 N \alpha_j E_L \qquad (2.38)$$

where V is the volume. Substitute Eq. (2.34) to (2.38), we have

$$P = \frac{Ne^2}{m} \frac{1}{\omega_0^2 - \omega^2 - i\gamma\omega} E_L \qquad (2.39)$$

From **P**$=\varepsilon_0 \chi$**E**$_L$, Eq. (2.39) can be rewritten as

$$\chi(\omega) = \frac{Ne^2}{\varepsilon_0 m} \frac{1}{\omega_0^2 - \omega^2 - i\gamma\omega} \qquad (2.40)$$

Then plasma frequency can be defined as $\omega_p^2 = Ne^2/(\varepsilon_0 m)$, and Eq. (2.40) can be rewritten as

$$\chi(\omega) = \frac{\omega_p^2}{\omega_0^2 - \omega^2 - i\gamma\omega} \qquad (2.41)$$

From Eq. (2.41), frequency dependence ω on χ can be deduced as

$$\varepsilon = 1 + \chi = 1 + \frac{\omega_p^2}{\omega_0^2 - \omega^2 - i\gamma\omega} \qquad (2.42)$$

Its complex form can be extend as

$$\varepsilon' + i\varepsilon'' = 1 + \chi' + \chi'' = 1 + \frac{\omega_p^2}{\omega_0^2 - \omega^2 - i\gamma\omega} \qquad (2.43)$$

ε' and ε'' can be expressed accordingly as

$$\varepsilon' = 1 + \chi'(\omega) = 1 + \frac{\omega_p^2 (\omega_0^2 - \omega^2)}{(\omega_0^2 - \omega^2) + \gamma^2 \omega^2} \qquad (2.44)$$

$$\varepsilon'' = \chi''(\omega) = \frac{\omega_p^2 \gamma\omega}{(\omega_0^2 - \omega^2) + \gamma^2 \omega^2} \qquad (2.45)$$

Realistic atoms have many resonances. Resonances occur due to motion of the atoms (low ω) and electrons (high ω) at

$$\chi = \sum_k \frac{N_k e^2}{\varepsilon_0 m} \frac{1}{\omega_k^2 - \omega^2 - i\gamma\omega} \tag{2.46}$$

where N_k is the density of the electrons/atoms with a resonance at ω_k. $n' > 1$ indicates presence high ω oscillators.

Now we consider solid media. Atom "feels" field from: (1) incident light beam; and (2) induced dipolar field from other atoms, p_i, and hence the local field is $\mathbf{E}_L = \mathbf{E}_0$ (field without matter) + $\mathbf{E}_I$ (field induced dipolar field from all the other atoms). Induce dipolar field, *e.g.*, for cubic symmetry, $\mathbf{E}_I = \mathbf{P}/(3\varepsilon_0)$. For polarization of a solid, the solid consists of atom type j at a concentration N_j can be expressed

$$P = \varepsilon_0 \sum_j N_j \alpha_j E_L = \varepsilon_0 \sum_j N_j \alpha_j (E_0 + \frac{P}{3\varepsilon_0}) \tag{2.47}$$

We get expression of susceptibility χ as function $\mathbf{P}$ below

$$\chi = \frac{P}{\varepsilon_0 E} = \frac{\sum_j N_j \alpha_j}{1 - \frac{1}{3}\sum_j N_j \alpha_j} \tag{2.48}$$

Limit case of low atomic concentration, *e.g.*, pretty good for gasses and glasses, or weak polarizability, $\chi \approx \sum_j N_j \alpha_j$;

dielectric properties of solids related to atomic polarizability can be described by Clausius-Mossotti equation

$$\frac{\varepsilon - 1}{\varepsilon + 2} = \frac{1}{3\varepsilon_0} \sum_j N_j \alpha_j \tag{2.49}$$

It is a general expression.

2.3.3 Optical properties of bulk and nano-structured materials

For bulk and nano-structured materials, outermost electrons interact and form band, and electrons in inner shells do not interact, and thus do not form bands. Classification of matter is done on the basis of band structure: insulator has valence band, semiconductor has semi-conduction band, and metal has conduction band. Insulators with a large $\mathbf{E}_{GAP}$ should not show absorption. Ion beam irradiation or x-ray exposure results in beautiful colors due to formation of color (absorption) centers (peak wavelength). For semiconductors, with $\hbar\omega$(eV) increasing, photon excites phonon and corresponding spectrum experience the processes of lattice resonance absorption, multi-phonon regime, Urbach tail, exciton line, and absorption edge. We cite H-atom as an example here. Electron orbit around a hole is similar to the electron orbit around an H-core. Niels Bohr put forth his opinion in 1913 that electron restrictes to well-defined orbits. Electron binding energy is

$$E_B = -\frac{me^4}{2(4\pi\varepsilon_0 hn)^2} = -\frac{13.6}{n^2} eV \text{ , n=1,2,3,...} \tag{2.50}$$

where m_e is the Electron mass, ε_0 the permittivity of vacuum, h the Planck's constant, and n the energy quantum number/orbit identifier. For binding energy of an electron to hole, electron orbit is expected to be qualitatively similar to an H-atom. Use reduced effective mass instead of m_e as $1/m^* = 1/m_e + 1/m_h$, the binding energy of electrons becomes

$$E_B = \frac{m^*}{m_e} \frac{1}{\varepsilon^2} 13.6 eV \text{ , n=1,2,3,...} \tag{2.51}$$

Typical value for semiconductors is E_B=10 *me*V~100 *me*V. Exciton Bohr radius is ~ 5 nm for many lattice constants.

Current induced by a time varying field can be deduced in the following steps:

Consider a time varying field, $\mathbf{E}(t)=\text{Re}\{\mathbf{E}(\omega)\exp(-i\omega t)\}$, equation of motion electron in a metal is

$$m\frac{d^2x}{dt^2} = \frac{d\mathbf{v}}{dt} = -m\frac{\mathbf{v}}{\tau} - eE \tag{2.52}$$

where τ is the relaxation time $\sim 10^{-14}$ s. Steady state solution of Eq. (2.52) is $v(t) = \text{Re}\{v(\omega)\exp(-i\omega t)\}$. Substituting $v(t)$ into motion equation Eq. (2.52), we have

$$\mathbf{v}(\omega) = \frac{-e}{m(1/\tau - i\omega)}E(\omega) \tag{2.53}$$

The current density is defined as $\mathbf{J}(\omega) = -nev$, whereas n is the electron density. It thus follows

$$J(\omega) = \frac{ne^2/m}{1/\tau - i\omega}E(\omega) \tag{2.54}$$

From definition of conductivity $\mathbf{J}(\omega) = \sigma(\omega)\mathbf{E}(\omega)$, and Eq. (2.53), $\sigma(\omega)$ conductivity can be written as

$$\sigma(\omega) = \frac{\sigma_0}{1 - i\omega t} \tag{2.55}$$

where $\sigma_0 = ne^2\tau/m$.

Both bound electrons and conduction electrons contribute to ε. From the curl equation (2.13) and a time varying field $E(t) = \text{Re}\{E(\omega)\exp(-i\omega t)\}$, we have

$$\nabla \times H = \frac{\partial \varepsilon_B E(t)}{\partial t} + J = -i\omega\varepsilon_B(\omega)E(\omega) + \sigma(\omega)E(\omega) = -i\omega\varepsilon_0[\varepsilon_B(\omega) - \frac{\sigma(\omega)}{i\omega\varepsilon_0})]E(\omega) \tag{2.56}$$

where ε_B represents bound electrons, and the term $\varepsilon_0\omega$ indicates conduction electrons, thus we define ε_{EFF} as

$$\varepsilon_{eff} = \varepsilon_B(\omega) - \frac{\sigma(\omega)}{i\omega\varepsilon_0} = \varepsilon_B(\omega) + i\frac{\sigma(\omega)}{\omega\varepsilon_0} \tag{2.57}$$

Since $\omega_{visible}\tau \gg 1$, dielectric constant at $\omega \approx \omega_{visible}$, conductivity $\sigma(\omega)$ can be approximated as

$$\sigma(\omega) = \frac{\sigma_0}{1 - i\omega t} = \frac{\sigma_0(1 + i\omega t)}{(1 + \omega^2\tau^2)} \approx \frac{\sigma_0}{\omega^2\tau^2} + i\frac{\sigma_0}{\omega\tau} \tag{2.58}$$

It follows that

$$\varepsilon_{eff} = \varepsilon_B(\omega) + i\frac{\sigma(\omega)}{\omega\varepsilon_0} = \varepsilon_B(\omega) + i\frac{\sigma_0}{\varepsilon_0\omega^3\tau^2} - \frac{\sigma_0}{\varepsilon_0\omega^2\tau} \tag{2.59}$$

Define $\omega_p^2 = \sigma_0/\varepsilon_0\tau = ne^2/(\varepsilon_0 m)$ (≈ 10 eV for metals), then the above equation can be rewritten as

$$\varepsilon_{eff} = \varepsilon_B - \frac{\omega_p^2}{\omega^2} + i\frac{\omega_p^2}{\omega^3\tau} \tag{2.60}$$

where the last two items represent free electrons.

For example, for Al, $\varepsilon_B \approx 1$, the corresponding ε_{eff} becomes

$$\varepsilon_{eff,Al} = 1 - \frac{\omega_p^2}{\omega^2} + i\frac{\omega_p^2}{\omega^3\tau} \tag{2.61}$$

For Ag, it show interesting feature in reflection which is caused by interband transition. Both conduction and bound electrons contribute to ε_{eff}, $\varepsilon_B = \varepsilon_{ib} \neq 0$, thus ε_{eff} can be expressed as

$$\varepsilon_{eff,Ag} = \varepsilon_{ib} - \frac{\omega_p^2}{\omega^2} + i\frac{\omega_p^2}{\omega^3\tau} \tag{2.62}$$

Similarly, the ε_{eff} for other metal materials can be deduced by the same way in respect to different ε_B.

2.4 Surface plasmon polaritons

2.4.1 Metal optics

Detailed information regarding optics of metals can be seen from Ref. 2 in chapter 13. In this section, we introduce something relevant to surface plasmon issues only.

Majority of optical components are designed on the basis of dielectric materials for performances of high speed, and high bandwidth (ω) etc. However, they do not scale well while need the design for large scale integration. But they still have problems such as bending loss, diffraction limit, and some other fundamental problems. To solve these problems, metals are considered due to its unique feature — surface plasmon (SP). What is a plasmon? A plasmon is a collective excitation of the electronic "fluid" in a piece of conducting material, like ripples on the surface of a pond are a collective mode of the water molecules of the liquid. The similarity here is not too far off, because like water, the electronic fluid in a metal is pretty close to incompressible. If you push down on the surface of a pond somewhere with a float, the density of the water doesn't change; instead the water elsewhere is displaced, because the water molecules have finite volume and push each other out of the way. The electronic fluid acts similarly, not because of any finite size or even the Coulomb repulsion of the electrons, but mostly because of the Pauli exclusion principle, which tends to keep the electrons out of each others' way. Comparing electron gas in a metal and real gas of molecules, metals are expected to allow for electron density waves: plasmons. There are bulk plasmon and surface plasmon. For bulk plasmon, metals allow for EM wave propagation above the plasma frequency ω_p. Metals become transparent at this frequency. Surface plasmon, sometimes is called a surface plasmon-polaritons (SPPs) which is a strong coupling to EM field for TM mode illumination only, as shown in Fig. 2-2. It has characteristics of strong localization of the EM field and high local field intensities which is easy to be obtained. SPP has applications of guiding of light below diffraction limit (near-field optics), non-linear optics, sensitive optical studies of surfaces and interfaces, bio-sensors, and study film growth etc. For instance, array of 50 nm diameter Au particles spaced by 75 nm has function of guiding electromagnetic energy at optical frequency below diffraction limit, enabling communication between nanoscale devices, and information transportation at speeds and densities exceeding current electronics.[3,4] Another typical example is the bull-eye-like structure. [5] It is composed of Ag film with a 440 nm diameter hole surrounded by circular grooves with transmission enhancement of $10 \times$ compared to that of a bare hole. The transmission is 3 folds more than the light directly impingent on hole due to excitation of plasmon-polaritons.

Firstly, we discuss optical properties of an electron gas (metal) with dielectric constant of a free electron gas (no interband transitions). Consider a time varying field: $E(t)=Re\{E(\omega)exp(-i\omega t)\}$, from Equation of motion electron (no damping) $md^2r/dt^2= -e\mathbf{E}$ and dipole moment electron $\mathbf{p}(t)= -er(t)$, we have a harmonic time dependence $\mathbf{p}(t) = \operatorname{Re}\{\mathbf{p}(\omega)e^{-i\omega t}\}$. Substitution $\mathbf{p}$ into motion equation, we get

$$\mathbf{p}(\omega) = -\frac{e^2}{m}\frac{1}{\omega^2}\mathbf{E}(\omega) \tag{2.63}$$

The dielectric constant is

$$\varepsilon_r = 1 + \chi = \frac{N\mathbf{p}(\omega)}{\varepsilon_0 \mathbf{E}(\omega)} = 1 - \frac{Ne^2}{\varepsilon_0 m}\frac{1}{\omega^2} = 1 - \frac{\omega_p^2}{\omega^2} \tag{2.64}$$

Now we discuss determination of dispersion relation for bulk plasmons. The wave equation is given by

$$\frac{\varepsilon_r}{c^2}\frac{\partial^2 \mathbf{E}(r,t)}{\partial t^2} = \nabla^2 \mathbf{E}(r,t) \tag{2.65}$$

Investigate solutions of the form: $E(r,t)=Re\{E(r,\omega)exp(ikr-i\omega t)\}$, we get dielectric constant by $\varepsilon_r=1-\omega_p^2/\omega^2$, and $\omega^2\varepsilon_r=c^2k^2$ as

$$\omega^2(1-\frac{\omega_p^2}{\omega^2}) = \omega^2 - \omega_p^2 = c^2 k^2 \tag{2.66}$$

From dispersion relation of $\omega=(\omega_p^2+c^2k^2)^{1/2}$, the solutions is above light line ($\omega=ck$), and thus there is no allowed propagating modes. For metals: $\hbar\omega_p \approx 10$ eV; and semiconductors $\hbar\omega_p < 0.5$ eV (depending on dopant concentration). In order to know dispersion relation of SPPs, we need to solve Maxwell's equations with boundary conditions. For the structures shown in Fig. 2-2, we have **E** and **H** fields for dielectric and metal expressed mathematically as follows

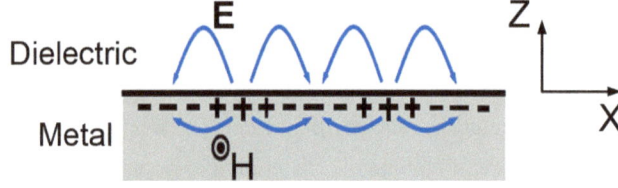

Fig. 2-2 Physical model of evanescent SPP wave propagation at interface of dielectric/metal.

$$z<0 \begin{cases} \boldsymbol{H}_d = (0,H_{yd},0)\exp i(k_{xd}x+k_{zd}z-\omega t) \\ \boldsymbol{E}_d = (E_{xd},0,E_{zd})\exp i(k_{xd}x+k_{zd}z-\omega t) \end{cases}$$

$$z>0 \begin{cases} \boldsymbol{H}_m = (0,H_{ym},0)\exp i(k_{xm}x+k_{zm}z-\omega t) \\ \boldsymbol{E}_m = (E_{xm},0,E_{zm})\exp i(k_{xm}x+k_{zm}z-\omega t) \end{cases}$$

$$(2.67)$$

Maxwell's Equations in medium i (i = metal or dielectric) are listed as follows

$$\nabla \cdot \varepsilon_i \boldsymbol{E} = 0$$

$$\nabla \cdot \boldsymbol{H} = 0$$

$$\nabla \times \boldsymbol{E} = -\mu_0 \frac{\partial \boldsymbol{H}}{\partial t}$$

$$\nabla \times \boldsymbol{H} = \varepsilon_i \frac{\partial \boldsymbol{E}}{\partial t}$$

$$(2.68)$$

At the boundary, we have $E_{x,m}=E_{x,d}$, $\varepsilon_m E_{z,m}=\varepsilon_d E_{z,m}$, $H_{y,m}=H_{y,d}$. Start with curl equation for **H** in medium i (as we did for EM waves in vacuum) $\nabla\times H_i=\varepsilon_i\partial E_i(t)/\partial t$, where $H_i=(0,H_{yi},0)\exp i(k_{xi}x+k_{zi}z-\omega t)$, $E_i=(E_{xi},0, E_{zi})\exp i(k_{xi}x+k_{zi}z-\omega t)$, the extension form is

$$\left(\frac{\partial H_{zi}}{\partial y}-\frac{\partial H_{yi}}{\partial z}, \frac{\partial H_{xi}}{\partial z}-\frac{\partial H_{zi}}{\partial x}, \frac{\partial H_{yi}}{\partial x}-\frac{\partial H_{xi}}{\partial y} \right) = \left(ik_{zi}H_{yi},0,ik_{xi}H_{yi} \right) = \left(-i\omega\varepsilon_i E_{xi},0,i\omega\varepsilon_i E_{zi} \right)$$

$$(2.69)$$

We will use that $k_{zi}H_{yi}=-\omega\varepsilon_i E_{xi}$, and get $k_{zm}H_{yi}=-\omega\varepsilon_m E_{xm}$ and $k_{zd}H_{yd}=-\omega\varepsilon_d E_{xd}$, combine with continuous across boundary of $E_\parallel$: $E_{x,m}=E_{x,d}$, the following relation can be obtain

$$\frac{k_{zm}}{\varepsilon_m} H_{ym} = \frac{k_{zd}}{\varepsilon_d} H_{yd}$$

$$(2.70)$$

Similarly, for continuous across boundary of $H_\parallel$: $H_{y,,m}=H_{y,d}$, combining with Eq. (2.69), we have $k_{zm}/\varepsilon_m=k_{zd}/\varepsilon_d$. Now, we see the relations regarding **k** vectors. Relation for k_x (Continuity $E_{//}$, $H_{//}$) is $k_{xm}=k_{xd}$. For any EM wave, $k_x^2+k_{zi}^2=\varepsilon_i(\omega/c)^2$, we have k_{SP} both in the dielectric and metal

$$k_{SP} = k_x = \sqrt{\varepsilon_i\left(\tfrac{\omega}{c}\right)^2 - k_{zi}^2}$$

$$(2.71)$$

and dispersion relation

$$k_x = \frac{\omega}{c}\sqrt{\frac{\varepsilon_m\varepsilon_d}{\varepsilon_m+\varepsilon_d}}$$

$$(2.72)$$

For low ω, Eq. (2.72) can be changed to

$$k_x = \frac{\omega}{c} \lim_{\varepsilon_m \to -\infty} \sqrt{\frac{\varepsilon_m \varepsilon_d}{\varepsilon_m + \varepsilon_d}} \approx \frac{\omega}{c} \sqrt{\varepsilon_d} \qquad (2.73)$$

At $\omega = \omega_{sp}$ (when $\varepsilon_m = -\varepsilon_d$), $k_x \to \infty$, then solution of Eq. (2.72) lies below the light line. Higher index medium on metal results in lower ω_{sp}.

If we excite SPPs with electrons, $k_{electron} \gg k_{light}$, there is stringent requirement on divergence of e-beam column. The ever-increasing need for faster information processing and transport is undeniable. Electronic components are running out of steam due to the issues with RC-delay times. As data rates and component packing densities increase, electrical interconnects become progressively limited by RC-delay. Electronics are aspect-ratio limited in speed. However, the bit rate in optical communications is fundamentally limited only by the carrier frequency: $B_{max} < f \sim 100$ Tbit/s, but light propagation is subjected to diffraction. Photonics device is diffraction-limited in size. In contrast, SP wavelengths can reach nanoscale at optical frequencies. SPPs are "x-ray waves" with optical frequencies. Plasmonic will enable an improved synergy between electronic and photonic devices. Plasmonics naturally interfaces with similar size electronic components and similar operating speed of photonic networks.

2.4.2 Surface plasmon polaritons excitation

In this subsection, we discuss dispersion relation for SPPs. Higher index medium on metal results in lower ω_{SP}, when $\varepsilon_m = -\varepsilon_d$, $\omega = \omega_{sp}$, we have $\omega = \omega_p/(1+\varepsilon_d)^{1/2}$. Some problems may occur when SPPs modes lie below the light line, as shown in Fig. 2-3. There is no coupling of SPPs modes to far field and vice versa (reciprocity theorem). A method is required

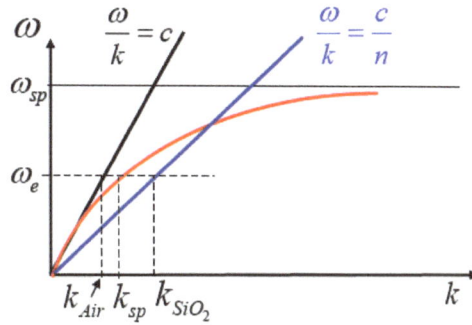

Fig. 2-3 Plot of dispersion relation for SPP wave propagation. Energy and momentum are necessary for the SPP excitation.

to excite modes below the light line. One method is excitation from a high index medium. Excitation of SPPs at a metal/air interface from a high index medium $n = n_h$, *i.e.*, SPPs at metal/air interface can be excited from a high index medium. To realize this, one approach is the well known Kretchmann geometry with dispersion relation of

$$k_{\parallel, SiO_2} = \sqrt{\varepsilon_d} \frac{\omega}{c} \sin\theta = k_{SP} \qquad (2.74)$$

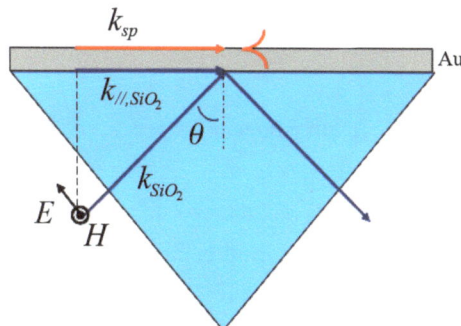

Fig. 2-4 Configuration of Kretchmann geometry. It consists of an Au layer coated on top surface of glass.

where θ is the incident angle. In this configuration, making use of SiO_2 prism creates evanescent wave by total internal reflection (TIR), and enables excitation surface plasmons at the Air/Metal interface, as shown in Fig. 2-4. There is a strong coupling when $k_{//,SiO_2}$ is coupled to k_{sp}. To quantitatively describe the coupling to SPPs, we need to calculate reflection coefficient firstly. For a plane polarized light, it can be expressed as

$$R = \left| \frac{E_r^p}{E_0^p} \right|^2 = \left| \frac{r_{01}^p + r_{12}^p \exp(2ik_{z1}d)}{1 + r_{01}^p r_{12}^p \exp(2ik_{z1}d)} \right|^2 \tag{2.75}$$

where d is the thickness of metal film, r_{ik}^p (p means planar polarized light) are the amplitude reflection coefficients, also known as Fresnel coefficients, can be written as

$$r_{ik}^p = \frac{\dfrac{k_{zi}}{\varepsilon_i} - \dfrac{k_{zk}}{\varepsilon_k}}{\dfrac{k_{zi}}{\varepsilon_i} + \dfrac{k_{zk}}{\varepsilon_k}} \tag{2.76}$$

Light intensity reflected from the back surface depends on the film thickness. There a thin film thickness exists for perfect coupling (destructive interference between two reflection beams) when light coupled in perfectly, all the EM energy dissipated in the film. Width of the resonance relates to damping of the SPPs. Light escapes prism below critical angle for total internal reflection. This technique can also be used to determine the thickness of metallic thin films.

In the resonating system, when $|\varepsilon'_m| \gg 1$, ω will be well below ω_{SP}; and when $|\varepsilon''_m| \ll |\varepsilon'_m|$, the resonance will lead to low loss. Reflection coefficient has Lorentzian line shape (characteristic of resonators)

$$R = 1 - \frac{4\Gamma_i \Gamma_{rad}}{(k_x - k_x^0)^2 + (\Gamma_i + \Gamma_{rad})^2} \tag{2.77}$$

where Γ_i is the damping due to resistive heating, Γ_{rad} is the damping due to re-radiation into the prism, and k_x^0 the resonance wave vector (maximum coupling). R goes to zero when $\Gamma_i = \Gamma_{rad}$. Coupling angle strongly depends on the film thickness of the metal film.[6]

The second excitation method is the usage of grating structures. For the grating coupling geometry, periodic dielectric constant couples waves for which the k-vectors differ by a reciprocal lattice vector **G**. Bloch boundary condition is necessary for computational numerical calculation. The strong coupling occurs when $k_{||,SiO_2}=k_{SP}\pm m\mathbf{G}$, where $|\mathbf{G}|=2\pi/\mathbf{P}$, and m is the mode orders. Then the dispersion relation changes to

$$k_{||,SiO_2} = |k_e| = \sqrt{\varepsilon_d} \frac{\omega}{c} \sin\theta \tag{2.78}$$

$$k_{SP} = \frac{\omega}{c} \sqrt{\frac{\varepsilon_m \varepsilon_d}{\varepsilon_m + \varepsilon_d}} \tag{2.79}$$

where k_e is incident wave vector. Corresponding wavelength λ_{SP} for classical normalized wave vector at a metal-dielectric interface can be expressed as

$$\lambda_{SP} = \frac{\Lambda}{m} \{ \mathrm{Re}[\sqrt{\frac{\varepsilon_1 \varepsilon_2}{\varepsilon_1 + \varepsilon_2}}] \pm \sin(\theta) \} \tag{2.80}$$

where Λ is the period of the structure, m is the relative integer for periods of grating-like indentation, ε_1 is the dielectric, ε_2 is the metal constant, and θ is the incident angle.

The third excitation method is metallic nanodots and nanoparticles. The strong coupling occurs at $k_{||,SiO_2}=k_{SP}\pm\Delta k_{dot}$. [7] Spatial Fourier transform of the dot contains significant contributions of Δk_{dot} values up to $2\pi/d$, whereas d is the diameter of dots. Dipole radiation is in direction of charge oscillation because plasmon wave is in direction of longitudinal. For the metal nanoparticles embedded in biological environment, Eq. (2.81) is Lorentz-Loenz model describing refractive index change due to the SP-supported radiation of the nanoparticles.[8]

$$\Delta n = -\frac{1}{6n}(n^2+2)^2\left(\frac{n^2-1}{n^2+2} - \frac{n_w^2-1}{n_w^2+2}\frac{V_P}{V}\right)\frac{\Delta d}{d} \tag{2.81}$$

where Δd is the change in protein thickness, Δn is the refractive index change, n_w is the refractive index of water, n is the refractive index of protein, V_w is the volume of water, V_p is the volume of protein, and V is the volume of protein layer, here $V=V_p+V_w$.

In addition, some other structures for the SPP excitation such as single strip [9], short cylinder [10], launch pad diversity [11], two-dimensional metallic-dielectric photonic crystals[12], and oil-imersion objective lens with large numerical aperture [13, 14], were reported in recent years.

2.5 Summary

Electromagnetic wave propagation and Maxwell equations are introduced in this chapter. EM wave propagation is described in detail; including light interaction with matters, dispersion in materials, and optical properties of bulk and nano-structured materials. With the wave propagation background, surface plasmon is highlighted as a main topic being described in detail, including propagation of SPP wave in both dielectric and metals. Several SPP excitation methods are illustrated on the basis of EM theory and dispersion relation. This chapter introduces theoretical issues of the SPP only. SPP relevant applications will be described as an important part in the following chapters of this book.

Reference

[1]　Edward J. Rothwell, Michael J. Cloud., *Electromagnetics*, CRC Press LLC, 2001, Chapt. 1.

[2]　Max Born and Emil Wolf, Edt.: *Principles of Optics*, Camrbidge University Press, 1980, Chapt. 1 and 13.

[3]　Mark L. Brongersma, John W. Hartman, and Harry A. Atwater, "Electromagnetic energy transfer and switching in nanoparticle chain arrays below the diffraction limit", Phys. Rev. B 62, R16356. (2000).

[4]　S. A. Maier, M. L. Brongersma, P. G. Kik, S. Meltzer, A. A. G. Requicha, H. A. Atwater, "Plasmonics - A Route to Nanoscale Optical Devices", Advanced materials 13, 1501 (2001).

[5]　Thio, Tineke; Pellerin, K M; Linke, R A; Lezec, H J; Ebbesen, T W, "Enhanced light transmission through a single subwavelength aperture", Optics Letters 26, 1972-1974 (2001).

[6]　Hiroshi Kano, *Near-field optics and Surface plasmon Polaritons*, Springer Verlag, 2001.

[7]　H. Ditlbacher, J. R. Krenn, N. Felidj, B. Lamprecht, G. Schider, M. Salerno, A. Leitner, and F. R. Aussenegg, "Fluorescence imaging of surface plasmon fields", Appl. Phys. Lett. 80, 404 (2002).

[8]　Englebienne, Patrick; Van Hoonacker, Anne; Verhas, Michel., Surface plasmon resonance: principles, methods and applications in biomedical sciences. *Spectroscopy* 17, 255-273 (2003).

[9]　Igor I. Smolyaninov, David L. Mazzoni, and Christopher C. Davis, "Imaging of Surface Plasmon Scattering by Lithographically Created Individual Surface Defects", Phys. Rev. Lett. 77, 3877 (1996).

[10]　J. R. Krenn, B. Lamprecht, H. Ditlbacher, G. Schider, M. Salerno, A. Leitner, and F. R. Aussenegg, "Non–diffraction-limited light transport by gold nanowires", Europhys.Lett. 60, 663-669 (2002).

[11]　S.C. Kitson, W. L. Barnes, and J. R. Sambles, "Full Photonic Band Gap for Surface Modes in the Visible", Phys Rev Lett. 77, 2670 (1996).

[12]　S.I. Bozhevolnyi, John Erland, Kristjan Leosson, Peter M. W. Skovgaard, and Jørn M. Hvam, "Waveguiding in Surface Plasmon Polariton Band Gap Structures", Phys Rev Lett. 86, 3008 (2001).

[13]　Hiroshi Kano and Wolfgang Knoll, "Locally excited surface-plasmon-polaritons for thickness measurement of LBK films". Opt. Commun. 153, 235-239 (1998).

[14]　Hiroshi Kano and Wolfgang Knoll, "A scanning microscope employing localized surface-plasmon-polaritons as a sensing probe". Opt. Commun. 182, 11-15 (2000).

3 ANALYSIS APPROACHES AND ALGORITHMS FOR PLASMONIC STRUCTURES

Abstract: Firstly, several stationary analyzing approaches: rigorous coupling wave analysis/Fourior model method, Mie theory, multiple multipole program, and discrete dipole approximation were presented. Then dynamic analyzing approaches: finite difference and time domain, and beam propagation method were described briefly in this chapter. Both the stationary and dynamic analyzing approaches were presented by theoretical description in respect of wave propagation through the subwavelength metallic structures.

3.1 Introduction

Subwavelength-optics design and modeling methods can be classified as following approaches:

1) Rigorous solution of Maxwell's equations;

2) Stationary analysis: band solver, rigorous coupling wave analysis (RCWA) / Fourior model method (FMM), multiple multipole program (MMP), Mie theory, and scattering codes such as discrete dipole approximation (DDA);

3) Dynamic analysis: finite difference and time domain (FDTD), beam propagation method (BPM), and custom methods.

Maxwell equations are essence for above analyses. Theoretical calculations start from Maxwell's equations with the following procedures:

1) Assuming light propagates in homogeneous isotropic media, electric field can be solved from Helmholtz equation;

2) Obtaining eigen solutions;

3) Aanalyzing dispersion relation.

The fundamental issues in the calculation are that light propagation in homogeneous volumes. Then the **E**-field wave propagates in the medium follows the relationships below

$$\Delta\mathbf{E}(\mathbf{r}, \omega) + \varepsilon^2/c^2\varepsilon(\omega)\mathbf{E}(\mathbf{r}, \omega) =0, \tag{3.1}$$

$$\mathbf{E}^{\wedge}(\mathbf{r}, \omega) =\mathbf{E}(\mathbf{k}, \omega)\exp(i\mathbf{kr}), \tag{3.2}$$

$$k^2=k_x^2+k_y^2+k_z^2=\varepsilon(\omega)\omega^2/c^2, \quad (\text{for } \mathbf{k}\Box\mathbf{E}(\mathbf{k}, \omega)) \tag{3.3}$$

These equations describe light propagation and dispersion relation, are essence of the electromagnetic wave analyses as well as sequential algorithms. The following sections give detailed illustrations regarding these analyses and corresponding algorithms.

3.2 Stationary analyzing approaches

3.2.1 RCWA / FMM

Rigorous coupling wave analysis (RCWA), also called Fourior model method (FMM), is an algorithm for solving the vector diffraction waves. It is ideally suited to simulate the optical response of two-dimensional (2D), periodic surface relief structures, and binary surfaces. It involves the expansion into spatial Fourier components of both the dielectric function and the associated fields, followed by a numerical solution of the boundary equations. The approach can yield both reflected and transmitted complex diffraction efficiencies, as well as the associated field distributions both inside and outside the structure. Initially, it originates from analyzing the gratings with binary structures. To analyze the response of subwavelength gratings, the vectorial nature of light must be taken into account through a resolution of the Maxwell equation. Fields and permittivities are decomposed on the Fourier basis and then matched at the grating layer boundaries to yield the diffraction order complex amplitudes. The RCWA algorithm is often used because of its good convergence and relatively simple implementation. The algorithm uses a layered structure to approximate the grating profile, but no approximations are used for the material properties. In this section, RCWA is introduced from one-dimensional (1D) structures firstly, after that we extended the theory to 2D structures.

In order to better understand the complexity of the grating problems, it is necessary to realize that nowadays gratings have complex profiles such as sinusoidal, rectangular or trapezoidal grooves and consist of all kinds of different materials. Moreover, in the visible region and for shorter wavelengths, the finite conductivity complicates the grating response and requires more complex mathematical models. Kraaij. *et. al.* put forth a modified RCWA method.[2] First of all, we discuss the gratings with 1D geometries, as shown in Fig. 3-1 which leads to Maxwell's equations: the basis of the RCWA algorithm. This is a standard model with a standard assumption: a linearly polarized electromagnetic field with angle ψ is obliquely incident at an arbitrary angle of incidence θ and an azimuthal angle ϕ upon a dielectric or lossy grating. The grating is assumed to be infinitely long in the periodic x-direction with a period Δ.

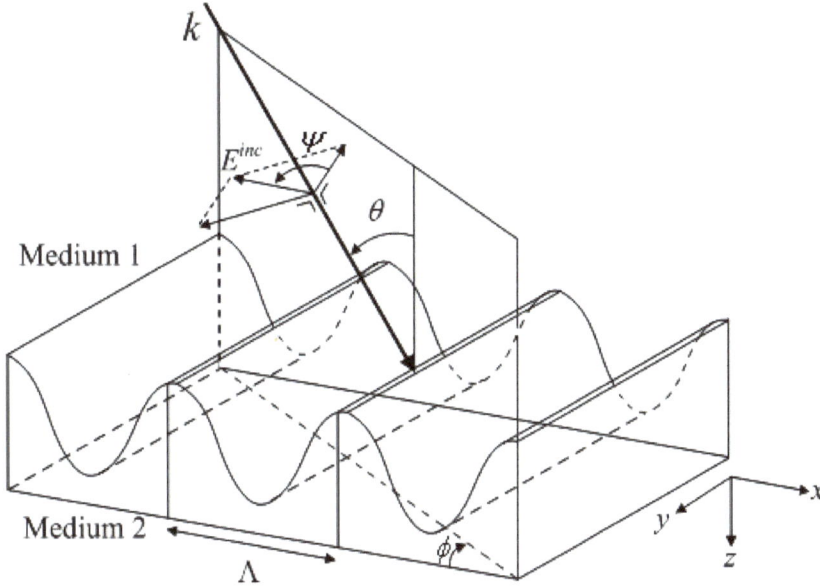

Fig. 3-1 Schematic diagram of one-dimensional periodic grating.

The grating grooves along y-direction are also assumed to be infinitely long. In the example below only two different media were present with refraction indices of n_I and n_{II}. Figure 3-2 shows how a general grating is approximated by a multilayered grating. Note that all calculations can be restricted to only one period. In each layer the material constants depend on the horizontal x-coordinate only and are independent of the vertical z-coordinate. Furthermore the different media are assumed to be homogeneous, linear and isotropic. Assuming that no primary or external sources are presented and considering only time-harmonic field quantities, Maxwell's equations applied to the discrete model in Fig. 3-2 and the constitutive relations can be reduced (only TM polarized incident light in a planar diffraction case is considered here. For other diffraction cases, readers can see Ref. 2.) to the following equations for each grating layer i:

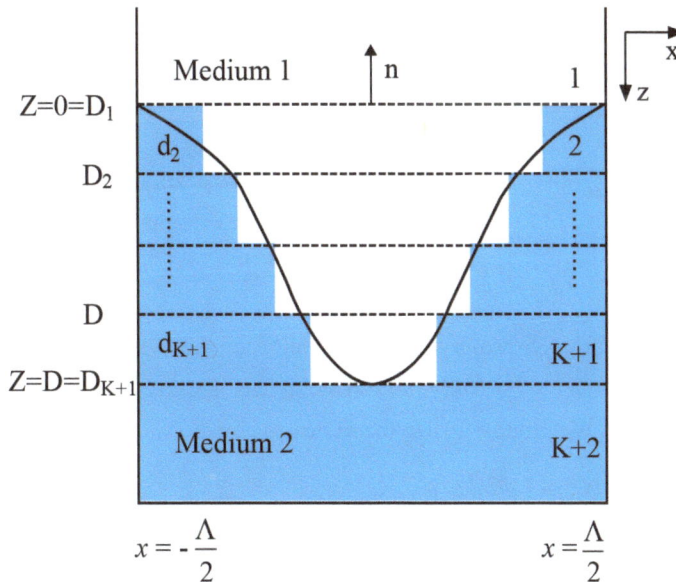

Fig. 3-2 Layered one-dimensional grating in $[-\Delta/2,\Delta/2]\times(-\infty,\infty)$.

$$\frac{\partial}{\partial z}E_{i,x}(x,z)=-j\omega\mu_0 H_{i,y}(x,z)+\frac{\partial}{\partial x}E_{i,z}(x,z),\qquad\qquad(3.13a)$$

$$\frac{\partial}{\partial z} H_{i,y}(x,z) = -j\omega\tilde{\varepsilon}_i(x)E_{i,x}(x,z),$$

(3.13b)

$$\frac{\partial}{\partial x} H_{i,y}(x,z) = j\omega\tilde{\varepsilon}_i(x)E_{i,z}(x,z),$$

(3.13c)

All the electric field components $\mathbf{E}_i$ can be eliminated so that only one equation for the y-component of the magnetic field $\mathbf{H}_i$ remains for each layer:

$$\frac{\partial^2}{\partial z^2} H_{i,y}(x,z) = -k_0^2 \frac{\tilde{\varepsilon}_i(x)}{\varepsilon_0} H_{i,y}(x,z) - \frac{\tilde{\varepsilon}_i(x)}{\varepsilon_0} \frac{\partial}{\partial x}\left(\frac{\varepsilon_0}{\tilde{\varepsilon}_i(x)} \frac{\partial}{\partial x} H_{i,y}(x,z)\right)$$

(3.14)

On the left and right of the domain, the pseudo-periodic boundary conditions are applied and above and below the grating, Rayleigh's radiation condition are used. Finally at the layer interfaces the continuity of the tangential electromagnetic field components are preserved. This can be reformulated in the following set of equations for the magnetic field:

$$H_{i,y}(x,D_i) = H_{i+1,y}(x,D_i)$$

(3.15a)

$$\frac{1}{\tilde{\varepsilon}_i(x)} \frac{\partial}{\partial z} H_{i,y}(x,D_i) = \frac{1}{\tilde{\varepsilon}_{i+1}(x)} \frac{\partial}{\partial z} H_{i+1,y}(x,D_i)$$

(3.15b)

The complex permittivity $\tilde{\varepsilon}_i(x)$ within each layer is expanded in a Fourier series. The incident magnetic field $\mathbf{H}_y^{inc.}$ is assumed to be one plane wave and the magnetic field expansions for each grating layer are given by Eq. (3.16):

$$H_y^{inc}(x,z) = n_1\sqrt{\frac{\varepsilon_0}{\mu_0}} e^{-jk_0 n_1 (x\sin\theta + z\cos\theta)}$$

(3.16a)

$$H_{1,y}(x,z) = n_1\sqrt{\frac{\varepsilon_0}{\mu_0}} \sum_n R_n e^{-j(k_{xn}x - k_{1,zn}z)} + H_y^{inc}(x,z)$$

(3.16b)

$$H_{i,y}(x,z) = n_1\sqrt{\frac{\varepsilon_0}{\mu_0}} \sum_n U_{i,n}(z) e^{-jk_{xn}x}$$

(3.16c)

$$H_{K+2,y}(x,z) = n_1\sqrt{\frac{\varepsilon_0}{\mu_0}} \sum_n T_n e^{-j(k_{xn}x + k_{K+2,zn}(z-D))}$$

(3.16d)

Note that these expansions already satisfy the pseudo-periodic boundary condition and Rayleigh's radiation condition. Substituting these expansions into Eq. (3.6) and truncating the equations results in:

$$\frac{d^2}{dz'^2} U_i(z') = \mathbf{E}_i(\mathbf{K_x}\mathbf{P_i}\mathbf{K_x} - \mathbf{I})U_i(z')$$

(3.17)

However Eq. (3.7) does not uniformly preserve the continuity of the appropriate field components across the discontinuities in one layer of the complex permittivity function. This is caused by the way in which the Fourier series are used in the truncated equations. We propose to use the truncated equations:

$$\frac{d^2}{dz'^2} U_i(z') = \mathbf{P}_i^{-1}(\mathbf{K_x}\mathbf{E}_i^{-1}\mathbf{K_x} - \mathbf{I})U_i(z')$$

(3.18)

This proposal is put forth on the basis of the theory in Ref. 3 which suggest that these truncations are better when there are discontinuities in one layer of the permittivity function. Equation (3.18) is not derived from Eq. (3.14) but from the basis of Eq. (3.13) after multiplying Eq.(3.13b) with $1/\tilde{\varepsilon}_i(x)$. Thus instead of firstly eliminating the electric field components, substituting the expansions and truncating the equations, we now start with substituting the expansions in

Eq. (3.13), truncating the equations and then eliminating the electric field components. Finally, the complex reflected and transmitted field amplitudes are determined by calculating eigenvalues and eigenvectors of Eq. (3.18) and using the boundary conditions in Eq. (3.15) at the layer interfaces. An enhanced transmittance matrix approach is used to calculate the reflected and transmitted field amplitudes in a stable way [4].

Now let's see the 2D structures, as shown in Fig.3-3. It shows the top view of a crossed grating. We introduced a nonrectangular coordinate system $Ox^1x^2x^3$ such that the x^1 and x^3 axes are parallel to x and z axes, respectively, but

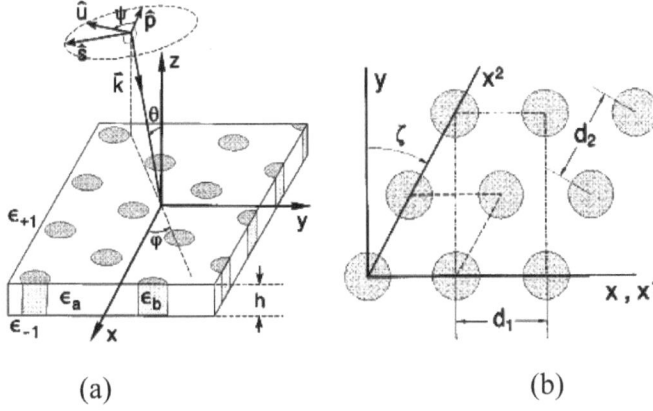

(a)   (b)

Fig.3-3 (a) Crossed grating illuminated by a plane wave. A rectangular Cartesian coordinate system is attached to the grating so that its x axis is along one of the periodic directions. (b) Top view of a crossed grating and a nonrectangular Cartesian coordinate system in the grating plane. The x^1 and x^2 axes are parallel to two periodic directions of the grating. The $x1$ axis is parallel to the x axis, and the $x2$ axis forms an angle z with the y axis. Reprinted with permission from Lifeng Li, J. Opt. Soc. Am. A 14 (10),2758-2767 (1997) with copyright of ©1997 Optical Society of America

the x^2 axis is parallel to another periodic direction of the grating. Among the denumerable periodic directions in the Oxy plane we identify x^1 and x^2 axes with those two directions that have the shortest periods and denote the two periods d_1 and d_2, respectively. The origin of z axis is assumed to be at the lower boundary of the grating layer. The angle formed by the positive directions of x^2 and y axes is ζ, $|\zeta|<\pi/2$. The parallelogram marked by dashed and solid lines and the rectangle marked by the dotted–dashed and solid lines are two possible unit cells of the periodic pattern. The transformation between the two coordinate systems is given by [5]

$$x^1 = x - y \tan\zeta, \quad x^2 = y \sec\zeta, \quad x^3 = z \tag{3.19}$$

For the nonrectangular coordinate system, the covariant and the contravariant basis vectors are

$$\mathbf{b}_1 = \hat{x}, \qquad\qquad \mathbf{b}_2 = \hat{x} \sin\zeta + \hat{y} \cos\zeta, \qquad \mathbf{b}_3 = \hat{z},$$

(3.20)

$$\mathbf{b}^1 = \hat{x} \sin\zeta - \hat{y} \tan\zeta, \quad \mathbf{b}^2 = \hat{y} \sec\zeta, \qquad\qquad \mathbf{b}^3 = \hat{z},$$

(3.21)

respectively. The two sets of basis vectors obey the reciprocal relation: $\mathbf{b}_\rho \cdot \mathbf{b}^\sigma = \delta_\rho{}^\sigma$, where $\delta_\rho{}^\sigma$ is the Kronecker delta symbol. In order to keep the flow of presentation there smooth, some unconventional notations for these matrices will be defined here. The 2D Fourier coefficients of $\varepsilon(x^1, x^2)$ are given by

$$\varepsilon_{mn} = \frac{1}{d_1 d_2} \int_0^{d_2} \int_0^{d_1} \varepsilon(x^1, x^2) \times e^{-i(mK_1 x^1 + nK_2 x^2)} dx^1 dx^2 \tag{3.22}$$

where $K_1 = 2\pi/d_1$ and $K_2 = 2\pi/d_2$. These coefficients can be turned into a matrix by splitting the indexing numbers. Introducing the notation $\lfloor\lfloor \cdot \rfloor\rfloor$, we have

$$\lfloor\lfloor \varepsilon \rfloor\rfloor_{mn,jl} = \varepsilon_{m-j,n-l} \tag{3.23}$$

Next we introduce the notations $\lceil \cdot \rceil$ and $\lfloor \cdot \rfloor$ by

$$\lceil\varepsilon\rceil_{mn} = \frac{1}{d_1}\int_0^{d_1}\varepsilon(x^1,x^2)\,e^{[-i(m-n)K_1x^1)]}dx^1 \tag{3.24}$$

$$\lceil\varepsilon\rceil_{mn} = \frac{1}{d_2}\int_0^{d_2}\varepsilon(x^1,x^2)\,e^{[-i(m-n)K_2x^2)]}dx^2 \tag{3.25}$$

Obviously $\lceil\varepsilon\rceil$ and $\lfloor\varepsilon\rfloor$ are still functions of x^2 and x^1, respectively. Finally, we introduce two more notations, $\lfloor\lceil\cdot\rceil\rfloor$ and $\lceil\lfloor\cdot\rfloor\rceil$, by

$$\lfloor\lceil\varepsilon\rceil\rfloor_{mn,\,jl} = \lfloor\{\lceil 1/\varepsilon\rceil^{-1}\}_{mj}\rfloor_{nl} = \frac{1}{d_2}\int_0^{d_2}\{\lceil 1/\varepsilon\rceil^{-1}\}_{mj}(x^2)\times e^{[-i(n-l)K_2x^2)]}dx^2 \tag{3.26}$$

$$\lfloor\lceil\varepsilon\rceil\rfloor_{mn,\,jl} = \lfloor\{\lceil 1/\varepsilon\rceil^{-1}\}_{nl}\rfloor_{mj} = \frac{1}{d_2}\int_0^{d_2}\{\lceil 1/\varepsilon\rceil^{-1}\}_{nl}(x^1)\times e^{[-i(m-j)K_1x^1)]}dx^1 \tag{3.27}$$

In the above, we have assumed that $\varepsilon(x^1, x^2)\neq0$ and the two inverse matrices exist, which is guaranteed on physical grounds. In terms of contravariant basis vectors the wave vector of the incident plane wave may be written as

$$\mathbf{k}= k_\sigma\mathbf{b}^\sigma =\alpha_0\mathbf{b}^1 + \beta_0\mathbf{b}^2 - \gamma_{00}^{(+1)}\mathbf{b}^3 \tag{3.28}$$

where

$$\alpha_0=k^{(+1)}\sin\theta\cos\varphi,$$
$$\beta_0=k^{(+1)}\sin\theta\cos(\varphi+\zeta),$$
$$\gamma_{00}^{(+1)}=k^{(+1)}\cos\theta, \tag{3.29}$$

where $k^{(+1)}= 2\pi[\varepsilon^{(+1)}\mu]^{1/2}/\lambda$, and λ is the vacuum wavelength. $k^{(-1)}$ will be similarly defined. The covariant components of the electric field vectors in regions ±1 can be represented by Rayleigh expansions as:

$$E_\sigma(r) = I_\sigma e^{i(\alpha_0x^1+\beta_0x^2-\gamma_{00}^{(+1)}x^3)} + \sum_{m,n} R_{\sigma mn}\times e^{i(\alpha_mx^1+\beta_nx^2+\gamma_{mn}^{(+1)}x^3)}, \ (x^3>h), \tag{3.30}$$

$$E_\sigma(r) = \sum_{m,n} T_{\sigma mn}\times e^{i(\alpha_mx^1+\beta_nx^2+\gamma_{mn}^{(-1)}x^3)}, \ (x^3<0), \tag{3.31}$$

where I_σ, $R_{\sigma mn}$, and $T_{\sigma mn}$ are the incident and diffracted field amplitudes,

$$\alpha_m=\alpha_0+mK_1, \qquad \beta_n=\beta_0+nK_2, \tag{3.32}$$

and $\gamma_{mn}^{(\alpha)}$ is to be solved from

$$\sec^2\zeta(\alpha_m^2+\beta_n^2-2\alpha_m\beta_n\sin\zeta)+ \gamma_{mn}^{(\alpha)2}=k^{(\alpha)2}, \quad (\alpha=\pm1) \tag{3.33}$$

The sign of $\gamma_{mn}^{(\alpha)}$ should be chosen so that

$$\mathrm{Re}[\gamma_{mn}^{(\alpha)}]+\mathrm{Im}[\gamma_{mn}^{(\alpha)}]>0, \ (\alpha=\pm1) \tag{3.34}$$

The coefficients α_m, β_n, and $\gamma_{mn}^{(\alpha)}$ are the covariant components of the wave vector of the (m, n)th diffracted order in medium α,

$$\mathbf{k}_{mn}^{(\pm1)}=\alpha_m\mathbf{b}^1+\beta_n\mathbf{b}^2\pm\gamma_{mn}^{(\pm1)}\mathbf{b}^3 \tag{3.35}$$

The projection of tips of the above two k vectors onto the Ox^1x^2 plane for all possible m and n forms the reciprocal lattice of the diffraction orders. In Fig. 3-3, each intersection point of the dotted lines corresponds to a diffraction order. The cross marks the origin of the space that contains the reciprocal lattice (the k space), and the black dot marks the location of the $(0, 0)$th order. The lattice constants along directions $\mathbf{b}^1$ and $\mathbf{b}^2$ are $K_1\sec\zeta$ and $K_2\sec\zeta$, respectively. Each smallest parallelogram formed by the dotted lines is a unit cell of the reciprocal lattice. The set of propagating orders in medium a is given by

$$\mathbf{U}^{(\alpha)} = \{(m, n)|\mathrm{Im}[\gamma_{mn}^{(\alpha)}]=0, \ \forall m, \ \forall n\} \tag{3.35}$$

Graphically, $U^{(a)}$ is represented by the lattice points that falls within the solid circle in Fig. 3-4, whose radius is $k^{(a)}$. It is

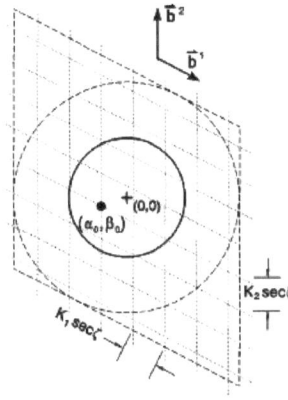

Fig.3-4 Section of the reciprocal lattice. The diffraction orders are located at crossings of the grid lines, which are parallel to the contravariant basis vectors. The solid circle indicates the disk of propagating orders. The dashed circle and parallelogram show two truncation schemes. Reprinted with permission from Lifeng Li, J. Opt. Soc. Am. A 14 (10), 2758-2767 (1997) with copyright of ©1997 Optical Society of America

easy to verify that, when $d_1 d_2 / \lambda_2 \gg 1$, the number of elements in set $U^{(a)}$ is approximately given by

$$\left[\frac{\pi\, k^{(\alpha)2}}{K_1 K_2 / \cos\zeta} \right] = \left[\frac{\pi(d_1 d_2 \cos\zeta)}{\lambda^2 / \varepsilon^{(\alpha)} \mu} \right] \tag{3.36}$$

where $[\cdot]$ means "the integral part of·" Thus the number of propagating orders is proportional to the area of the unit

cell in real space and inversely proportional to that in the k space. We now pause to address one of the advantages of being able to use a nonrectangular coordinate system. In FMM the dimension of the truncated matrix for numerical computation is essentially the number of lattice points that fall within an area in the reciprocal lattice that contains the disk of propagating orders. Since the density of lattice points in k space is proportional to the area of the unit cell in real space, we should choose one of the smallest unit cells in order to minimize the matrix dimension. If a larger unit cell is chosen, then some of the lattice points will correspond to diffraction orders whose amplitudes are identically zero. This point was clearly recognized by the authors of Ref. 6. However, when they analyzed a periodic pattern as sketched in Fig. 3-3(b), they were forced to use the rectangular unit cell, because their formulation was done based on a rectangular coordinate system. By using the parallelogramic unit cell, one can reduce the matrix dimension and the computation time by a factor of 2 and 8, respectively, because the area of the parallelogramic unit cell is half that of the rectangular one. This section concludes with the formula for calculating diffraction efficiencies. If the incident plane wave is normalized such that

$$\frac{1}{\gamma_{00}^{(+1)}} [(k^{(+1)2} - \beta_0^2) |I_1|^2 + (k^{(+1)2} - \alpha_0^2) |I_2|^2 + (\alpha_0 \beta_0 - k^{(+1)2} \sin\zeta)(I_1 \bar{I}_2 + I_2 \bar{I}_1)] = 1, \tag{3.37}$$

then the (m, n)th-order diffraction efficiency in medium α is given by

$$\eta_{mn}^{(\alpha)} = \frac{1}{\gamma_{mn}^{(\alpha)}} [(k^{(\alpha)2} - \beta_n^2) |E_{1mn}^{(\alpha)}|^2 + (k^{(\alpha)2} - \alpha_m^2) |E_{2mn}^{(\alpha)}|^2 + (\alpha_m \beta_n - k^{(\alpha)2} \sin\zeta) \times (E_{1mn}^{(\alpha)} \overline{E}_{2mn}^{(\alpha)} + \overline{E}_{1mn}^{(\alpha)} E_{2mn}^{(\alpha)})] \tag{3.38}$$

where $E_{\alpha mn}^{(+1)} = R_{\sigma mn}$ and $E_{\alpha mn}^{(-1)} = T_{\sigma mn}$.

In the grating layer the electromagnetic field can be expanded into Floquet–Fourier series,

$$\Psi(x^1, x^2, x^3) = \sum_{m,n} \Psi_{mn}(x^3) e^{i\alpha_m x^1 + i\beta_n x^2} \tag{3.39}$$

where Ψ stands for any one of the six components of the electric and magnetic vectors.

After solving the above formulations, the model solution in the grating layer can be found. The model expansion of the total field in the grating layer can be written as

$$E_\sigma(\mathbf{r}) = \sum_{m,n,q} [u_q e^{i\gamma_q x^3} + d_q e^{-ir_q x^3}] \times e^{i(\alpha_m x^1 + \beta_n x^2)} E_{\sigma mnq} \tag{3.40}$$

$$H_\sigma(\mathbf{r}) = \sum_{m,n,q} [u_q e^{i\gamma_q x^3} - d_q e^{-ir_q x^3}] \times e^{i(\alpha_m x^1 + \beta_n x^2)} H_{\sigma mnq} \tag{3.41}$$

where $\sigma = 1, 2$, and u_q and d_q are the unknown amplitudes of the upward and downward propagations or decaying modal fields. If the grating region is approximated by several thin layers, then the field expressions in Eq. (3.39) are valid for each layer. Considering length of the section, we do not give the detailed deducing process for the solving solutions here. More information regarding the solution and numerical calculations can be seen in Ref. 5, 7, and 8.

For the plasmonic structures with metallic binary periodic grooves, RWCA algorithm can be used for transmission calculation on the basis of the dispersion relation of the dielectric media described by Drude model. In this case, only one layer is enough for the theoretical model. For more information, readers can see Chapt. 11 and Ref. 1 in Chapt. 13 of this book.

3.2.2 Mie theory: a brief introduction

Mie theory, also called Lorenz-Mie theory or Lorenz-Mie-Debye theory, is an analytic solution of Maxwell's equations for the scattering of electromagnetic radiation by spherical particles (also called Mie scattering). Mie theory provides rigorous analytic solutions for light scattering by an isotropic sphere embedded in a homogeneous medium. Extensions of Mie theory include solutions for core/shell spheres and gradient-index spheres. Although these theories are restricted to the case of a perfect sphere, the results have provided insight into the scattering and absorption properties for a wide variety of pigment systems, including non-spherical pigments such as elliptical particles. In most paper applications, where TiO_2 concentrations are relatively low (<15% by volume), theoretical calculations predict the relative effects of particle size, particle composition, composition of the surrounding medium and wavelength of light. These trends correlate well with experimental data. In applications with TiO_2 concentrations greater than ~15% by volume, near-field optical interactions between neighboring particles become significant and can dramatically impact macroscopic optical properties. The optical theories applied in the present study describe the light scattering properties of an isolated spherical particle and therefore cannot be applied to systems in which the particles are crowded together and near-field interactions between particles are significant.

The concepts of geometrical optics (refraction by lenses and reflection by mirrors) that are familiar in the macroscopic world do not adequately describe the interactions of particles with light when the particle size is comparable to the wavelength of the light. Rigorous optical theories such as Mie theory addresses the full complexity of vector electromagnetic quantities interacting with a particle. The mathematics of Mie theory is straightforward but tedious, requiring the computation of a potentially large number of series expansions. Digital computers are ideally suited to this task. In the present study, the computer codes BHMIE and BHCOAT provided by Bohren and Huffman have been used to compute the scattering results presented, after slight modification to incorporate computation of the asymmetry parameter. When the light wavelength is similar to the particle diameter, light interacts with the particle over a cross-sectional area larger than the geometric cross section of the particle. The Mie calculation output provides this scattering cross section, C_{sca}. For more information, readers can refer to Ref. 1.

3.2.3 Multiple Multipole Program

The multiple multipole program (MMP) was firstly proposed in 1980 by Christian Hafner.[9] It is a semi-analytic method for numerical field computations that has been applied to electromagnetic fields and to acoustics. Essentially, the field is expanded by a series of basis fields. Each of the basis field is an analytic solution of the field equations within a homogeneous domain. The amplitudes of the basis fields are computed by a generalized point matching technique that is efficient, accurate, and robust. Meanwhile, MMP 'knows' many different sets of basic fields, but multipole fields were still considered to be most useful. Due to its close relations to analytic solutions, MMP is very useful and efficient when accurate and reliable solutions are desired. MMP belongs to a class of boundary discretization methods, also known as generalized multipole technique (GMT). On the basis of some extensions of Mie-Vekua theory, it has close links to the method of moments, the method of auxiliary (or fictitious) sources, and the method of discrete sources etc.

The MMP is a compromise between a pure analytical and a pure numerical approach.[10] It is a well established technique for solving Maxwell equations in arbitrarily shaped, isotropic, linear, and piecewise homogeneous media. In the MMP technique, the infinite space is divided into subdomains D_i. The interface between the individual properties but fictitious boundaries may be defined as well. In every D_i the scalar fields f may be approximated by a series expansion

$$f^{(i)}(\mathbf{r}) \approx \sum_j a_j^{(i)} f_j(\mathbf{r}), \tag{3.42}$$

in which the basic functions f_j cover any of the known analytical solutions of the Helmholtz equation $(\nabla^2 + k^2)f(\mathbf{r}) = 0$. The general solution is obtained by combining the TE and TM solutions. The MMP models a single scatter using multiple multipoles for both the outer and inner domains. Such a multiple multipole approach achieves a better convergence for boundaries deviating considerably from spherical surfaces. Usually several multipoles are set along the boundary of the domain in which the field is expanded. To avoid numerical dependences, the origins have to be sufficiently separated. The highest possible degree and orders of an individual multipole are limited by a spatial sampling criterion that depends on the boundary discretization and on the proximity of the multipole to the boundary. If the result is not accurate enough, additional or more appropriate basic functions have to be included in the series expansion Eq. (3.42). For more information, please read Ref. 11.

3.2.4 DDA algorithm

Efforts to infer the size, morphology, and composition of solid particles in general rely on comparison between observations of extinction, absorption, scattering, or emission of electromagnetic radiation by the particles. However, current ability to theoretically compute scattering and absorption cross sections for solid particles is extremely limited. If the particles are spherical, homogeneous, and isotropic, the rigorous solutions obtained by Mie scattering theory can be employed to compute any desired optical property of the grain provided only that the sphere is not extremely large compared to the wavelength. Sometimes, optical properties of non-spherical particles are investigated. Moreover, some candidate grain materials have anisotropic optical properties. For these particles Mie theory would not be applicable even if the particles were to be homogeneous spheres. For non-spherical particles, rigorous analytic scattering solutions exist only for special cases: particles are very small compared to the wavelength, infinite cylinders and spheroids. The interested readers can referred to the book by Bohren and Huffman [12] for a review of these solutions. The "T-matrix" technique [13] involves an exact expansion of the exterior radiation field in terms of spherical harmonics; the expansion is truncated, and the expansion coefficient is determined numerically by casting the scattering problem in integral form. While the method can in principle be applied to homogeneous particles in arbitrary shape and size. It appears that considerable effort is involved in modifying the treatment for each new shape.

Purcell *et. al.* developed the "discrete dipole approximation" (DDA) algorithm, a very flexible and general technique for calculating the optical properties of particles of arbitrary shapes.[14] The DDA replaces the solid particles by an array of N point dipoles, with the spacing among the dipoles small compared to the wavelength. Each dipole has an oscillating polarization in response to both an incident plane wave and the electric fields due to all of the other dipoles in the array; the self-consistent solution for the dipole polarizations can be obtained as the solution to a set of coupled linear equations. They presented solutions with $N \leq 256$; and reported that under some circumstances they encountered difficulties with convergence of the interative methods employed to obtain solutions. Considering this, Draine et. al.

gave some corrections for radiative reaction which were incorporated into the formalism. Formulae used to extract the desired cross sections from a discrete-dipole solution were presented in Ref. 15. DDA has the following validity criteria.

(1) Granularity of surface

On the basis of a limited number of numerical calculations for spheres, it can be estimated that in order to attain a fractional error less than Δ (in the zero-frequency limit) the particle should satisfy the criterion $N>N_{\min}\approx60|m-1|^3(\Delta/0.1)^{-3}$. This estimates for $N_{\min}$ applies to spheres; $N_{\min}$ for other convex shapes will differ by factors of order unity. The strong dependence on $|m-1|$ implies that discrete dipole calculations for the materials with large refractive indices must either accept relatively large fractional errors Δ, or else employ very large values of N.

(2) Granularity versus wavelength and skin depth

It is a second necessary condition for the dipole array to provide an accurate representation of a homogeneous grain. The condition is that the length scale for variation of the electric field within the grain must be large compared to the nearest-neighbor distance between two adjacent dipoles d ($d=(4\pi/3N)^{1/3}a_{eq.}$, $a_{eq.}=(3V/4\pi)^{1/3}$, where N is grains number, V the volume of a grain. For specified $ka_{eq.}$ and refractive index, it is useful to have a criterion with which to estimate how large N must be in order to provide an accurate approximation for the discrete dipole model. The validity criterion depends on the complex refractive index m of the materials. One necessary condition is that the small d is not just compared to the wavelength in vacuum $2\pi/k$ but also with the wavelength $2\pi/\mathrm{Im}(m)k$ for the electromagnetic wave. A simple criterion suggests itself: $kd|m|<\beta$, where β is some constants of order unities; obviously the value of β will depends on the desired degree of calculation accuracy.

(3) Granularity: neglect of magnetic dipole absorption

The above criterion is necessary, but not sufficient. The discrete dipole method replaces a cubic volume of grain material, of volume d^3, by a point electric dipole-magnetic dipole effects which are neglected. Even for non-magnetic materials, the induced magnetic moment associated with volume d^3 of the grain material may not be negligible if the material is sufficiently conductive. In particular, "magnetic-dipole" absorption may be comparable to, or greater than the 'electric dipole" absorption. To estimate the relative importance of magnetic dipole absorption, consider a spherical volume, of radius $r=(3/4\pi)^{1/3}d$, having the same volume as the unit cell. For this sphere the ratio of magnetic dipole absorption to electric dipole absorption is

$$\frac{C_{abs}^{m}}{C_{abs}^{e}}\approx\frac{(kr)^2}{90}[(\varepsilon_1+2)^2+\varepsilon_2] \tag{3.43}$$

where ε_1 and ε_2 is the dielectric constants of real part and imaginary part, respectively. If we require $C_{abs}^{m}/C_{abs}^{e}<\Delta$, one obtains the condition $N\geq(ka_{eq})^3|m|^6\Delta^{-3/2}/90$, where we have replaced $(\varepsilon_1+2)^2+\varepsilon_2^2$ by $|m|^4$, since magnetic dipole effects are important only for the case of $|\varepsilon|>>1$.

The DDA is a flexible terming of radiation by the particles in arbitrary shape. The method has been extended to incorporate the effects of radiative reaction and allow for possible anisotropy of the dielectric tensor of the materials,

3.3 Dynamic analyzing approaches

3.3.1 FDTD algorithm

The finite-difference time-domain (FDTD) method [16, 17] has recently become a start-of-the-art method for solving Maxwell's equations in complex geometries. Being a direct time and space solution, it offers the user a unique insight into all types of problems in electromagnetics and photonics. It is necessary to use FDTD algorithm when feature sizes are on the order of a wavelength. In addition, FDTD can also obtain the frequency solution by exploiting Fourier transforms, thus a full range of useful quantities can be calculated, such as the complex Poynting vector and the transmission / reflection of light. This section will introduce the basic mathematical and physics formalism behind the FDTD algorithm, starting from the linear Maxwell's equations. The simulator can be used for advanced research and development or as an ideal teaching and learning environment in photonics, optics, and electromagnetics.

Finite-difference and time-domain (FDTD) algorithm treats Maxwell equations as a set of finite difference equations in both time and space. The model space considered here includes both the probe and the sample surface and consists of an aggregation of cubic cells with each cell having its own complex dielectric constant. The finite difference equations can be written as:

$$\frac{H_{z(t;x,y+\Delta y,z)}-H_{z(t;x,y-\Delta y,z)}}{2\Delta y}-\frac{H_{y(t;x,y,z+\Delta z)}-H_{y(t;x,y,z-\Delta z)}}{2\Delta z}=\tilde{\varepsilon}(x,y,z)\frac{E_{x(t+\Delta t;x,y,z)}-E_{x(t-\Delta t;x,y,z)}}{\Delta t},$$

$$\frac{H_{x(t;x,y,z+\Delta z)}-H_{x(t;x,y,z-\Delta z)}}{2\Delta z}-\frac{H_{z((t;x+\Delta x,y,z)}-H_{z(t;x-\Delta x,y,z)}}{2\Delta x}=\tilde{\varepsilon}(x,y,z)\frac{E_{y(t+\Delta t;x,y,z)}-E_{y(t-\Delta t;x,y,z)}}{\Delta t}, \qquad (3\text{-}44)$$

$$\frac{H_{y(t;x+\Delta x,y,z)}-H_{y(t;x-\Delta x,y,z)}}{2\Delta x}-\frac{H_{x(t;x,y+\Delta y,z)}-H_{x(t;x,y-\Delta y,z)}}{2\Delta y}=\tilde{\varepsilon}(x,y,z)\frac{E_{z(t\_\Delta t;x,y,z)}-E_{z(t-\Delta t;x,y,z)}}{\Delta t}.$$

where $\mathbf{E}=\mathbf{E}(E_x,E_y,E_z)$ and $\mathbf{H}=\mathbf{H}(H_x,H_y,H_z)$ are the electric field and the magnetic induction vectors, respectively, and $2\Delta x$, $2\Delta y$, $2\Delta z$ are increments along the three coordinate directions, respectively; Δt is the unit time increment, and

$\tilde{\varepsilon}$ (x, y, z) is the complex dielectric constant of the medium at that point. Equation (3.44) can be simultaneously solved to determine the component values at the time $t+\Delta t$.

In perfect conductor, the scattered field may be written as [18]

$$\frac{\varepsilon}{\sigma}\frac{\partial E^{scat}}{\partial t}=-E^{scat}-E^{inc}-\frac{(\varepsilon-\varepsilon_0)}{\sigma}\frac{\partial E^{inc}}{\partial t}+\frac{1}{\sigma}(\nabla\times H^{scat}) \qquad (345)$$

We now difference the free space scattered field equations. In essence, finite differencing replaces derivatives differences:

$$\frac{\partial f}{\partial t}=\lim_{\Delta t\to 0}\frac{f(x,t_2)-f(x,t_1)}{\Delta t}\approx\frac{f(x,t_2)-f(x,t_1)}{\Delta t} \qquad (3.46)$$

$$\frac{\partial f}{\partial x}=\lim_{\Delta x\to 0}\frac{f(x_2,t)-f(x_1,t)}{\Delta x}\approx\frac{f(x_2,t)-f(x_1,t)}{\Delta x} \qquad (3.47)$$

where Δt and Δx are finite rather than infinitesimal in the above approximation. In short, calculus becomes algebra. Some critical issues aside from this algebraic replacement include:

(1) What form the differencing takes: we use an explicit central difference scheme here that only retains first order term. The $\mathbf{E}$ and $\mathbf{H}$ fields are interleaved spatially and temporally because of the central differencing. What results is often referred to a "leapfrog" scheme.

(2) Stability: only for Δt given by the Courant stability condition, $\Delta t\leq(\Delta x)/c\sqrt{3}$ for cubical cells, is the formulation stable.

To start FDTD algorithm, $\mathbf{E}$ and $\mathbf{H}$ are discretized in both time and space. For the time discrete, $\mathbf{E}(t)\to\mathbf{E}^{n\Delta t}$, and $\mathbf{H}(t)\to\mathbf{H}^{(n+1/2)\Delta t}$. The basic FDTD time-stepping relation is

$$\boldsymbol{E}^{n+1}=\boldsymbol{E}^{n}+\alpha\nabla\times\boldsymbol{H}^{n+1/2} \qquad (3.48)$$

$$\boldsymbol{H}^{n+3/2}=\boldsymbol{H}^{n+1/2}+\beta\nabla\times\boldsymbol{E}^{n+1} \qquad (3.49)$$

The time sequence follows the orders of $\mathbf{E}^0\to\mathbf{H}^{1/2}\to\mathbf{E}^1\to\mathbf{H}^{3/2}\to\ldots$. For instance, for the 2nd order accurate in time, the corresponding error is $\sim\Delta t^2$. For the space discrete, the analyzed object volume space is discretized by means of dividing numerous unit cells, called Yee cell.[19] The above notation shows $\mathbf{E}$ at time corresponding to $n=N$ updated from its prior value at time $n=N-1$ and the curl of $\mathbf{H}$ at time $n=N-1/2$. Next $\mathbf{H}$ is evaluated at $n=N+1/2$ from its earlier value at $n=N-1/2$ and the curl of $\mathbf{E}$ at $n=N$. This interleaves $\mathbf{E}$ and $\mathbf{H}$ temporally and results in a centered difference or approach of "leapfrog in time". After each update of $\mathbf{E}$ and $\mathbf{H}$, the index N is increased by 1 and the process is repeated.

The spatial indices in the curl calculations are determined by the Yee cell geometry. The curl calculations are also center differenced, and represent the nearest-neighbour interactions. It spatially staggers the vector components of the **E**-field and **H**-field about rectangular unit cells of a Cartesian computational grid. The total simulation time can be estimated as $\sim V(\lambda/\mathrm{d}x)^4$, and $\sim A(\lambda/\mathrm{d}x)^3$ for 3D and 2D simulation respectively, whereas V and A is the volume and area of the simulated objects, respectively. Total memory requirements for computers can be estimated as $\sim V(\lambda/\mathrm{d}x)^3$, and $\sim A(\lambda/\mathrm{d}x)^2$ for 3D and 2D simulation respectively.

Here we introduced a commercial professional software: FDTD Solution [20]. It can solve two and three dimensional Maxwell equations in both linear and non-linear dispersive media, where the user can specify arbitrary geometric structures and various input excitation sources. The two dimensional FDTD simulator solves the TE and/or TM Maxwell equations. FDTD is a time domain technique, meaning that the electromagnetic fields are solved as a function of time. In general, FDTD Solution is commonly used to calculate the electromagnetic fields as a function of frequency or wavelength by means of performing Fourier transforms during the computational numerical simulation. This allows it to obtain complex-valued fields and other derived quantities such as the complex Poynting vector, normalized transmission, and far field projections as a function of frequency or wavelength. The field information can be returned in two different normalization states. Dispersive materials with tabulated refractive index (n, k) data as a function of wavelength can be solved using the multi-coefficient model with auto-fitting. Alternatively, specific theoretical models such as Plasma (Drude), Debye or Lorentz can be used. Boundary conditions are very important in electromagnetics and simulation techniques. FDTD Solution supports a range of boundary conditions, such as PML, periodic, and Bloch. Sources are another important component of a numerical simulation. FDTD Solution supports a number of different types of sources such as point dipoles, beams, plane waves, a total-field scattered-field (TFSF) source, a guided-mode source for integrated optical components, the light source with different polarization states (*e.g.*, linear/circuar/elliptical/radial polarizations, and vector cylindrical beam) and an imported source to interface with external photonic design software as well as incoherent light source.

The advantages of FDTD Solution can be summarized as its ability to work within a wide range of frequencies, stimuli, objects, environments, response locations, and computers. To further list, it can be added that the advantages of computational efficiency is an important issue in comparison to other techniques such as the method of moments, especially when broadband results are required. Furthermore, the FDTD code, while inherently volumetric, has successfully treated thin plates and thin wire antennas. Its accuracy, using a sufficiency of cells, can be made as high as desired. Conversely, engineering estimates of a few decibels' accuracy can be made with surprisingly few cells. Finally, powerful visualization tools are being developed to enhance the user's understanding of the essential physics underlying the various processes that FDTD can model, simulate, and analyze. Presently, FDTD algorithm is a crucial tool for simulating and analyzing optical behaviours of the plasmonic structures and nanophotonic devices at both near- and far-field regions.

Currently, there are several commercial softwares for the FDTD algorithm-based electromagnetic field analysis: FDTD Solution, XFDTD, OptiFDTD, CST, and Ansoft/Rsoft. Amongst, FDTD Solution (from *Lumerical Inc.* in Canada) is the most commonly used software in near-field optics and nanophotonics because of its unique characteristics and most specific functions in these areas.[20]

For more information regarding FDTD algorithm, readers can see Ref. 17.

3.3.2 Beam propagation method

Beam Propagation Method (BPM) is one of the modules: OlympIOs which offers for simulating the propagation of light through planar waveguide structures. For contrast from low to medium index, the BPM module allows fast and accurate 2D and 3D simulations. Users can define waveguide structures of their choice using the extensive optical element library. The library can be further extended to include user defined, custom elements (external elements). Complete parameterization means various design options can be analyzed quickly and automatically. Once the basic waveguide structure is optimized, the basic waveguide component is easily incorporated within a complete mask layout, as the mask module uses the same method of structure definition. It has the following features:

1). 2D and 3D BPM simulations

2). Wide angle of 2D-BPM (2nd and 4th order Padé approximation)

3). Novel 5-points BPM (less sensitive to lateral discretization)

4). Field analyses and manipulation capabilities

5). Second order non-linearity

BPM is generally formulated as a solution to Helmholtz equation in a time-harmonic case, [21]

$$(\nabla^2 + k_0^2 n^2)\psi = 0 \tag{3.50}$$

with the field written as, $E(x,y,z,t) = \psi(x,y,x)exp(-j\omega t)$. Now the spatial dependence of this field is written according to any one TE or TM mode polarization as $\psi(x,y) = A(x,y)exp(+jk_o vy)$, with the envelope $A(x,y)$ following a slowly varying approximation,

$$\frac{\partial^2(A(x,y))}{\partial y^2} = 0 \tag{3.51}$$

Now the solution when replaced into the Helmholtz equation follows,

$$\left[\frac{\partial^2}{\partial x^2} + k_0^2(n^2 - v^2)\right] A(x,y) = \pm 2jk_0 v \frac{\partial A_k(x,z)}{\partial z} \tag{3.52}$$

With the aim to calculate the field at all points of space for all times, we only need to compute the function $A(x,y)$ for all space, and then we are able to reconstruct $\psi(x,y,z)$. Since the solution is for the time-harmonic Helmholtz equation, we only need to calculate it over one time period. We can visualize the fields along the propagation direction, or the cross section waveguide modes.

The master equation is discretized using various centralized difference, crank nicholson scheme etc. and rearranged in a causal fashion. Through iteration the field evolution is computed, along the propagation direction.

BPM is a quick and easy method for solving fields in integrated optical devices. It is typically used only in solving for intensity and modes within shaped (bent, tapered, and terminated) waveguide structures, as opposed to scattering problems. It is not suited as a generalized solution to Maxwell equations, like the FDTD or FEM methods.

3.4 Summary

Approaches for electromagnetic analysis: stationary analysis and dynamic analysis were illustrated respectively in this chapter. To give readers an outline regaring the relevant numerical methods in subwavelength optics, we collected the relevant analysis methods and numerical algorithms as more as possible. Readers will have a complete and clear frame regarding subwavelength optics from theory, concepts, to fabrication, characterization and applications after reading this book. Considering scope of this book, only brief introduction was given for each approach and algorithm. Currently, there are numerous relevant books regarding electromagnetic field analysis and Maxwell equations. And many technical papers discussed various relevant computational numerical algorithms. Thus for more detailed information regarding a certain analysis method or algorithm, the readers can refer to the relevant books and reference papers.

Reference

[1] Wax Born and Emil Wolf, *Principle of Optics*, (7th Edition), Cambridge University Press/Pergamon Press, 1980.

[2] M.G.M.M.v.Kraaij. A more Rigorous Coupled-Wave Analysis (MSolver). Master's thesis, Technische Universiteit Eindhoven, 2004.

[3] Lifeng Li. "Use of Fourier series in the analysis of discontinuous periodic structures". J. Opt. Soc. Am. A, 13(9):1870-1876 (1996).

[4] M. G. Moharam, Drew A. Pommet, and Eric B. Grann. "Stable implementation of the rigorous coupled-wave analysis for surface-relief gratings: enhanced transmittance matrix approach". J. Opt. Soc. Am. A, 12(5):1077-1086 (1995).

[5] Lifeng Li, "New formulation of the Fourier model method for crossed surface-relief gratings", J. Opt. Soc. Am. A 14, 2758-2767

(1997).

[6]   E. Noponen and J. Turunen, ''Eigenmode method for electromagnetic synthesis of diffractive elements with three-dimensional profiles,'' J. Opt. Soc. Am. A 11, 2494–2502 (1994).

[7]   Lifeng Li, "Use of Fourier series in the analysis of discontinuous periodic structures", J. Opt. Soc. Am. A 13, 1870-1876 (1996).

[8]   Lifeng Li, Formulation and comparison of two recursive matrix algorithms for modeling layered diffraction gratings, J. Opt. Soc. Am. A 13, 1024-1035 (1996).

[9]   C. Hafner and R. Ballisti, "The multiple multipole method (MMP)", *COMPEL - The International Journal for Computation in Electrical and Electronic Engineering*, 2(1), 1983.

[10] C. Hafner, *The Generalized Multiple Multipole Technique for Computational Electromagnetics*, Boston: Artech (1990).

[11] Lukas Novotny and Bert Hecht, *Principles of Nano-Optics*, Cambridge University Press, 2007, Chapter 15.

[12] Bohren, C.F., and Huffman, D.R., 1983, *Absorption and Scattering of light by Small Particles* (New York: Wiley).

[13] Geller, P E; Tsuei, T G; Barber, P W , "Information content of the scattering matrix for spheroidal particles", Applied Optics 24, pp.2391-2396 (1985).

[14] Purcell,E.M., and Pennypacker,C.R., Ap. J., 186, 705 (1973).

[15] Bruce T. Draine, "The discrete dipole approximation and its application to interstellar graphite grains", The Astrophysical Journal 333, 848-872 (1988).

[16] Dennis M. Sullivan, *Electromagnetic Simulation Using The FDTD Method*. New York: IEEE Press Series, (2000).

[17] Allen Taflove, *Computational Electromagnetics: The Finite-Difference Time-Domain Method*. Boston: Artech House, 2005.

[18] Karls Kunz, Raymond J. Luebbers, *The Finite Difference and Time Domain for Electromagnetic*, CRC press LLC, 1993.

[19] Kane Yee, "Numerical solution of initial boundary value problems involving Maxwell's equations in isotropic media". *Antennas and Propagation, IEEE Transactions on* 14: 302–307 (1966).

[20] Professional commercial software for FDTD algorithm from Lumerical Inc. in Canada with website http://www.lumerical.com.

[21] Okamoto K., *Fundamentals of Optical Waveguides*. San Diego, CA: Academic, 2000.

4 NEAR-FIELD SCANNING OPTICAL MICROSCOPE AND APPLICATIONS

Abstract: Literature review of near-field scanning optical microscopy was given firstly. Then introduction of the near-field probing and corresponding systems were presented in detail. Finally, some typical applications of the near-field probing systems in different areas were described briefly combining with some concrete examples.

Near-field scanning optical microscope (NSOM), also named scanning near-field optical microscope (SNOM), appeared at end of last century, is a revolution in optics because it is a breakthrough of conventional diffraction limit. By this approach, a spatial resolution reaches to nanoscale level was realized. Its working principle follows technique of scanning probe microscope such as atomic force microscope (AFM) and scanning tunneling microscope (STM). Hence it provides one more option for the metrology in nanoscale level after AFM and STM, especially for optical characterization of nanophotonic devices, life cell/molecular, and light emission devices. Photoluminescent properties with tiny structures and localized in small area can be easily investigated experimentally using the NSOM method. Benefitting from invention of the NSOM, nanosciences and nanotechnologies including life science, nano-optics, nano-materials, semiconductor, nanomedicine, and nano-mechanical-electronic system (NEMS) have rapidly developed since its appearance. A brief outline about NSOM is given in this chapter for the purpose of fast understanding the basic working principle of NSOM and its main applications after a quick browse in a short period for the readers interested in the near-field probing technique.

4.1 Literature review of NSOM/SNOM

E.H. Synge proposed the idea of using a small aperture to image a surface with sub-wavelength resolution using optical light in 1928 and 1932.[1, 2] For the small opening, he suggested using either a pinhole in a metal plate or a quartz cone that is coated with a metal except for at the tip. He discussed his theories with A. Einstein, who helped him developing his ideas. In 1956, J.A. O'Keefe, a mathematician, proposed the concept of near-field microscopy without knowing about Synge's earlier papers. However, he recognized the practical difficulties of near field microscopy and wrote the following statements about his proposal: "the realization of this proposal is rather remote, because of the difficulty providing for relative motion between the pinhole and the object, when the object must be brought so close to the pinhole". In the same year, Baez performed an experiment that acoustically demonstrated the principle of near field imaging. At a frequency of 2.4 kHz ($\lambda = 14$ cm),[3] he showed that an object (his finger) smaller than the wavelength of the sound can be resolved. In 1972, E.A. Ash and G. Nichols demonstrated $\lambda / 60$ resolution in a scanning near field microwave microscope using 3 cm radiation.[4] In 1984, The first paper on the application of NSOM/SNOM appeared. This paper is the first article showing that NSOM/SNOM is a practical possibiltity, spurring the growth of this new scientific field.[5] In the early 1870s, Ernst Abbe formulated a rigorous criterion for being able to resolve two objects in a light microscope: $d > \lambda / (2\sin\theta)$, where d is the distance between the two neighbored objects, λ the wavelength of the incident light, and 2θ the angle through which the light is collected. According to this equation, the best resolution achievable with optical light is about 200 nm. With the introduction of NSOM, this limitation no longer exists, and optical resolution of less than 50 nm can be achieved.

Schematic diagram of Fig. 4-1 shows the basic principle of near-field optics: light passes through a sub-wavelength diameter aperture and illuminates a sample that is placed within its near field, at a distance much less than the wavelength of the incident light. The resolution achieved is far better than that which conventional optical microscopes can attain.

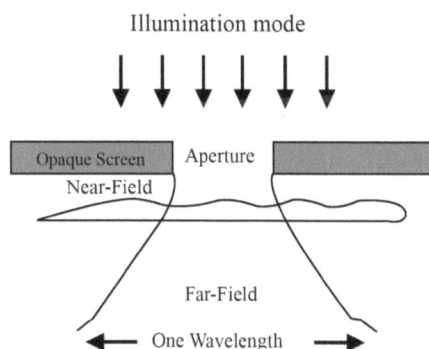

Fig.4-1 Schematic diagram of fiber probe scanning for optical intensity distribution ranging from near-field to far-field.

4.2 The Basic Setup of NSOM/SNOM
4.2.1 Basic system setup

In order to make an NSOM/SNOM experiment, a point light source must be brought near the surface that will be imaged (within nanometers). The point light source must then be scanned over the surface, without touching it, and the optical signal from the surface must be collected and detected.

1) There are a few different ways to obtain a point light source:

- One can use pulled or etched optical fibers (tapered optical fibers) that are coated with a metal thin film (~100 nm in thickness) except for at an aperture at the fiber's tip. The light is coupled into the fiber and is then emitted at the sub-wavelength (50 nm or larger) aperture of the fiber.
- Or, one can use a standard AFM cantilever with a hole in the center of the pyramidal tip. A laser is focused onto this hole, which is the dimension in sub-wavelength scale.
- Finally, the tip of a tapered pipette can be filled with a light emitting compound, which can then be excited either by light or by applying a voltage. It is also possible to use chemical luminescence.

The resolution of a NSOM/SNOM measurement is defined by the size of the point light source used (typically 50-100 nm).

2) The distance between the point light source and the sample surface is usually controlled through a feedback mechanism that is unrelated to the NSOM/SNOM signal. Currently, most instruments use one of the following two types of feedback:

- Normal force feedback (the standard feedback mode used in AFM), which enables one to perform experiments in contact and intermittent contact mode. This feedback mechanism is only possible with the cantilevered, tapered optical fibers and the AFM cantilevers with holes.
- Shear force feedback, or tuning fork feedback (most commonly used approach in NSOM systems). The straight tip is mounted to a tuning fork, which is then oscillated at its resonance frequency. The amplitude of this oscillation is strongly dependent on the tip-surface distance, and it can be effectively used as a feedback signal. Shear force imaging is not understood very well, and there are a lot of artifacts in the topographical images that one obtains using the method.

3) There are four possible modes of operation with NSOM/SNOM systems, as shown in Fig. 4-2:

- Transmission mode imaging: the sample is illuminated through the probe, and the light passing through the sample is collected and detected.
- Reflection mode imaging: the sample is illuminated through the probe, and the light reflected from the sample surface is collected and detected.
- Collection mode imaging: the sample is illuminated with a macroscopic light source from the top or bottom, and the probe is used to collect the light from the sample surface.
- Illumination/collection mode imaging: the probe is used for both the illumination of the sample and for the collection of the reflected signal.

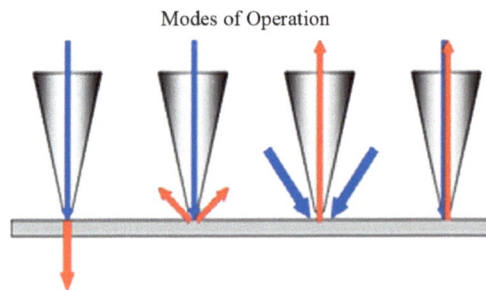

Fig.4-2 Four different working modes of the NSOM system (from left to right): transmission mode, reflection mode, collection mode, and illumination mode.

Detecting the collected light can be achieved with a wide variety of instruments: an avalanche photo diode (APD), a photomultiplier tube (PMT), a CCD, or a spectrometer. The signals obtained by these detectors are then used to create a NSOM/SNOM image from the surface. In NSOM/SNOM, changes in light intensity are usually used to create the image. However, it is also possible to use changes in the polarization of the light as a contrast mechanism or even the dependence of the light intensity from the wavelength used to illuminate. There are several properties of a sample that can give contrast in the NSOM/SNOM image:

- Changes in the index of refraction
- Changes in the reflectivity
- Changes in the transparency
- Changes in polarization
- Stress at certain points of the sample that changes its optical properties
- Magnetic properties, which can change the optical properties
- Fluorescent molecules
- Molecules excited through a Raman shift, SHG, or other effects
- Changes in the material

Essentially, the NSOM systems can capture the messages with optical resolution beyond diffraction limit because it collects the optical signals at near-field region instead of far-field region. The evanescence wave involving high spatial frequency is collected by the objective lens at near-field before it decays away. Because the signal is a tiny optical point only, line scanning is necessary to obtain more signals involving the high spatial frequency message in a larger area *e.g.*, several square microns. Then by aid of a technique of computational data processing, optical intensity distribution in a 3D volume can be reconstructed and presented in the forms of both 2D profiles and 3D topographies. Moreover, the intensity distribution in a 3D volume in free space can be realized by means of multilayer scanning and reconstruction of the multiple single layer images together through post data processing. The approach is similar to that of the 3D reconstruction of a confocal optical microscope.

4.2.2 Fabrication of NSOM Probes

The basis of the NSOM is short-range electromagnetism between two antennas, a probe antenna, which is much smaller than the wavelength of the driving field. It is apparent that fabricating and manipulating small antennas are the most important factors in the successful development of a NSOM system. One of the most realistic and commonly used methods for preparing a small antenna is sharpening an optical fiber to a very small apex. By employing the scanning technique already established in a scanning tunneling microscope (STM) and an atomic force microscope (AFM), the antenna at the apex of a sharpened fiber works as a probe on a small surface under precise distance control. [10]

A conventional optical fiber is sharpened by chemical etching in buffered hydrofluoric acid. An apex (diameter $d=2a$), which is the antenna of the fiber probe, can easily be made small, and a minimum apex size of only a few nanometers has already been achieved. In NSOM, only the apex works as an antenna, which interacts electromagnetically with the sample. The tapered part has a simple conical shape, and the cone angle θ is clearly defined. The apex radius and the cone angle can be varied by controlling the etching conditions. The exterior surface of the probe is coated with opaque metal, such as aluminum or gold in thickness of ~100 nm, to avoid the illumination of excitatory light or the detection of the background signal through the sidewall of the tapered part. The very end of the metal coating is removed to allow the sharpened glass part, including the apex, to protrude. The small aperture with a diameter (also called the foot diameter) makes an important contribution on the probing results.

The most crucial part of the NSOM probe is a subwavelength sized aperture perforated at the apex of the tip. The aperture can be naturally formed during metal film coating meanwhile rotating the probe, or drilled using the technique of focused ion beam directly milling. Currently, the tapered optical fiber probes are most widely used in NSOM systems. Although improvements have been achieved in fabricating such an optical fiber NSOM probe, problems still remain at present. The fiber tip is very fragile. The shape of the tip and the size of the metallic aperture at the fiber tip are not reproducible. The opening angle of the fiber tip is small. Therefore, most of the light is absorbed on the metal wall, which leads to low optical transmission efficiency. Contrarily, a Si-based NSOM probe can be easily fabricated in a batch process. It is easy to combine the microfabricated NSOM probe with a well-developed AFM as an AFM/NSOM combination system. [11, 12] Several technological approaches have been developed to create the aperture at the apex of the tip such as coating and selective etching metal at the apex of a Si_3N_4 tip, [13] metal molding with an aperture at the pyramidal etched pit on an Si cantilever, [14] creating the metallic aperture at the apex of a metal coated SiO_2 tip on a Si cantilever by field evaporation, [15] and low temperature oxidation and a selective etching technique [16]. In this section, another approach, focused ion beam (FIB) milling, for directly fabricate the aperture on the apex of the fiber is presented in detail. Undoubtedly, this is an effective way for which no other alternative approach can be found so far.

In the past decade, NSOM has successfully demonstrated its capability of subwavelength optical resolution by breaking the diffraction limit of conventional optical microscopy.[17,18] The optical near field of a source exists within a distance of less than one optical wavelength from it. By following Synge's [19] original suggestion, involving the use of a tiny aperture (diameter $<<\lambda$) in a metal screen that is placed in near-field proximity to a sample, optical microscopy can achieve a resolving power below the classical Abbe´ limit. Practical implementation of a tiny aperture in a conductive screen is accomplished by tapering down an optical fiber to subwavelength diameter using the heating–pulling technique followed by metallic over-coating of the sides of the fiber to define an aperture. The major drawback of pulled the near-field tips is their low optical throughput, which is a key parameter influencing their better utilization in applications such as optical spectroscopy,[20] nanofabrication,[21] and high-density data storage.[22] To overcome this problem, many researchers [23–26] have attempted to fabricate NSOM probes of a desired shape and with higher throughput. For example, Zeisel *et al.*[27] used a chemical etching technique similar to that developed earlier by Pangaribuan *et al.*[24] for a photon scanning tunneling microscopy. The resultant fiber tips had a much shorter taper region and higher cone angle; they provided a three orders of magnitude higher throughput than normally obtained with pulled fiber probes. In another experiment, Islam and co-workers fabricated hybrid pull-etch tapered probes with throughput four orders of magnitude higher than that of the pulled fibers. The process of applying a metallic overcoat to the tip is an equally important aspect of fabricating high-resolution NSOM probes.[20] Through the addition of an opaque metallic coating such as aluminum that covers the sidewalls of the tapered fiber, a tiny aperture at the end of the fiber is created. The usual process for coating the tip is through vacuum deposition of aluminum by thermal evaporation while the tip is rotated and held at an angle to the source. The exposed glass fiber end resulting in this manner is generally located inside the metallic aperture, is partially obstructed by rough aluminum grains situated near the aperture boundary, and often contains aluminum protrusions, as evidenced by numerous scanning electron microscope (SEM) images of fiber tips.[28,29] This may contribute to the poor polarization capabilities of the majority of fiber tips prepared by angled evaporation.[30] A distance larger than optimal separation between the fiber tip and sample and the absence of a clean opening results in lower throughput of the probe and poor optical resolution of the NSOM. Furthermore, the angle-evaporation procedure is very time consuming and may not result in the fabrication of a batch of reproducible and well-defined apertures of the desired diameter. In addition, other problems associated with the angle evaporation technique are the fabrication of a special holder to accommodate many fibers with adjustable tilt angle, the loading of the probes, rotation of the tips about their axis during the coating operation, and keeping the fibers untangled throughout the coating procedure. It is the aim of this work to develop a method for fabricating the clean, well-defined, and highly reproducible subwavelength apertures for NSOM applications using a FIB milling technique. Fiber probes were produced through the use of a heating–pulling process and by adjusting the parameters described by Valaskovic *et al.*[20] for obtaining probes of optimum shapes and throughputs. The pulled fiber probes were then uniformly coated all over with a 100–150 nm aluminum film in batches of 10–15 fibers with a sputtering unit (Edwards Xenosput XE200). Usually, inspection of the metal-coated probe specimens is performed by scanning electron microscopy (SEM). The coating is smooth, and exhibits no leakage or pin holes upon injection of laser light into such fiber probes. In order to use the tips in a near-field microscope, a subwavelength opening at the apex of the probe must be formed. We have used FIB milling to fabricate an aperture of desired diameters at the very end of the metal coated fibers. Figure 4-3 is an example for probe drilling using FIB milling. A FIB system used from FEI company with model of Dualbeam 620 FIB/SEM workstation (30 KeV Ga$^+$ ions, minimum spot size 5 nm) was employed. The electron beam of the system images the profile and diameter of each individual tip in the batch containing 5–10 fibers positioned inside the chamber. To remove the metal coating and thus fabricate the aperture, the ion beam was raster scanned in a rectangular pattern at the apex of the fiber, as represented by the shaded region in Fig. 4-3. In contrast to simply cutting off the end of the

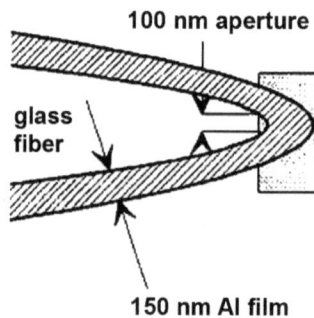

Fig.4-3 Schematic diagram of the FIB milling procedure to produce a near-field aperture. The rectangular shaded region represents the area in which the ion-beam raster scans over the tip from right to left in order to remove material cleanly. Reprinted with permission from Saeed Pilevar *et. al.*, Appl. Phys. Lett. 72, 3133 (1998) with Copyright ©1998 of American Institute of Physics.

fiber in one sweep with the FIB, which will result in redeposition of material onto the aperture, this procedure completely removed the material at the end of the fiber. By varying the size of the rectangular scanned region and with knowledge of the coating thickness and fiber end diameter, we can precisely make a clean-cut aperture with the desired diameter and optical throughput suitable for a variety of NSOM applications. Figure 4-4 (a) shows a SEM image of a tapered fiber probe coated with 150 nm in thickness of aluminum. The aperture is then formed with the aforementioned FIB milling procedure creating an extremely well-defined 100 nm diameter opening, as shown in Fig. 4-4 (b). Figure 4-4 (c) shows another aluminized fiber tip milled by the FIB to make a 300 nm diameter aperture. The entire FIB milling time for each tip is between 2 and 3 min. As revealed in Fig. 4-5 (a) and (b), the fabricated near-field apertures are, the best-defined and clean-cut ones ever fabricated. In contrast to the angled metal evaporation, the FIB milling procedure removes any rough aluminum grains at the tip end surrounding the aperture.

The advantages of this technique are twofold: the optical throughput is improved and the aperture can give the true near-field optical contrast of a sample without the influence of tip protrusions. The limitations of fabrication technique of the FIB tip aperture involve the coating uniformity, tip geometry, tip charging under ion imaging, and the ion-beam profile. An RF sputterer was used to achieve a smooth and uniform coating over the tip so that for a known coating thickness, one can cut the aperture to a desired diameter. Clearly, non-uniformity of the coating can lead to errors in the desired aperture diameter. Tip geometry is also a factor since a high aspect ratio tip will allow for more precise control of the final aperture diameter (this is especially important for fabrication of the apertures smaller than 100 nm). A tradeoff must be struck, however, because the tips with high aspect ratio have fewer throughputs because the evanescent wave propagates through a longer region. The FIB has a Gaussian profile with a minimum 5 nm spot size, allowing for precise aperture milling of even ultrahigh aspect ratio of the tips. A final important consideration is the grounding of the tip to avoid charging while imaging. This is accomplished using a small amount of silver paint, which is then easily removed by acetone. FIB fabricated fiber probes were tested in the near-field system developed in laboratory. Near-field distance regulation was maintained through the use of shear force phase-locked tuning fork feedback. [31] The imaged sample was a series of lines of various widths (1 μm~100 nm) milled by the FIB through a 50 nm aluminum film coated on a glass substrate.

(a)

(b)

(c)

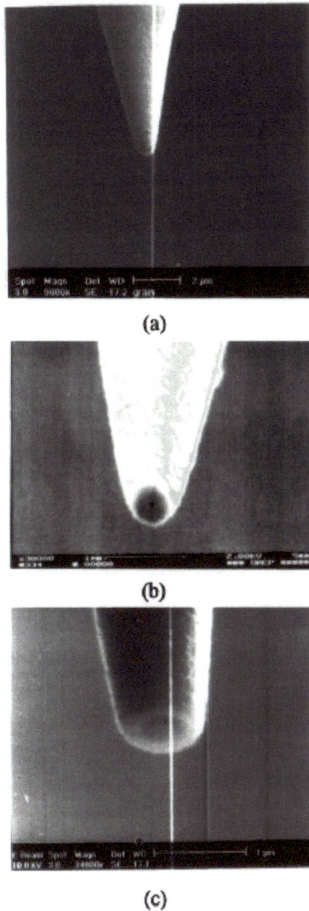

Fig.4-4 SEM photographs of the near-field optical fiber probe (a) before FIB milling (bar=2μm), (b) after FIB fabrication of a 100 nm diameter aperture (bar=1μm), and (c) after FIB fabrication of a 300 nm diameter aperture (bar=1 μm). Reprinted with permission from "Saeed Pilevar *et. al.*, Focused ion-beam fabrication of fiber probes with well-defined apertures for use in near-field scanning optical microscopy, Appl. Phys. Lett. 72, 3133 (1998)" with Copyright © 1998 of American Institute of Physics.

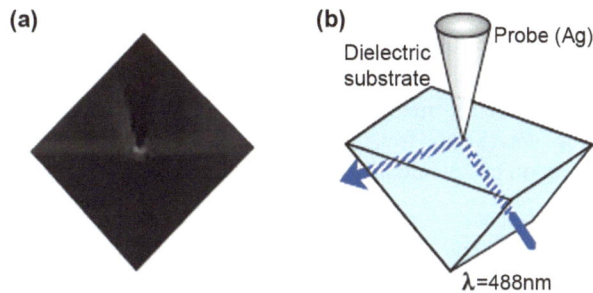

Fig.4-5 (a) Enhanced electric-field intensity localized at a silver probe tip, calculated by the FDTD method. (b) Schematic of the tip probing on prism. Reprinted with permission from S. Kawata Ed., *Near-Field Optics and Surface Plasmon Polaritons*, Topics Appl. Phys. 81, 29–48 (2001) with copyright ©2001of Springer-Verlag Berlin Heidelberg.

4.2.3 Apertureless fiber probe [32]

The NSOM described here is so-called apertureless NSOM, and the resolution is determined by radius of the tip. In this chapter, we will show the principle and features of the NSOM systems using an apertureless metallic probe, as well as several near-field images obtained by the systems. Furthermore, we will discuss near-field nanospectroscopy for chemical analysis of organic materials and molecules with the aid of the experimental data. The apex of a metallic probe, such as the tip of a scanning tunneling microscope (STM), functions as a scatterer by enhancing electric field locally. What is shown in Fig. 4-6 (a) is the intensity of a light field scattered at a tip, obtained from numerical analysis. Figure 4-6 (b) shows the model used for the analysis. A silver metallic tip of radius 20 nm is placed in contact with a glass substrate (refractive index is 1.5). The silver tip is illuminated by a plane wave traveling from the substrate. The working wavelength of the incident field is 488 nm and its polarization is in the TM mode (*i.e.* p-polarization). The incidence angle is 45°. An evanescent field is generated over the surface of the glass substrate because the incident angle satisfies the condition of total internal reflection. The finite-differential time-domain (FDTD) method was employed in the calculation [33]. A localized and enhanced field spot is observed around the tip in the figure. The size of the beam spot is as small as ~30 nm, which corresponds approximately to the radius of the tip. The peak intensity of the small spot is enhanced by a factor of 80 compared to the intensity of the incident field. If, instead, the tip is a glass probe with the same tip radius of 20 nm, the enhancement factor is 7 only when the illumination is in the TM mode [34]. No enhancement of the electric field is induced, when the TE mode (i.e. s-polarization) is used for illumination in the calculation [34, 35]. These analyses show that strong enhancement of the localized field requires the use of a metallic tip and TM mode illumination for an apertureless probe, and that light field of the spot in small-scale strongly enhanced at the tip is caused by excitation of a localized SPPs. The basis of the metallic-tip NSOM is the scattering of the evanescent field localized around the tip due to the sample structure or the scattering of the evanescent field localized around the sample structure. The function of the metallic tip is the same as that of an aperture probe, which is used for the generation of an evanescent field or conversion of an evanescent field into a propagating field at the aperture. A difference between the metallic tip and the aperture probe is whether it functions as a waveguide for illumination/detection or not.

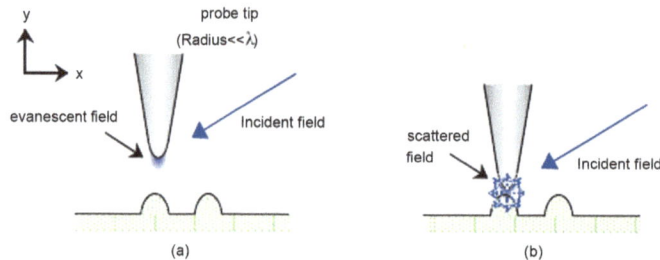

Fig. 4-6 Principle of an apertureless NSOM, (a) a small spot is generated at the tip, but does not interact with the sample, because the tip is far from the sample surface. (b) The localized spot is rescattered by the sample structure as the tip is in the vicinity of the sample. Reprinted with permission from S. Kawata Ed., *Near-Field Optics and Surface Plasmon Polaritons* Topics Appl. Phys. 81, 29–48 (2001) with copyright ©2001of Springer-Verlag Berlin Heidelberg.

Compared to the aperture probe, the metallic tip has several features, and is listed as follows:

1. Higher resolution is obtainable by making the tip sharper, because the resolution of an NSOM with a metallic tip is determined by the radius of the tip. Atomic resolution is achievable, provided that the tip apex is processed at the atomic level. So far, the accomplishment of imaging with 1 nm resolution has been reported [36, 37]. On the other hand, the resolution of the NSOM with an aperture probe is limited to several tens of nanometers because the size of the aperture is restricted by the skin depth of the metal coated around the fiber [38].

2. Collection of the light field scattered at a metallic tip is efficient if external collection optics are used, while light field is strongly absorbed during waveguide transmission by the metal film coated around the fiber tip during waveguide transmission. The collection efficiency is getting better as higher *NA* lenses are employed.

3. A metallic tip can produce a more intensive scattered field than that of a tip made from other materials [39] because a metal can scatter the light field more efficiently than other materials as mentioned above.

4. A metallic tip can be sharpened more finely and easily than an aperture probe. An aperture probe has larger effective diameter than a pure metal tip because the former has metal coating that has to be thicker than the skin depth of light. Hence, the metallic tip can be scanned over steep areas of a sample when the tip is regulated by AFM or STM. Furthermore, atomic contact between the tip and sample or single-channel transport of tunneling electrons is possible with a sharp metallic tip, while a light field is scattered at the tip [40]. This means that scattering center of the evanescent field coincides with the contact point or electron channel. Thus, the near-field optical image corresponds to the topography obtained by an AFM or STM. In contrast, an aperture probe does not always provide a single contact point or electron channel, owing to thickness of the probe, as shown in Fig. 4-7. This means that the probe–sample distance may not be controlled and regulated precisely and reliably. As a result, the near-field optical image and the topography of the same sample surface can exhibit different characteristic responses [41].

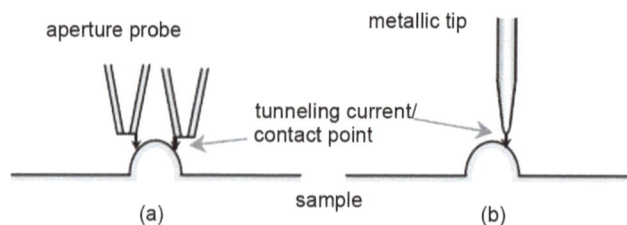

Fig. 4-7 Operation of NSOM probe tips as STM/AFM probes: (a) a fiber probe with a coated metal film, and (b) a metallic tip. Reprinted with permission from S. Kawata Ed., *Near-Field Optics and Surface Plasmon Polaritons*, Topics Appl. Phys. 81, 29–48 (2001) with copyright ©2001of Springer-Verlag Berlin Heidelberg.

5. The spectral responses of near-field detection with a metallic tip ranges from the ultraviolet to the infrared because of the use of external optics (*e.g.* a Cassegrain objective mirror or a lens made of an appropriate material), while the spectral response of an aperture probe is limited by the material of the waveguide. Because the scattering efficiency of a metal is higher in the infrared region than in the visible, then the use of metallic probe tip can be beneficial in infrared micro-spectroscopy [42, 43].

4.3 Applications of NSOM

There are many applications for NSOM such as the fields in nanometrology, biomedical, data storage, semiconductor, and nanomaterials. It can give the message of surface topography, and near-field optical intensity distribution. Therefore, it is particularly suitable to measure the photoluminescent (PL) and the samples with light emission. Moreover, the aqueous biosamples such as molecule and DNA also can be measured using the NSOM. The following are some examples of the NSOM measurement.

The excitation of dielectric-loaded surface plasmon polaritons (SPPs) by a CdS nanostripe placed on an Ag layer is observed by using SNOM, as shown in Fig. 4-8. [44] The far-field optical image of this excited CdS nanostripe is shown in Fig. 4-8 (a). The emitted fluorescent intensity at position C (left end) is lower than that of at position A (right end). There is an intensity gradient at the boundary of the Ag layer, as shown in Fig. 3 (a). Figure 3 (b) is a near-field optical image of the part of the nanostripe corresponding to the region in Fig. 4-8 (a). The dashed lines in both Figs. 4-8 (a) and (b) denote the boundary of the Ag layer. The position of the laser focus is unchanged during the scanning and collection measurements, as shown in Fig. 4-8 (b). This shows a spatial map of the minimum emission intensities at the end of CdS nanostripe placed on the Ag layer. Figure 4-8 (c) shows the locally detected near-field PL spectra recorded by the SNOM collection tip at positions A, B, and C, respectively. The near-field PL at site of full width and half maximum (FWHM) of the spectra peak is about 0.7 eV, much smaller than that of the far field, which is about 1.5 eV. Both the energy peak position and the fluorescent intensity change significantly when the SNOM tip is changed among positions A, B, and C, while the incident laser spot is kept unchanged in the middle of the stripe "position B". The changes in photon energy show that the emission bands exhibit a red-shift of about 30 meV from A (2.404 eV) to C (2.374 eV) while keeping the waveguide propagation distance equal (AB=BC=3 μm).

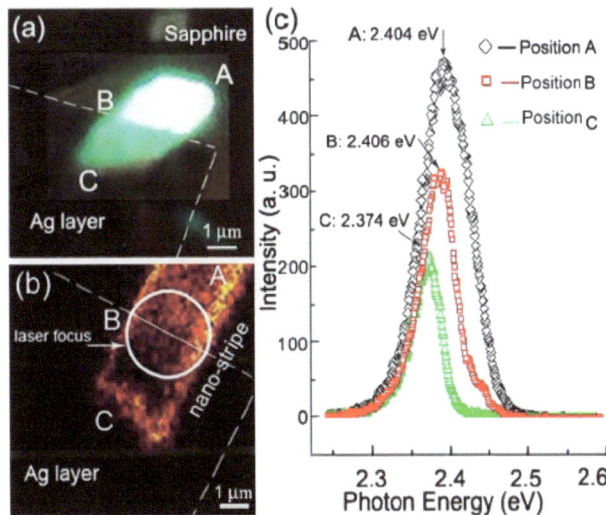

Fig. 4-8 (a) Far-field PL image of CdS nanostripe excited by the 442 nm laser. (b) Near-field optical image of CdS nanostripe corresponding to the region of (a). The position of the laser focus is unchanged during the measurement. (c) Near-field PL spectra detected with the SNOM collection tip at positions A (rhombus), B (square), and C (triangle), respectively. Reprinted with permission from "Zheyu Fang, Xuejin Zhang, Dan Liu, and Xing Zhu, Excitation of dielectric-loaded surface plasmon polariton observed by using near-field optical microscopy, Appl. Phys. Lett. 93, 073306 (2008)" with copyright © 2008 of American Institute of Physics.

Figure 4-9 presents both computed and experimental field patterns for this fibre. In panels (a) and (d), the distribution of the time-averaged Poynting vector's z-component at the fibre surface shows concentration of light in the central bore.[45] For shorter wavelengths (700 nm, upper row), the hole is too large to show strong effects and most of the light is in the annular silica region. However, at longer wavelengths (1,050 nm, lower row), the hole is well below the diffraction limit and so fills with light, and the material contrast ensures that the maximum intensity occurs inside the bore. To visualize the near field directly, we used a SNOM working in collection mode. Figure 4-9 (b) and (e) show experimental images recorded at two different wavelengths, 700 nm and 1,050 nm, respectively. The near-field profiles of the light at the output end of the fibres were imaged using a white-light source and the NSOM (Nanonics Multiview

2000^{TS}). The white-light source was an SC generated using a microchip laser and a PCF, filtered to the required wavelength by using interference filters. The fibre end was mounted vertically in the SNOM and an optical fibre tip (multi-mode fibre with a 50-nm diameter, and Au-coated tip) was lowered onto the fibre-end face. The collected light was detected using a silicon avalanche photodetector. Each image is a 1.6×1.6 mm square scan with a resolution of 170 points per micrometre.

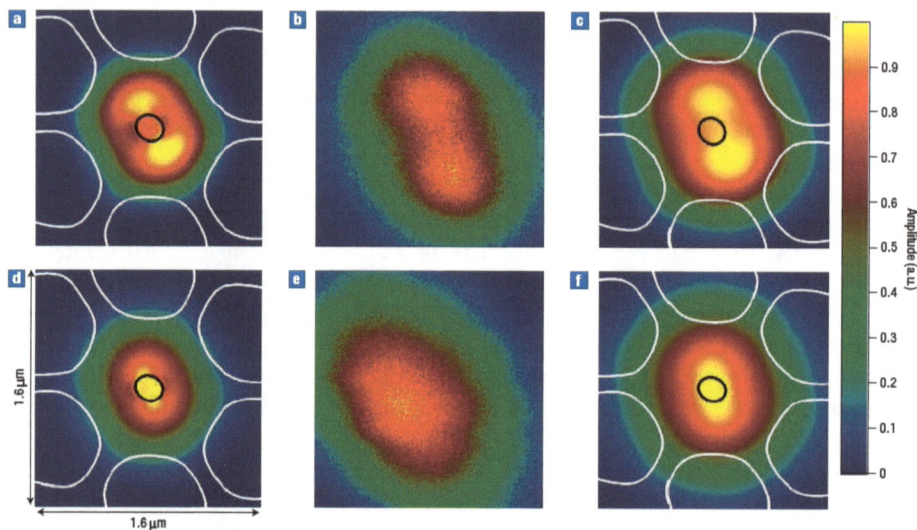

Fig. 4-9 Near-field mode patterns for PCF 3. The three patterns in the upper row are for l ¼ 700 nm and those in the lower row for l ¼ 1,050 nm. a, d, Patterns in the left column represent the z-component of the calculated time-averaged Poynting vector at the fibre surface. b, e, The middle column are SNOM experimental images. c, f, Patterns in the right column represent the z-component of the calculated time-averaged Poynting vector at a distance of 30 nm from the fibre surface. The white and black solid lines represent the contours of the fibre geometry. Reprinted with permission from "G. S. Wiederhecker, C. M. B. Cordeiro, F. Couny, F. Benabid, S. A. Maier, J. C. Knight, C. H. B. Cruz And H. L. Fragnito, Nature Photonics 1 Feb. 2007" with copyright © 2007 of Nature Publishing Group.

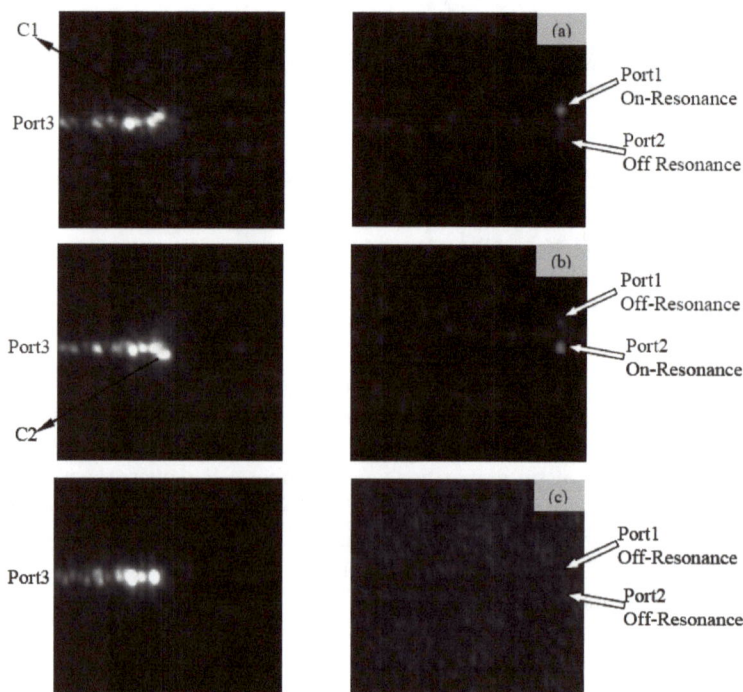

Fig. 4-10 Near-field images observed at the wavelengths of (a) 1529.5 nm (b) 1531 nm and (c) 1520 nm, respectively. Reprinted with permission from "Cheng Ren, Jie Tian, Shuai Feng, Haihua Tao, Yazhao Liu, Kun Ren, Zhiyuan Li, Bingying Cheng, and Daozhong Zhang, Opt. Express 14, 10014-10020 (2006)" with copyright (C) 2006 of Optical Society of America.

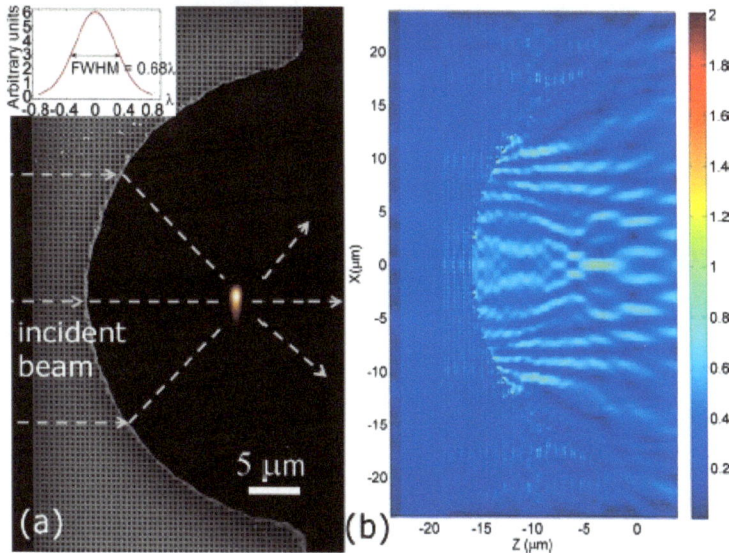

Fig. 4-11 (a) NSOM reconstructed image of the transmitted field in the air cavity at 1530 nm. Focusing is observed ~12 μm away from the concave face. The FWHM of the beam spot size is in the order of ~0.68λ (shown in the inset picture). The white dashed lines indicate the beam path. Note that (a) is a composite picture consisting of a SEM image of the lens superposed on a NSOM scan. (b) Electric field intensity plot of FDTD simulations results of the planoconcave microlens. For better visual illustration the radius of the concave face in the Z plane has been truncated to 14 μm, while the transverse size (X plane) is still maintained at 40 μm. The focusing effect occurs 12 μm away from the concave face matching the experimental results. The FWHM of the focused beam is 0.67λ. Reprinted with permission from "B. D. F. Casse, W. T. Lu, Y. J. Huang, and S. Sridhar, Nano-optical microlens with ultrashort focal length using negative refraction,Appl. Phys. Lett. 93, 053111(2008)" with copyright © 2008 of American Institute of Physics.

To directly see the propagation behaviour of the waveguides and cavities, the light transmission from top of the sample recording by use of an infrared camera was carried out.[46] Figure 4-10 shows the image of the filter performance, as light enters the filter from port 3. It is clearly seen in Figs 4-10 (a) and (b) that light in the input waveguide drops upwards and downwards to the cavity C1 and C2 when the input light wavelengths were set at 1529.5 nm and 1531 nm, respectively. Meanwhile, one bright spot will appear at port 1 and port 2, respectively. On the other hand, in the off-resonant case, *i.e.* the input light wavelength was set neither at 1529.5 nm nor at 1531 nm, the light in the input waveguide cannot couple to C1 and C2 and strong reflection occurs, as shown in Fig. 4-10 (c). Correspondingly, the bright spots at port 1 and port 2 disappear simultaneously. It turns out that this scattering-light analysis technique can become a powerful supplemental tool to probe the optical properties of photonic crystal functional devices to the usual quantitative transmission spectrum technique. [46]

Figure 4-11 is a most recently reported results regarding NSOM characterization of a negative refractive index-based microlens for superfocusing.[47] An ultrashort focal length a plano-concave microlens in an InP/InGaAsP semiconductor two-dimensional (2D) photonic crystal with negative refraction index of −0.7 was experimentally realized. At λ=1.5 μm, the lens exhibits ultrashort focal lengths of 12 μm (~8λ) and numerical aperture close to unity. The focused beam has a near diffraction-limited spot size of 1.05 μm (~0.68λ) at full width at half maximum. The negative refractive index and focusing properties of the microlens were confirmed by 2D finite-difference and time-domain simulations. Such refractive negative-index-based nano-optical microlenses can be integrated into existing semiconductor heterostructures and platforms for next-generation optoelectronic applications. Figure 4-11(a) is a composite picture consisting of a scanning electron microscopy (SEM) image of the lens superposed on a NSOM scan. The results were further confirmed using 2D FDTD simulations. A 40 μm wide plane parallel TE-polarized Gaussian beam was chosen as incident field excitation for the microlens in order to match the experimental conditions. The FDTD simulation results of the microlens for the electric field intensity are shown in Fig. 4-11(b). The calculated FWHM from simulations is 0.67λ.

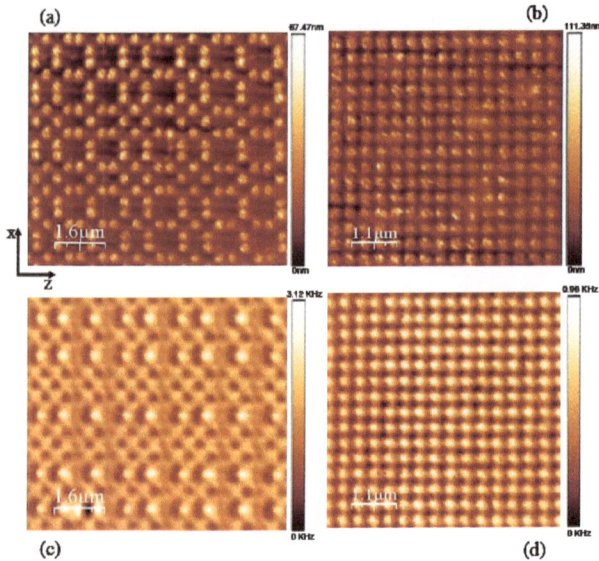

Fig. 4-12 (a) and (b) show topographical AFM images on the Fibonacci and Periodic arrays respectively, (c) and (d) are NSOM images obtained from Fibonacci and Periodic nanoparticle arrays obtained under identical conditions. Reprinted with permission from "Ramona Dallapiccola1, Ashwin Gopinath2, Francesco Stellacci1, and Luca Dal Negro, Opt. Express 16, 5544-5555 (2008)" with copyright of © 2008 of Optical Society of America.

The last example is topography of a 2D Fibonacci plasmonic lattices fabricated by electron-beam lithography on transparent quartz substrates measured using NSOM.[48] The NSOM images shown in Fig. 4-12 clearly demonstrated the presence of hot spots corresponding to the resonant excitation of localized nanoparticle plasmons (LNPs) both in the periodic and Fibonacci samples. The direct comparison to the topographic AFM images shown in Fig. 4-12 (a) and Fig. 4-12 (b) allows us to conclude that the LNPs are localized at interstitial positions with respect to the nanoparticles sites. However, while for the periodic structures the observed LNPs are periodically arranged into a simple square lattice, the question arises on how the LNPs are arranged in a Fibonacci lattice. A closer look at Fig. 4-12 (c) suggests that the LNPs are localized in between vertically-coupled dimmer structures in the Fibonacci array, which are efficiently excited by the vertical polarized (x-direction) incident field.

There are still many other helpful applications reported before. It is too difficult to involve all of them due to length limitation of this chapter. The examples cited here just play a role of a guideline for readers. Some new applications will need further exploring in the future.

4.4 Summary

Relevant literature review, working principle, system configuration, and typical applications of NSOM systems were described in this chapter. Fabrication issue of NSOM probes was highlighted. In addition, apertureless probe was introduced. Although the NSOM has been developed near 20 years till now, there are still some unknown problems and unrevealed phenomena existing in the systems. Strictly speaking, it is still too early to say that it is a mature technology already although three commercial producers grown so far: Nanonics Inc. in Israel, WiTECK in Germany, and ND-MDT in Russia, respectively. NSOM performances, *e.g.*, repeatability, automation, and operation difficulty etc. are still need to be improved for current commercial systems in future. It is reasonable to believe that there will be more producers appear in the market with high-technologies in the near future. Different models from the different producers have their own characteristics and specific performances. Therefore, it is too difficult to address a certain model which can involve all the customers required functions. Thus many home-built systems were extensively setup in universities and institutes in recent years so as to meet their own specific requirements.

Reference

[1]   E.H. Synge, A suggested method for extending the microscopic resolution into the ultramicroscopic region, Phil. Mag. 6, 356 (1928);

[2]   E.H. Synge, An application of piezoelectricity to microscopy, Phil. Mag., 13, 297 (1932).

[3]   J.A. O'Keefe, Resolving power of visible light, J. of the Opt. Soc. of America, 46, 359 (1956).

[4]   E.A. Ash and G. Nichols, Super-resolution aperture scanning microscope, Nature 237, 510 (1972).

[5]   A. Lewis, M. Isaacson, A. Harootunian and A. Murray, Ultramicroscopy 13, 227 (1984); D.W. Pohl, W. Denk and M. Lanz, APL 44, 651 (1984).

[6]   N. F. van Hulst, M. H. P. Moers, O. F. J. Noordman, R. G. Tack, F. B. Segerink, and B. Bölger, Near-field optical microscope using a silicon-nitride probe. Appl. Phys. Lett. 62, 461 (1993).

[7]   Science 277, 637 (1997).

[8]   S. Shalom et al., An Optical Submicrometer Calcium Sensor with Conductance Sensing Capability, Analytical Biochem. 244, 256 (1997)

[9]   United States Patent Number 5, 264,698, 1993.

[10]   M.Ohtsu and K.Sawada, in *Nano-Optics* edited S.Kawata,M.Ohtsu and M.Irie, Springer (2002), chap. 3, p.61.

[11]   A.G.T.Ruter, M.H.P.Moers, N.F. van Hulst, and M. de Boer, Microfabrication of near-field optical probes. J. Vac. Sci. Technol. B 14, 597 (1996).

[12]   D.Drews, W.Ehrfeld, M.Lacher, K.Mayr, W. Noell, S.Schmitt, and M.Abraham, Nanostructured probes for scanning near-field optical microscopy. Nanotechnology 10, 61 (1999).

[13]   C.Mihalcea, W.Scholz, S.Werner, S.Munster, E.Oesterschulze, and R.Kassing, Multipurpose sensor tips for scanning near-field microscopy. Appl. Phys. Lett. 68, 3531 (1996).

[14]   P.N. Minh, T.Ono, and M.Esashi, *Nat. Conf. Phys. Sensors*, Tokyo, Japan, November 1998.

[15]   P.N.Minh, T.Ono, and M.Esashi, Proc. of 11th IEEE Int. Conf. on *Micro Electro. Mech. Syst.*, Orlando, Sens. Actuators A 80, 163 (1999).

[16]   P.N.Minh, T.Ono, and M.Esashi, Nonuniform silicon oxidation and application for the fabrication of aperture for near-field scanning optical microscopy. Appl. Phys. Lett. 75, 4076 (1999).

[17]   D. W. Pohl, W. Denk, and M. Lanz, Optical stethoscopy: Image recording with resolution λ/20. Appl. Phys. Lett. 44, 651 (1984).

[18]   E. Betzig, J. K. Trautman, T. D. Harris, J. S. Weiner, and R. L. Kostelak, Breaking the Diffraction Barrier: Optical Microscopy on a Nanometric Scale. Science 251, 1468 (1992).

[19]   E. H. Synge, A suggested method for extending the microscopic resolution into the ultramicroscopic region. Philos. Mag. 6, 356 (1928).

[20]   G. A. Valaskovic, M. Holton, and G. H. Morrison, Fourth-order optical aberrations and phase-space transformation for reflection and diffraction optics. Appl. Opt. 34, 125 (1995).

[21]   W. P. Ambrose, P. M. Goodwin, J. C. Martin, and R. A. Keller, Alterations of Single Molecule Fluorescence Lifetimes in Near-Field Optical Microscopy. Science 265, 364 (1994).

[22]   A. Shchemelinin, M. Rudman, K. Liberman, and A. Lewis, A simple lateral force sensing technique for near-field micropattern generation. Rev. Sci. Instrum. 64, 3538 (1993).

[23]   E. Betzig, J. K. Trautman, R. Wolfe, E. M. Gyorgy, and P. L. Finn, Near-field magneto-optics and high density data storage. Appl. Phys. Lett. 61, 142 (1992).

[24]   T.Pangaribuan, K. Yamada, S.Jiang, H.Ohsawa, and M. Ohtsu, Reproducible Fabrication Technique of Nanometric Tip Diameter Fiber Probe for Photon Scanning Tunneling Microscope. Jpn. J. Appl. Phys., Part 1 31, 1302 (1992).

[25]   D. Zeisel, S. Nettesheim, B. Dutoit, and R. Zenobi, Pulsed laser-induced desorption and optical imaging on a nanometer scale with scanning near-field microscopy using chemically etched fiber tips. Appl. Phys. Lett. 68, 2491 (1996).

[26]   R. U. Maheswari, S. Mononobe, and M. Ohtsu, Deducing structural variations of the apex of probes used in near-field optical microscopy through simultaneous measurement of shear force and evanescent intensity. Appl. Opt. 35, 6740 (1996).

[27]   M. N. Islam, X. K. Zhao, A. A. Said, and C. F. Vail, Conference on Lasers and Electro-Optics (Optical Society of America), Washington, DC, 1997, Vol. 11, p. 70.

[28]   E. Betzig and J. K. Trautman, Near-Field Optics: Microscopy, Spectroscopy, and Surface Modification Beyond the Diffraction Limit . Science 257, 189 (1992).

[29]   C. Obermuller and K. Karrai, Far field characterization of diffracting circular apertures. Appl. Phys. Lett. 67, 3408 (1995).

[30]   K. P. Muller and H. C. Petzold, "Microstructuring of gold on x-ray masks with focused Ga+ ion beams", Electron-Beam, X-Ray, and Ion-Beam Technology: Submicrometer Lithographies IX, Proc. SPIE Vol.1263, 12 (1990).

[31]   B. Hecht, H. Bielefeldt, Y. Inouye, D. W. Pohl, and L. Novotny, Facts and artifacts in near-field optical microscopy, J. Appl.Phys. 81, 2492 (1997).

[32]   S. Kawata Ed., *Near-Field Optics and Surface Plasmon Polaritons*, Topics Appl. Phys. 81, 29–48 (2001), Springer-Verlag Berlin Heidelberg, 2001.

[33]   H. Furukawa, S. Kawata, Analysis of image formation in a near-field scanning optical microscope: effects of multiple scattering, Opt. Commun. 132, 170–178 (1996).

[34]   H. Furukawa, S. Kawata, Local field enhancement with an apertureless near-field-microscope probe. Opt. Commun. 148, 221–224 (1998).

[35]   O. J. F. Martin, C. Girard, Controlling and tuning strong optical field gradients at a local probe microscope tip apex, Appl. Phys. Lett. 70, 705–707 (1998).

[36]   F. Zenhausern, Y. Martin, H. K. Wickramasinghe, Scanning Interferometric Apertureless Microscopy: Optical Imaging at 10 Angstrom Resolution. Science 269, 1083– 1085 (1995).

[37]   J. Koglin, U. C. Fischer, H. Fuchs, Formation of Mn-derived impurity band in III-Mn-V alloys by valence band anticrossing . Phys. Rev. B 55, 7977–7984 (1997).

[38]   J. P. Fillard, *Near Field Optics and Nanoscopy*, World Scientific, Singapore 1996.

[39]   F. Zenhausern, M. P. O'Boyle, H. K. Wickramasinghe, Apertureless near-field optical microscope. Appl. Phys. Lett. 65, 1623–1625 (1994).

[40]   Y. Inouye, S. Kawata, J. Microsc. 178, 14–19 (1995).

[41]   K. Lieberman, A. Lewis, Simultaneous scanning tunneling and optical near-field imaging with a micropipette. Appl. Phys. Lett. 62, 1335–1337 (1993).

[42]   A. Lahrech, P. Bachelot, P. Gleyzes, A. C. Boccara, Infrared-reflection-mode near-field microscopy using an apertureless probe with a resolution of λ/600. Opt. Lett. 21, 1315–1317 (1996).

[43]   B. Knoll, F. Keilmann, Near-field probing of vibrational absorption for chemical microscopy. Nature 399, 134–137 (1999).

[44]   Zheyu Fang, Xuejin Zhang, Dan Liu, and Xing Zhu, Excitation of dielectric-loaded surface plasmon polariton observed by using near-field optical microscopy, Appl. Phys. Lett. 93, 073306 (2008).

[45]   G. S. Wiederhecker, C. M. B. Cordeiro, F. Couny, F. Benabid, S. A. Maier, J. C. Knight, C. H. B. Cruz And H. L. Fragnito, Field enhancement within an optical fibre with a subwavelength air core, Nature Photonics Vol. 1, Feb. 2007, 115-118.

[46]   Cheng Ren, Jie Tian, Shuai Feng, Haihua Tao, Yazhao Liu, Kun Ren, Zhiyuan Li, Bingying Cheng, and Daozhong Zhang, High resolution three-port filter in two dimensional photonic crystal slabs, Opt. Express 14, 10014-10020 (2006).

[47]   B. D. F. Casse, W. T. Lu, Y. J. Huang, and S. Sridhar, Nano-optical microlens with ultrashort focal length using negative refraction, Appl. Phys. Lett. 93, 053111 (2008).

[48]   Ramona Dallapiccola1, Ashwin Gopinath2, Francesco Stellacci1, and Luca Dal Negro, Quasi-periodic distribution of plasmon modes in two-dimensional Fibonacci arrays of metal nanoparticles , Opt. Express 16, 5544-5555 (2008).

5 METAMATERIALS

Abstract: Aim of this chapter is that readers can obtain an outline about metamaterials from concept to principle after reading. Concept introduction of negative refraction index was given firstly, including literature review, history regarding the negative refraction index, and illustration of energy and momentum in negative refractive index materials. Then three types of metamaterials: electronic metamaterials, magnetic metamaterials, and double negative-index metamaterials were described in this chapter.

5.1 Introduction

"Meta" is a Greek word, means "beyond". Metamaterial is an arrangement of artificial structural elements, designed to achieve advantageous and unusual electromagnetic properties. Metamaterials are artificial periodic structures with "lattice constants" that are still smaller than the working wavelength of light. Again, the light field "sees" an effective homogeneous material. The "atoms," however, are not real atoms but are rather artificial nanostructures composed of many atoms. This allows for tailoring their properties in a way which is not possible with normal atoms. Indeed, Pendry [1] showed that a combination of "magnetic atoms" and "electric atoms" (*i.e.*, split rings and metallic wires) with negative permeability μ and permittivityε, respectively, can lead to materials with a negative refraction index n. These materials open a whole new chapter of photonics connected with novel concepts and potential applications. Metamaterials typically use metallic structures to provide a negative permittivity and use resonant structures (equivalent to inductor-capacitor tank circuits) with a scale much smaller than the wavelength to provide a negative permeability leading to negative refraction.

The concept of negative $\mu_{eff}(\omega)$ is of particular interest, not only because this is a regime not observed in ordinary materials, but also because such a medium can be combined with a negative $\varepsilon_{eff}(\omega)$ to form a "left-handed" material (*i.e.*, **E**×**H** lies along the direction of wave vector **k** for propagating plane waves). In 1967, Veselago [2] theoretically investigated the electrodynamic consequences of a medium having both ε and μ negative and concluded that such a medium would have dramatically different propagation characteristics stemming from the sign change of the group velocity, including reversal of both the Doppler shift and Cherenkov radiation.

J. B. Pendry was the first scientist to imagine a practical way to make a left-handed metamaterial (LHM). "Left-handed" in this context means a material in which the "right-hand rule" is not obeyed, allowing an electromagnetic wave to convey energy in the opposite direction to wave propagation. Pendry's initial idea was that metallic wires aligned along propagation direction could provide a metamaterial with negative permittivity ($\varepsilon<0$). Note however that natural materials (such as ferroelectrics) were already known to exist with negative permittivity. The challenge is to construct a material that also shows negative permeability ($\mu<0$). In 1999, Pendry demonstrated that an open ring ("C" shape) [1] with axis along the propagation direction **k** could provide a negative permeability. In the same paper, he showed that a periodic array of wires and ring could give rise to a negative refractive index.

Snell law and Doppler effects are still available for the negative refraction. The first metamaterial was experimentally verified in microwave frequency (see schematic diagram of Fig. 5-1). Microwave antenna related metamaterials on the basis of design theory of transmission line are one of hot topics of the metamaterials. Considering scope of this book, the following sections in this chapter mainly address the metamaterials for optical frequency. The materials with $\varepsilon<0$ and $\mu>0$ are called electronic matematerials, $\varepsilon>0$ and $\mu<0$ are called magnetic matematerials, and $\varepsilon<0$ and $\mu<0$ are called double negative (DNG) metamaterials. They were introduced respectively in the following forthcoming sections. Applications of metamaterials in microwave antenna and optical imaging will be illustrated separately in chapters 10 and 13 of this book.

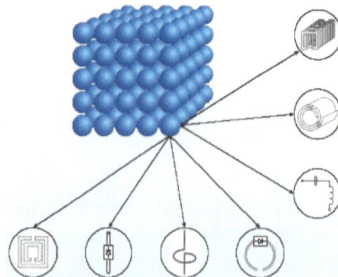

Fig.5-1 Schematic diagram of metamaterials with possible structures and equivalent physical models. Reprinted with permission from J. B. Pendry, A. J. Holden, D. J. Robbins, and W. J. Stewart, IET Microwaves, Antennas & Propagation, 1(1), pp. 3-11(2007) with copyright © 2005 of IEEE.

5.2 Electronic metamaterials

An approach was introduced that uses a pair of metal layers separated by a dielectric to provide resonant interactions (*e.g.*, distributed inductance or capacitance) along with a periodic array of holes through the film stack to facilitate interaction with the surface plasma waves of the composite structures.[3] The structure consists of a glass substrate with two metallic films (30 nm thick Au) separated by a dielectric layer (60 nm thick Al_2O_3) with a two-dimensional, square periodic array of circular holes (period 838 nm; hole diameter ~360 nm) perforating the entire multilayer structure. A schematic and a scanning electron microscopy (SEM) picture of the fabricated structure are shown in Fig. 5-2. The incident magnetic field interacts primarily with the hatched regions. The top and bottom metal films form a loop (inductor) and, whereas the induced current is interrupted by the holes, also form a capacitor between the two films giving a resonant permeability response very similar to that recently reported for Au ''staple'' structures [4]. In the vicinity of the resonance, this tank circuit provides a magnetization field opposite to that of the incident wave and, therefore, a reduced permeability. The dark regions dominate the electric field response forming a wire-grid polarizer that cancels the field in the metal and results in a negative permittivity.

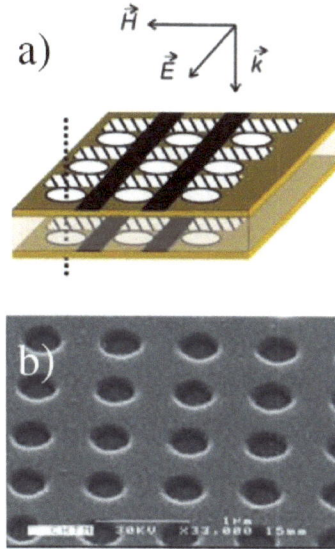

Fig. 5-2 (a) Schematic of the multilayer structure consisting of an Al_2O_3 dielectric layer between two Au films perforated with a square array of holes (838 nm pitch; 360 nm diameter) atop a glass substrate. For the specific polarization and propagation direction shown, the active regions for the electric (dark regions) and magnetic (hatched regions) responses are indicated. (b) SEM picture of the fabricated structure. Reprinted with permission from "Shuang Zhang, Wenjun fan, N. C. Panoiu, K. J. Malloy, R. M. Osgood, S. R. J. Brueck, Phys. Rev. Lett. 95, 137404 (2005)" with copyright © 2005 of The American Physical Society.

A simple Drude model was used for the dielectric constant of gold, $\varepsilon(\omega)=1-\omega_p^2/[\omega(\omega+i\omega_c)]$, where $\omega_p=1.37\times10^{16}$ Hz is the plasma frequency and $\omega_c = 4.08\times10^{13}$ Hz is the scattering frequency for bulk gold [5]. The thin metallic film (30 nm) in our device has more surface and boundary scattering than bulk material and, therefore, likely exhibits a higher scattering frequency. We can use the transmission and reflectance amplitude and phase measurements to evaluate the

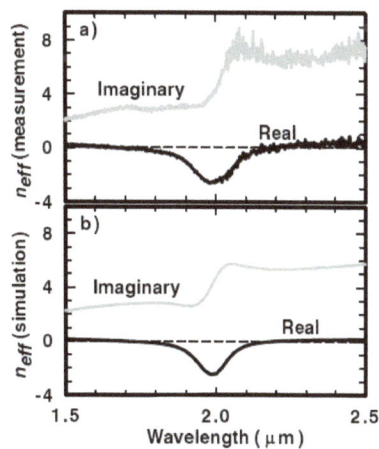

Fig. 5-3 The effective refractive index extracted from measurement (a) and from modeling (b) showing a resonance and a negative real part at ~2.0 μm. Reprinted with permission from "Shuang Zhang, Wenjun fan, N. C. Panoiu, K. J. Malloy, R. M. Osgood, S. R. J. Brueck, Phys. Rev. Lett. 95, 137404 (2005)" with copyright © 2005 of The American Physical Society.

metamaterial refractive index ($n=(\varepsilon\mu)^{1/2}$) and impedance ($\zeta=(\mu/\varepsilon)^{1/2}$). The results from both measurement and modeling are shown in Figs. 5-3 (a) and (b), respectively. The real parts are negative over a short range of wavelengths around 2 μm, where the imaginary part undergoes a strong modulation. The minimum value of the real part is about -2, while the imaginary part is larger than 3, which means that the negative-index material exhibits significant loss associated with electron scattering in the thin metal films. From the results for n and ζ, the effective permeability μ_{eff} and permittivity ε_{eff} can be evaluated (not shown). The real part of ε_{eff} shows a strong modulation, and the imaginary part has a peak characteristic of a strong magnetic resonance in the wavelength region where $n<0$. However, the minimum value of μ_{eff} is not negative (for both the measurement and the modeling results) as a result of the large scattering loss. The refractive index is expressed in terms of the permittivity and permeability as $n=\pm[(\varepsilon_1\mu_1-\varepsilon_2\mu_2)+i(\varepsilon_1\mu_2+\varepsilon_2\mu_1)]^{1/2}$ and Im(n)> 0, where $\varepsilon=\varepsilon_1+i\varepsilon_2$ and $\mu=\mu_1+i\mu_2$. To achieve a negative Re(n) with $\varepsilon_1 < 0$ and $\mu_1 > 0$, $-\varepsilon_1\mu_2$ must be larger than $\varepsilon_2\mu_1$, as is the case near the metamaterial resonance.

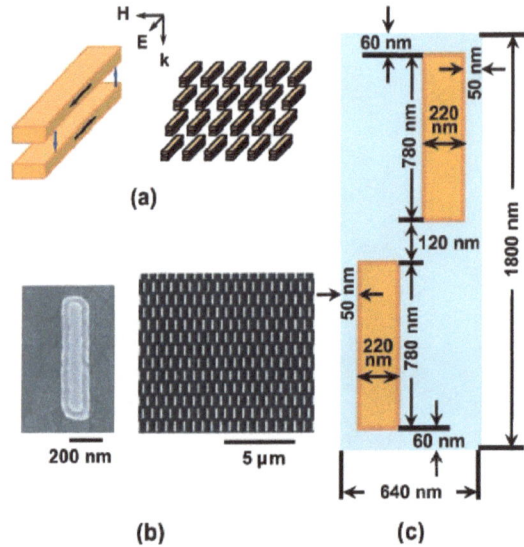

Fig. 5-4 (a) Schematic diagram for the array of nanorod pairs. (b) Field-emission scanning electron microscope images. (c) Elementary cell. Reprinted with permission from "Vladimir M. Shalaev, Wenshan Cai, Uday Chettiar, Hsiao-Kuan Yuan, Andrey K. Sarychev, Opt. Lett. 30, 3356-3358 (2005)" with copyright © 2005 of Optical Society of America.

The other electric matematerials was reported by Vladimir *et. al.* in Ref. 6. Figure 5-4 shows a negative refractive index material for the optical range, specifically for the wavelengths close to 1.5 μm (200 THz frequency), accomplished with a metal–dielectric composite. For normally incident light with the electric field polarized along the rods and the magnetic field perpendicular to the pair, the electric and magnetic responses both can experience resonant behavior at certain frequencies. Above the resonance frequency, the circular current in the pair of rods can lead to a

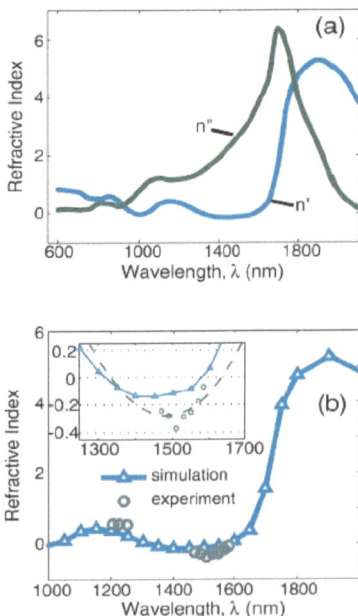

Fig. 5-5 (Color online) (a) Real and imaginary parts of the refractive index retrieved from simulations. (b) Real part of the refractive index retrieved from simulations (triangles) and experiments (circles). The inset in (b) is a magnified view of the region of negative refraction; the dashed curve shows the quadratic least-squares fitting for the experimental data. Reprinted with permission from "Vladimir M. Shalaev, Wenshan Cai, Uday Chettiar, Hsiao-Kuan Yuan, Andrey K. Sarychev, Opt. Lett. 30, 3356-3358 (2005)" with copyright © 2005 of Optical Society of America.

magnetic field opposing the external magnetic field of the light. The excitation of plasmon resonances for both the electric and the magnetic light components results in the resonant behavior of the refractive index, which can become negative above the resonance as previously predicted.[7, 8] This resonance can be thought of as a resonance in an optical LC circuit, with the metal rods providing the inductance L and the dielectric gaps between the rods acting as capacitive elements C. Note that coupling between metal rods may lead to other interesting optical properties.[9] Figure 5-5 shows the retrieved refractive index and demonstrates excellent agreement between measurements and simulations. The obtained phase shift of −61° in the light transmittance at λ=1.5 µm is well below the phase shift in air ϕ_0 =−40° at 1.5 µm; and thus the negative phase acquired in the sample is $\phi \approx$ −21°. The refractive index is negative between 1.3 and 1.6 µm, with n'=−0.3±0.1 at λ=1.5 µm. Note a rather high transmittance of $T \approx$ 25% and relatively low absorption of $A \approx$ 10%. The imaginary part of the refractive index also shows a resonant behavior and it is large near the resonance. The calculations show that by optimizing the system (*e.g.*, by matching impedances), the ratio of the real and imaginary parts of the refractive index can be significantly increased.

5.3 Magnetic metamaterials

Negative μ_{eff} (ω) was shown to be possible when a polariton resonance exists in the permeability, such as in the antiferromagnets such as MnF_2 and FeF_2 [10, 11], or certain insulating ferromagnets. Arrays of gold split rings with a minimum feature size of 50-nm and an LC resonance of 200 THz in frequency at 1.5 µm wavelength over an structured area of 100 µm^2 were reported, as shown in Fig.5-6.[12] It can realize negative permeability μ at 1.5 µm wavelength.

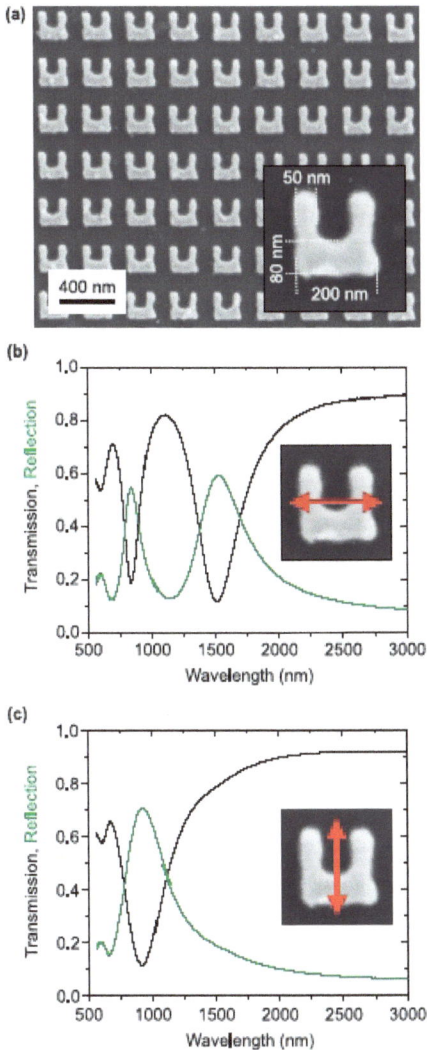

Fig. 5-6 Electron micrograph of a split ring array with a total area of 100 µm^2. The lower right-hand side inset shows the dimensions of an individual split ring. The corresponding measured normal-incidence transmission and reflection spectra for horizontal and vertical polarization are shown in (b) and (c), respectively. For (b), one can couple to the fundamental magnetic mode at 1:5 µm wavelength via the electric-field component of the incident light; for (c), one cannot. Reprinted with permission from "C. Enkrich, M. Wegener, S. Linden, S. Burger, L. Zschiedrich, F. Schmidt, J. F. Zhou, Th. Koschny, and C. M. Soukoulis,Phys. Rev. Lett. 95, 203901 (2005)" with copyright ©2005 of The American Physical Society.

The structures were fabricated using standard electron-beam lithography on a 1 mm thick glass substrate coated with a 5 nm thin film of indium-tin-oxide (ITO), in order to avoid charging effects of the poly(methyl methacrylate) resist layer (PMMA 950 k) during the exposure. The gold film thickness is 30 nm. To increase the resonance frequency at a given minimum feature size and simplify the nanofabrication, we almost eliminate the tiny upper arms of the SRR, leading to more ''U''-shaped structures [see Fig. 5-6 (a)]. Intuitively, these U structures$_s$ correspond to 3/4 of one winding of a magnetic coil with inductance L. The end of the U-shaped wires forms the capacitor C. If the incident light is polarized horizontally, as shown in Fig.5-6 (b), the electric field can be coupled to the capacitor of the SRR. The coupling induces a circulating current in the coil and leads to a magnetic-dipole moment normal to the SRR plane. The magnetic resonance disappears for vertical incident polarization [Fig. 5-6(c)], and leaves behind the Mie resonance of the SRR only around 950 nm wavelength.

Grigorenko *et. al.* reported another magnetic metamaterial for the optical frequency.[13] Figure 5-7 shows an example of their devices that illustrates the basic idea behind the experiments. The prepared structures are large arrays of Au pillars fabricated by high-resolution electron-beam lithography on a glass substrate and grouped in tightly spaced

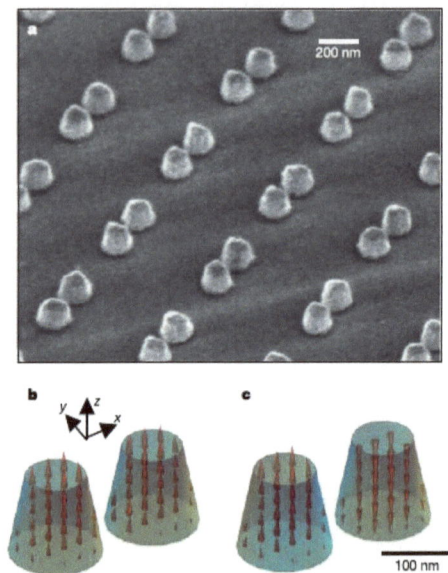

Fig. 5-7 Nanofabricated medium with magnetic response at optical frequencies. (a) Scanning electron micrograph (viewed at an angle) of an array of Au nanopillars. (b) and (c), Numerical simulation of the distribution of electric currents (arrows) inside a pair of such pillars for the symmetric and antisymmetric resonant z-modes, respectively. The non-cylindrical shape of pillars is important to provide an efficient coupling to incident light, and was intentionally introduced in our design through a choice of microfabrication procedures. Reprinted with permission from "A. N. Grigorenko, A. K. Geiml, H. F. Gleeson1, Y. Zhang, A. A. Firsov, I. Y. Khrushchev & J. Petrovic, Nature 438, 335-338 (2005)" with copyright © 2005 of Nature Publishing Group.

pairs (except for the reference samples consisting of similar but isolated Au pillars). The structures typically covered an area of ~0.1 mm^2 and contained ×10^6 pillars over the area. The lattice constant a, for periodic arrays is down to 400 nm—that is, smaller than the wavelength λ of visible light. Heights h of Au pillars (80–90 nm) and their diameters d<100 nm were chosen through numerical simulations so that the plasmon resonance in the reference samples appeared at red-light wavelengths, λ< 670 nm. A number of different structures were studied with diameter d between 80 nm and 140 nm, and the pair separation s between centers of adjacent pillars ranging from 140 nm to 200 nm; *i.e.*, the gap s-d between the neighboring pillars varies from 100 nm down to almost zero. (The best results were achieved for the two sets parameters of s=200 nm, d=140 nm, h=80 nm, and s=140 nm, d =110 nm, h=90 nm.) At these separations, electromagnetic interaction between the neighboring pillars within a pair is important and plasmon resonance observed for an individual pillar splits into two resonances for a pillar pair. These resonances are referred to as symmetric and antis-ymmetric, similar to the case of any classical or quantum system with two interacting parts and in agreement with the notation used for plasmon resonances in nanoparticles. Figure 5-8 is computational numerical simulations of optical response for interacting Au pillars: (a) and (b), are distribution of electric currents (red arrows) and magnetic field H$_y$ (color map measured in units of the magnetic field amplitude of the incident wave) for the pillars being illuminated by normal incident light of TM polarization at λ=500 nm (a) and λ=690 nm (b). Geometrical sizes were shown in meters at the bottom and the left of images. 'Dots' in the image refer to Au nanopillars. Figure 5-8 (c) and (d) are the spectral dependence of permittivity ε and permeability μ: (c) shows the absolute values and (d) the real parts. Figure 5-8 (e) and (f) are calculated reflection spectra.

5.4 Double negative-index metamaterials

Mathematically, the refractive index n is expressed as a function of the product of the permittivity and the permeability as $n = \sqrt{\varepsilon\mu}$. Hence, it is not immediately obvious that ε < 0 and μ < 0 imply n < 0. Like often in electromagnetics, one has to look at the lossy situation in order to extrapolate the conclusions to the lossless media. Upon including losses, the permittivity and the permeability are written in the polar coordinate system (in the complex plane) as [14]

The refractive index thus becomes

$$n = \sqrt{|\varepsilon||\mu|} \, e^{i\frac{1}{2}(\theta_\varepsilon + \theta_\mu)} \tag{5.2}$$

Our convention (using $i = \sqrt{-1}$ to denote the imaginary number) imposes that $\varepsilon_2 > 0$ (the imaginary part of the permittivity to be positive) and $\mu_2 > 0$, so that $\theta_\varepsilon \in [0, \pi]$ and $\theta_\mu \in [0, \pi]$. Consequently, the angle of the index of refraction is $(\theta_\varepsilon + \theta_\mu)/2 \in [0, \pi]$. This range is reduced to $[\pi/2, \pi]$ in the situation when $\varepsilon_1 < 0$ and $\mu_1 < 0$, so that the real part of the index of refraction is negative ($n_1 < 0$).

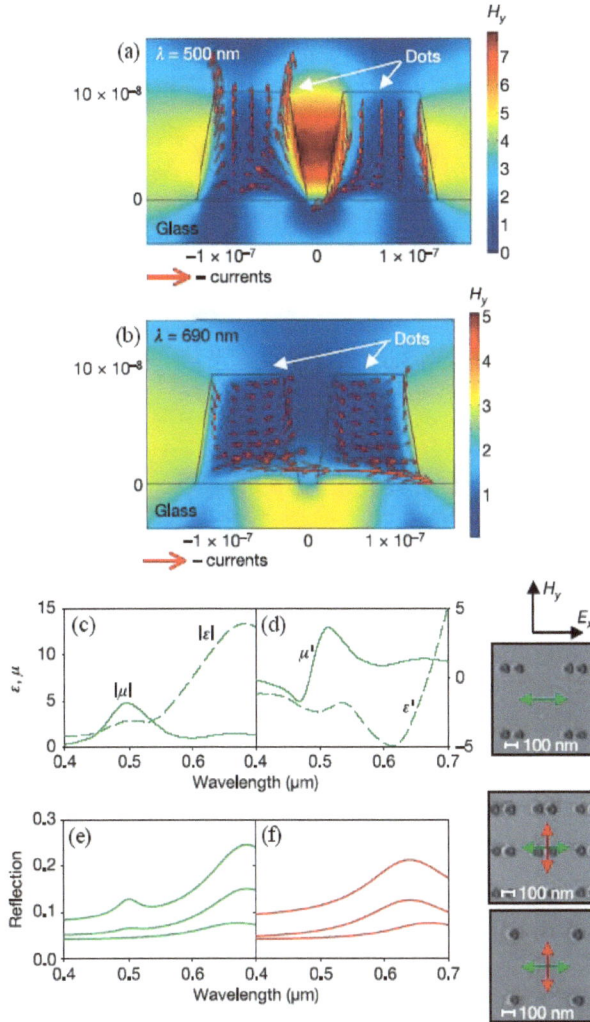

Fig.5-8 Numerical simulations of optical response for interacting Au pillars. (a) and (b), Distribution of electric currents (red arrows) and magnetic field Hy (colour map measured in units of the magnetic field amplitude of the incident wave) for the pillars being illuminated by normal incident light of TM polarization with wavelengths λ=500nm (a) and λ=690nm (b). Geometrical sizes are shown in metres at the bottom and the left of images. 'Dots' in the image refer to Au nanopillars. (c) and (d) The spectral dependence of permittivity 1 and permeability m: c shows the absolute values and d the real parts. (e) and (f) Calculated reflection spectra. Reprinted with permission from "A. N. Grigorenko, A. K. Geim1, H. F. Gleeson1, Y. Zhang, A. A. Firsov, I. Y. Khrushchev & J. Petrovic, Nature 438, 335-338 (2005)" with copyright © 2005 of Nature Publishing Group.

In particular, the lossless limit is obtained as $\theta_\varepsilon \to \pi$ and $\theta_\mu \to \pi$ which produces:

$$n = \sqrt{|\varepsilon||\mu|} \, e^{i\pi} = -\sqrt{|\varepsilon||\mu|} \tag{5.3}$$

Under the assumption of an isotropic medium, this refractive index n can be directly introduced into Snell's law and is seen to reverse the refraction direction, as already explained on the basis of phase matching of the wave-vector.

For propagating waves and $\varepsilon_1 < 0$ and $\mu_1 < 0$, the negative sign has to be chosen in order to yield waves of decaying amplitude with z if $[\varepsilon_1\mu_1 + \varepsilon_2\mu_2] < 0$. It is necessary to look at the limit of zero dissipation to obtain the physically sensible wave-vector in a non-dissipative medium. Obviously in a medium where ε_1 and μ_1 have opposite signs, the waves are all evanescent and we again have decaying waves into the medium. Since it is meaningless to associate a direction of propagation to evanescent waves, we do not discuss the choice of the sign for the square root for this case, except for remark that the sign should always be chosen such that the wave decays to zero at the infinity in a dissipative medium. Figure 5-9 presents graphically a few interesting cases of negative refraction. In the first case we have the usual negative refraction at an interface: denotes the opposite directions of the phase vector and the energy flow (ray). The second panel on the top shows a prism made of a negative material: the negative refraction causes the ray to deflect toward the apex, in contrast to the usual case of refraction toward the base in normal positive materials. The third panel (top) shows a flat lens or imaging device that can focus a source located on one side of the lens to an image on the other side. A convex lens, shown in the fourth panel, made of the negative materials behaves as a diverging lens and, by analogy, a concave lens causes rays from infinity to converge. One can see that as the angle of incidence changes

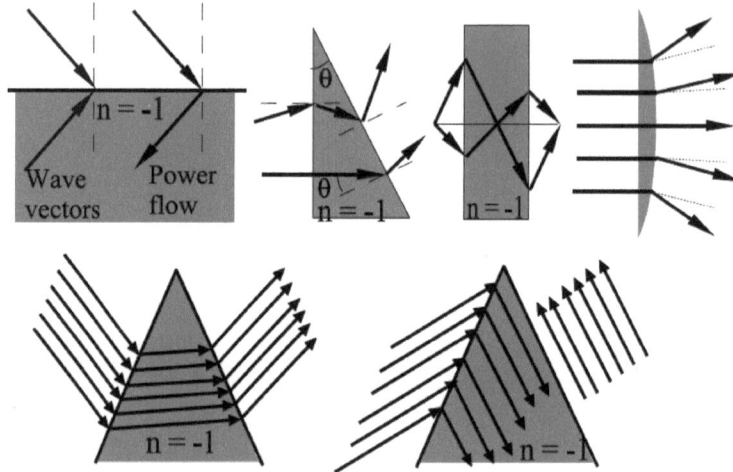

Fig. 5-9 Pictorial examples of the modified refraction process for a negative index medium where it is assumed $n = -1$ (Top, from left to right): A planar interface (the phase vectors are shown in grey), a prism, a flat slab and a plano-convex lens. Bottom: Refraction across a wedge. In the second case (right) which can be obtained by decreasing the angle in the first case (left), the rays become tangential to the second interface and the output rays appear as if associated with a source at infinity. Reprinted with permission from *Physics and Applications of Negative refractive index materials* with copyright © 2009 of Taylor & Francis Group, LLC.

beyond a critical point, the incident bundle of rays approaches the tangent to the lower interface. The rays on the other side of the interface appear as if they could be associated with another source located infinitely far away. This scenario arises because negative refraction at the other interface happens at a point infinitely far away from the apex of the wedge.

As examples, we give some double negative (DNG) index structures reported recently. Richard W. Ziolkowski reported an earlier DNG for microwave antenna use, as shown in Fig. 5-10.[15] DNG media are the materials in which the permittivity and permeability are both negative. Simulation and experimental results demonstrated the realization of DNG metamaterials matched to free space. The extraction of the effective permittivity and permeability for these metamaterials from reflection and transmission data at normal incidence is treated. It is shown that the metamaterials studied here exhibit DNG properties in the frequency range of interest.

Zhang *et. al.* numerically demonstrated a metamaterial with both negative ε and negative μ over an overlapping near-infrared wavelength range resulting in a low loss negative-index material, as shown in Fig.5-11.[16] To obtain negative refraction, both magnetic and electric responses (permeability and permittivity) need to be controlled over an overlapping wavelength range. Two reports demonstrated the fabrication and characterization of an electronic metamaterials [4, 5]. In Ref. [4], the negative index material consists of a pair of gold films separated by a dielectric layer with a two-dimensional, square-periodic array of circular holes perforating the entire multilayer structure. The magnetic resonance between the two gold films along with the electric response of the gold film to the external electrical field results in the negative refraction for the wave propagation normal to the films. The negative refractive index was obtained at a wavelength around 2 μm; the real part is as negative as -2 and the imaginary part about $+3$. The imaginary part of the index being larger than the real part indicates significant loss. The refractive index is expressed in terms of the real and imaginary parts of the permittivity and permeability as $n \equiv n_1 + i n_2 = \sqrt{(\varepsilon_1 \mu_1 - \varepsilon_2 \mu_2) + i(\varepsilon_1 \mu_2 + \varepsilon_2 \mu_1)}$

where $\varepsilon = \varepsilon_1 + i\,\varepsilon_2$ and $\mu = \mu_1 + i\,\mu_2$. To achieve a negative n_1, the imaginary part n_2 inside the square root needs to be $n_2 < 0$, which can be satisfied for a sufficiently large imaginary term without requiring both ε_1 and μ_1 to be negative. At the wavelength with negative index, ε_1 is negative and large, $\varepsilon_1 >> \varepsilon_2$, giving $n = \sqrt{\varepsilon_1 \mu_1 + i\varepsilon_1 \mu_2}$. Around the magnetic resonance, $\mu_2 > 0$, which indicates that the real part of n is negative even if $\mu_1 > 0$. However, the sign of μ_1 determines the relative magnitudes of the imaginary and real parts of the refractive index, with $n_1 > n_2$ for $\mu_1 < 0$; and is opposite for $\mu_1 < 0$. Thus, to achieve a negative index with a small imaginary part of the index, a negative permeability is required.

Figure 5-11(a) shows the schematic diagram of the staple structure exhibiting negative permeability. This structure is a L-C circuit with an inductance associated with both the loop and the electron inertia, and a capacitor formed by the staple footings. A simplified, higher-frequency version of the staple structure is just a pair of finite-width metal stripes which is parallel to the direction of magnetic field separated by a dielectric layer (*e.g.* distributed inductance/capacitance), as shown in Fig. 5-11 (b). It is well known that an array of thin metallic wires along the direction of electrical field can introduce a negative ε [see Fig. 5-11 (c)]. Combination of the structures shown in Figs. 5-11 (b) and (c) results in a negative index material with both magnetic and electric responses. A question is that how the incorporation of the thin wires affects the resonance of magnetic structure. This will be studied below by comparing the optical properties of structure shown in Fig. 5-11(d) (with thin metal wires along the electric field direction) with

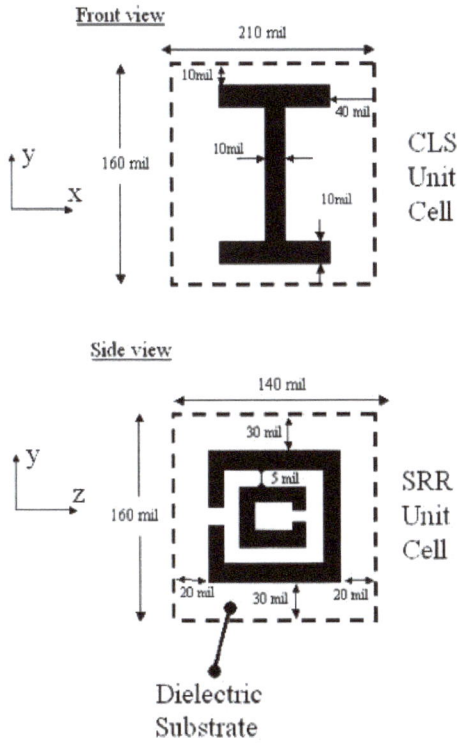

Fig.5-10 Nonplanar metamaterials geometry. Reprinted with permission from "Richard W. Ziolkowski, IEEE Transactions on Antennas and Propagation 51(7), 1516-1529 (2003)" with copyright © 2003 of IEEE.

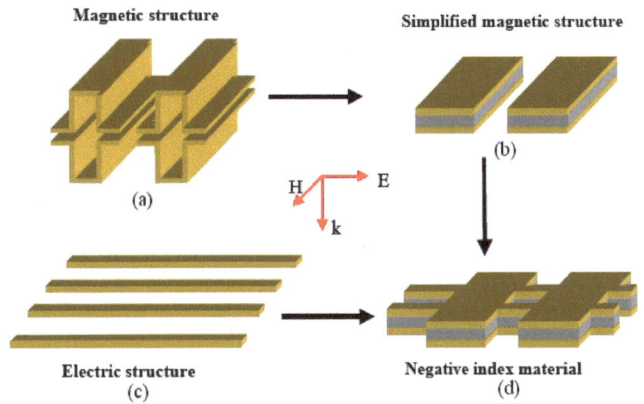

Fig. 5-11 Schematic of the NIM design (a) Staple structures for magnetic resonance, (b) A simplified structure for magnetic resonance. (c) Array of metallic wires along the electrical field direction for electrical response. (d) Combining (b) and (c) to get a negative index material. Reprinted with permission from "Shuang Zhang, Wenjun Fan, K. J. Malloy and S. R. J. Brueck, N. C. Panoiu and R. M. Osgood, Opt. Express 13, 4922-4930 (2005)" with copyright (C) 2005 of Optical Society of America.

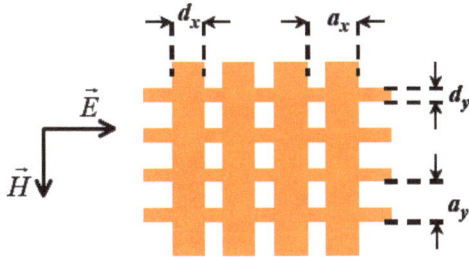

Fig. 5-12 Top view of the structure with geometrical parameters indicated. Reprinted with permission from "Shuang Zhang, Wenjun Fan, K. J. Malloy and S. R. J. Brueck, N. C. Panoiu and R. M. Osgood, Opt. Express 13, 4922-4930 (2005)" with copyright © 2005 of Optical Society of America.

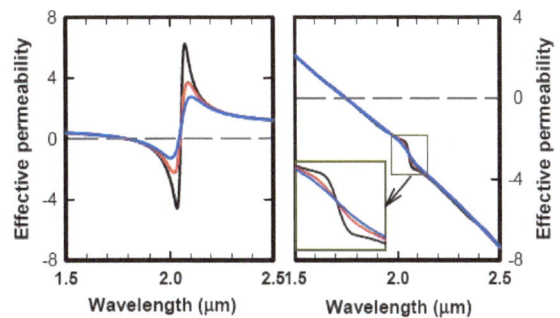

Fig. 5-13 (a) Effective permeability and (b) effective permittivity for different scattering loss parameters of Au. Reprinted with permission from "Shuang Zhang, Wenjun Fan, K. J. Malloy and S. R. J. Brueck, N. C. Panoiu and R. M. Osgood, Opt. Express 13, 4922-4930 (2005)" with copyright © 2005 of Optical Society of America.

that one in Fig. 5-11 (b) (without the metal wires). The geometrical parameters of the structure are indicated in the top view shown in Fig. 5-12. As an example, they designed a structure with the pitches of the 2D gratings a_x and a_y which were set both fixed to be 801 nm, less than the resonance wavelength of ~2 μm. The refractive index of the dielectric layer between the gold films was taken to be 1.5. The thicknesses of the Au/dielectric/Au layers were fixed at 30/60/30 nm, respectively. The authors claimed that refractive index can be realized by this structure. Rigorous coupled-wave analysis (RCWA), a commonly used algorithm to calculate the transmission and reflection of periodic structures, was used for the simulations. The effective permeability and permittivity were extracted and shown in Fig. 5-13. Clearly, the impact of the scattering loss depends mainly on the permeability, while the permittivity is slightly affected only. The range over which both the permeability and permittivity are negative. The NIM structure can be improved to exhibit simultaneous negative permeability and negative permittivity, *i.e.* to a double-negative material. In contrast to the works described in Refs. 8 and 9, the improved structure has a much lower loss, much better impedance matching, and thus a much higher transmission, which will lead to more extensive applications. Furthermore, the proposed structure has a minimum feature size of ~ 100 nm, and can be easily fabricated over a large-area by standard optical interferometric lithography techniques.

5.5 Energy and momentum in negative refractive index materials [14]

The study of energy propagation is deeply rooted in the study of left-handed media since negative constitutive parameters can lead to paradoxical conclusions if not considered carefully. The transfer of momentum from an electromagnetic wave to matter is formulated within the classical framework of the Maxwell stress tensor and the Lorentz force. Two physical subsystems are considered: the electromagnetic subsystem constituted by the propagating electric and magnetic fields governed by the Maxwell equations, and the matter subsystem described by the type of materials which the wave is propagating through.

Considering physical subsystems, one is often interested in knowing if energy or momentum is conserved. This is not to say, of course, that the energy is not conserved in general, but that it may not be when one restricts the consideration to specific subsystems. Mathematically, the equations of conservation of energy and momentum are cast in the most general form as

$$\nabla \cdot \mathbf{S} + \frac{\partial W}{\partial t} = -\phi \tag{5.3a}$$

$$\nabla \cdot \overline{\overline{T}} + \frac{\partial \mathbf{G}}{\partial t} = -\mathbf{f} \tag{5.3b}$$

where $\mathbf{S}$ and $\overline{\overline{T}}$ denotes the energy and momentum flow, respectively, while W and $\mathbf{G}$ denotes the energy density and momentum density, respectively. The quantities ϕ and $\mathbf{f}$, if non-zero, imply that the subsystem considered is open, *i.e.*, that energy and/or momentum is transferred to or from another subsystem that has not been considered.

The propagation of the energy of an electromagnetic wave is given by the Poynting vector, directly obtained from the Maxwell equations. The latter are here cast in the Chu form, where the material contributions are clearly identified by the polarization currents $\mathbf{P} = \mathbf{D} - \varepsilon_0 \mathbf{E}$ and $\mu_0 \mathbf{M} = \mathbf{B} - \mu_0 \mathbf{H}$ as

$$\nabla \times \mathbf{H}(t) - \varepsilon_0 \frac{\partial \mathbf{E}(t)}{\partial t} = \frac{\partial \mathbf{P}}{\partial t} \equiv J_e(t) \tag{5.4a}$$

$$\nabla \times \mathbf{E}(t) + \mu_0 \frac{\partial \mathbf{H}(t)}{\partial t} = -\mu_0 \frac{\partial \mathbf{M}}{\partial t} \equiv -J_m(t) \tag{5.4b}$$

$$\mu_0 \nabla \cdot \mathbf{H}(t) = -\nabla \cdot \mu_0 \mathbf{M} \equiv \rho_m(t) \tag{5.4c}$$

$$\varepsilon_0 \nabla \cdot \mathbf{E}(t) = -\nabla \cdot \mathbf{P} \equiv \rho_e(t) \tag{5.4d}$$

In this form, the fundamental electromagnetic fields are $\mathbf{E}(t)$ and $\mathbf{H}(t)$ and the electric and magnetic sources in the medium are $(\mathbf{J}_e, \rho_e)$, and $(\mathbf{J}_m, \rho_m)$. It is well known that the electromagnetic energy and momentum quantities can be derived solely for the electromagnetic subsystem without assuming specific models for $\mathbf{P}$ and $\mathbf{M}$, as is the case for lossless frequency dispersive media. Here we assume that the lossy dispersive medium can be modeled as an idealized assembly of independent oscillators where the motion of the electrons is described by

$$m(\frac{d^2 \mathbf{r}_i}{dt^2} + \gamma_i \frac{d\mathbf{r}_i}{dt} + \omega_i^2 \mathbf{r}_i) = -e\mathbf{E} \tag{5.5}$$

where $\mathbf{r}_i$ is the displacement vector of the N electrons and m is their mass. Defining the polarization vector $\mathbf{P} = -Ne\mathbf{r}$ where $-e$ is the charge of the electrons, the equation of the electrons motion under the action of an electric field can be rewritten as

$$\frac{\partial^2 \mathbf{P}}{\partial t^2} + \gamma_e \frac{\partial \mathbf{P}}{\partial t} + \omega_{eo}^2 \mathbf{P} = \varepsilon_0 \omega_{ep}^2 \mathbf{E} \tag{5.6a}$$

where ω_{ep} and ω_{eo} are the electric plasma and resonant frequencies respectively, and γ_e is the electric collision frequency which is responsible for losses in the medium. Similarly, the magnetization of the medium due to the series of split ring resonators is governed by

$$\frac{\partial^2 \mathbf{M}}{\partial t^2} + \gamma_m \frac{\partial \mathbf{M}}{\partial t} + \omega_{mo}^2 \mathbf{M} = F\omega_{mp}^2 \mathbf{H} \tag{5.6b}$$

where ω_{mp} and ω_{mo} are the magnetic plasma and resonant frequencies respectively, γ_m is the magnetic collision frequency, and F is the filling factor of the magnetic resonators. The energy conservation law of the entire system composed of the electromagnetic wave and of the material can be expressed as

$$\mathbf{S} = \mathbf{S}_{eh} = \mathbf{E} \times \mathbf{H} \tag{5.7a}$$

$$W = [\frac{\varepsilon_0}{2}\mathbf{E}\cdot\mathbf{E} + \frac{\mu_0}{2}\mathbf{H}\cdot\mathbf{H} + \frac{1}{2\varepsilon_0\omega_{ep}^2}(\frac{\partial\mathbf{P}}{\partial t}\cdot\frac{\partial\mathbf{P}}{\partial t} + \omega_{eo}^2\mathbf{P}\cdot\mathbf{P}) + \frac{\mu_0}{2F\omega_{mp}^2}(\frac{\partial\mathbf{M}}{\partial t}\cdot\frac{\partial\mathbf{M}}{\partial t} + \omega_{mo}^2\mathbf{M}\cdot\mathbf{M})] \tag{5.7b}$$

$$\phi = -\frac{\gamma_e}{2\varepsilon_0\omega_{ep}^2}\frac{\partial\mathbf{P}}{\partial t}\cdot\frac{\partial\mathbf{P}}{\partial t} - \frac{\mu_0\gamma_m}{2F\omega_{mp}^2}\frac{\partial\mathbf{M}}{\partial t}\cdot\frac{\partial\mathbf{M}}{\partial t} \tag{5.7c}$$

where W is the energy density.

The topic of momentum density in left-handed media is related to an important part of electrodynamics (electro-magneto-dynamics in fact), which relates electromagnetic effects to forces and motion of matter. A momentum conservation equation can be written as

$$\bar{\bar{T}} = [\frac{1}{2}(\mathbf{D}\cdot\mathbf{E} + \mathbf{B}\cdot\mathbf{H})\bar{\bar{I}} - \mathbf{DE} - \mathbf{BH}] + [\frac{1}{2}(\mathbf{P}\cdot\mathbf{E} + \mu_0\mathbf{M}\cdot\mathbf{H})\bar{\bar{I}}] + [\frac{1}{2\varepsilon_0\omega_{ep}^2}(\frac{\partial\mathbf{P}}{\partial t}\cdot\frac{\partial\mathbf{P}}{\partial t} - \omega_{eo}^2\mathbf{P}\cdot\mathbf{P})\bar{\bar{I}}] \tag{5.8a}$$

$$+[\frac{\mu_0}{2F\omega_{mp}^2}(\frac{\partial\mathbf{M}}{\partial t}\cdot\frac{\partial\mathbf{M}}{\partial t} - \omega_{mo}^2\mathbf{M}\cdot\mathbf{M})\bar{\bar{I}}]$$

$$\mathbf{G} = \mathbf{D}\times\mathbf{B} - \frac{1}{2\varepsilon_0\omega_{ep}^2}\nabla\mathbf{P}\cdot\frac{\partial\mathbf{P}}{\partial t} - \frac{\mu_0}{2F\omega_{mp}^2}\nabla\mathbf{M}\cdot\frac{\partial\mathbf{M}}{\partial t} \tag{5.8b}$$

$$\mathbf{f} = -\frac{\gamma_e}{\varepsilon_0\omega_{ep}^2}\nabla\mathbf{P}\cdot\frac{\partial\mathbf{P}}{\partial t} - \frac{\mu_0\gamma_m}{2F\omega_{mp}^2}\nabla\mathbf{M}\cdot\frac{\partial\mathbf{M}}{\partial t} \tag{5.8c}$$

Unlike the energy case in which the energy flow is unchanged and expressed by the Poynting vector, we see here that the medium contributes to the momentum flow with additive terms which is directly obtained from the polarization vectors $\mathbf{P}$ and $\mathbf{M}$. The momentum flow is therefore expressed as the sum of the momentum in non-dispersive media and a material contribution. Likewise, the momentum density $\mathbf{G}$ contains the Minkowski momentum $\mathbf{D} \times \mathbf{B}$ and terms due to the dispersion of the material. Finally, the force $\mathbf{f}$ is directly due to losses in the medium via γ_e and γ_m, and therefore represents dissipation. All these terms are expressed as function of the micro-structure of the medium, where the plasma frequency, the resonance frequency, and the damping rates appear explicitly.

The previous discussions did not suppose a specific form for the electromagnetic wave, but only a specific form of the medium micro-structure given by Eqs. (5.6). Here, we particularize the previous results to a monochromatic electromagnetic wave, where field quantities in the frequency domain are related to the field quantities in the time domain by $\mathbf{E} = \text{Re}\{\mathbf{E}(t)e^{-i\omega t}\}$, and similarly for all other quantities. In this regime, we can substitute the operator $\partial/\partial t$ by $-i\omega$ and ∇ by $i\mathbf{k}$ which yields the specific Lorentz model for the permittivity and permeability:

$$\varepsilon = \varepsilon_0(1 - \frac{\omega_{ep}^2}{\omega^2 - \omega_{eo}^2 + i\omega\gamma_e}) \tag{5.9a}$$

$$\mu = \mu_0(1 - \frac{F\omega_{mp}^2}{\omega^2 - \omega_{mo}^2 + i\omega\gamma_m}) \tag{5.9b}$$

Likewise, the electric and magnetic polarizabilities, as well as their derivatives, are expressed as

$$|\mathbf{P}|^2 = \frac{\varepsilon_0^2\omega_{ep}^4}{(\omega^2 - \omega_{eo}^2)^2 + \omega^2\gamma_e^2}|\mathbf{E}|^2 \tag{5.10a}$$

$$|\mathbf{M}|^2 = \frac{F\omega_{mp}^4}{(\omega^2 - \omega_{mo}^2)^2 + \omega^2\gamma_m^2}|\mathbf{H}|^2 \tag{5.10b}$$

$$\frac{\partial\mathbf{P}}{\partial t} = \omega^2|\mathbf{P}^2| \tag{5.10c}$$

$$\frac{\partial\mathbf{M}}{\partial t} = \omega^2|\mathbf{M}^2| \tag{5.10d}$$

Although left-handed media cannot be lossless, we can still suppose that in some regions of the frequency dispersion the real parts of the permittivity and permeability are much larger than their imaginary counterparts. These regions correspond to transparent regions, where we can assume that the losses can be negligible. The dissipation rate given by Eq. (6.21c) is directly related to the imaginary parts of the permittivity and permeability via

$$\phi = -\omega\varepsilon_2 \,|\, \mathbf{E} \,|^2 \,-\omega\mu_2 |\, \mathbf{H} \,|^2 \tag{5.11}$$

where

$$\varepsilon_2 = \varepsilon_0 \frac{\omega\gamma_e\omega_{ep}^2}{(\omega^2 - \omega_{eo}^2)^2 + \gamma_e\omega^2} \tag{5.12a}$$

$$\mu_2 = \mu_0 \frac{\omega\gamma_m F\omega_{mp}^2}{(\omega^2 - \omega_{mo}^2)^2 + \gamma_m\omega^2} \tag{5.12b}$$

Since the time average of the rate of energy change is zero, the energy conservation given by Eq. (6.1a) then reads

$$-\langle \nabla \cdot \mathbf{S} \rangle = [\omega\varepsilon_2|\mathbf{E}|^2 + \omega\mu_2|\mathbf{H}|^2]/2 \tag{5.13}$$

which is the complex Poynting theorem with $\langle S \rangle = 0.5\mathrm{Re}\{\mathbf{E} \times \mathbf{H}^*\}$.

Applying a similar procedure to the momentum density, the momentum flow due to the losses reduces to

$$\langle \overline{\overline{T}} \rangle = 0.5\mathrm{Re}\{0.5(\mathbf{D} \cdot \mathbf{E}^* + \mathbf{B} \cdot \mathbf{H}^*)\,\overline{\overline{I}}\,-\mathbf{D} \cdot \mathbf{E}^* - \mathbf{B} \cdot \mathbf{H}^*\} \tag{5.14}$$

The Wigner delay times [17] is defined as $\tau_\omega = \partial\phi/\partial\omega$, where the phase of the wave ϕ, is a popular measure for the time delay of a pulse with a slowly varying envelope and a well-defined carrier wave frequency of ω. The Wigner delay time is based on tracking a fiducial point on a wave packet that moves with the group velocity $v_g = \partial\omega/\partial k$, but is generalized to include the phase shifts which arise due to scattering as well. In a homogeneous medium, this arises from the superluminality or negativity of the group velocity

$$v_g = \left(\frac{\partial k}{\partial\omega} |_{\omega_c}\right)^{-1} = \frac{c}{n(\omega_c) + \omega\frac{\partial n}{\partial\omega} |_{\omega_c}} \tag{5.15}$$

where n is the refractive index, ω_c is the carrier frequency of the pulse, and the derivatives are evaluated at the carrier frequency. It is clear that when $\partial n/\partial\omega|_{\omega_c}$ is large and negative as happens in the case of anomalous dispersion, the group velocity can be superluminal and even negative. On the other hand, where this quantity is a large and positive number, the group velocity can become very small and gives rise to ultra-slow light where the pulse travels at terrestrial speeds of a few meters per second. It can also be seen that a positive group velocity is possible for media with $n < 0$ only when the dispersion in the refractive index is large enough. The transmission coefficient across the slab is given by

$$T(\omega) = \frac{tt'e^{(ik_{z2}d)}}{1 - r'^2 e^{(2ik_{z2}d)}} \tag{5.16}$$

where d represents the slab thickness and t, t', and r' represent the Fresnel coefficients of transmission and reflection relating the magnetic fields across the slab interfaces and are given by

$$t = \frac{2\frac{k_{z1}}{\varepsilon_1}}{\frac{k_{z1}}{\varepsilon_1} + \frac{k_{z2}}{\varepsilon_2}} \tag{5.17a}$$

$$t' = \frac{2\frac{k_{z2}}{\varepsilon_2}}{\frac{k_{z3}}{\varepsilon_3} + \frac{k_{z2}}{\varepsilon_2}} \tag{5.17b}$$

$$r' = \frac{\frac{k_{z2}}{\varepsilon_2} - \frac{k_{z3}}{\varepsilon_3}}{\frac{k_{z3}}{\varepsilon_3} + \frac{k_{z2}}{\varepsilon_2}} \tag{5.17c}$$

The Wigner delay time can be calculated using the phase of the transmission coefficient given above, and is given by

$$\tau = \frac{\partial\phi}{\partial\omega} = \frac{\frac{\partial p}{\partial\omega}\tan(k_{z2}d) + p\sec^2(k_{z2}d)\frac{\partial k_{z2}}{\partial\omega}d}{1 + p^2\tan^2(k_{z2}d)} \tag{5.18}$$

where

$$p = \frac{1}{2}\left(\frac{k_{z1}\varepsilon_2}{k_{z2}\varepsilon_1} + \frac{k_{z2}\varepsilon_1}{k_{z1}\varepsilon_2}\right) \tag{5.19}$$

for p-polarized radiation, and

$$p = \frac{1}{2}\left(\frac{k_{z1}\mu_2}{k_{z2}\mu_1} + \frac{k_{z2}\mu_1}{k_{z1}\mu_2}\right) \tag{5.20}$$

for s-polarized radiation. It is understood that all quantities in the above area evaluated at the carrier wave frequency of the pulse. Note that the expressions for p- and s-polarized light become identical at normal incidence.

In a comparatively uniform landscape of delay times, the resonant conditions for the slab surface plasmon polaritons (SPPs) stand out in stark contrast where the magnitude of the delay times are very large. The entire dispersion of the SPPs can be traced out by the regions of large magnitude of the Wigner delay times. Both the antisymmetric and symmetric SPPs modes of electric nature manifest for the p-polarized light, while similar SPP modes of magnetic nature manifest for s-polarized light.

5.6 Summary

Three types of metamaterials: electronic metamaterials ($\varepsilon < 0$ and $\mu < 0$), magnetic metamaterials ($\varepsilon > 0$ and $\mu < 0$), and DNG ($\varepsilon < 0$ and $\mu < 0$) were introduced. Considering many relevant publications appeared already, this chapter highlights fundamental and theoretical issues of the metamaterials applied in optical wavelength regime due to scope of this book, including energy and momentum in negative refractive index materials. Detailed applications such as superfocusing/imaging and in microwave antennas as well as fabrication issues will be illustrated separately in Chapt.6, Chapt.7, Chapt.10, and Chapt.13 respectively, of this book.

Reference

[1] J. B. Pendry, A. J. Holden, D. J. Robbins, and W. J. Stewart, "Magnetism from Conductors and Enhanced Non-Linear Phenomena", IEEE Trans. Microwave Theory Tech. 47, 2075 (1999).

[2] V. G. Veselago, "The electrodynamics of substances with simultaneously negative values of ε and μ", Sov. Phys. USPEKHI 10, 509 (1967).

[3] Shuang Zhang, Wenjun fan, N. C. Panoiu, K. J. Malloy, R. M. Osgood, S. R. J. Brueck, "Experimental demonstration of Near-Infrared Negative-Index Materials," Phys. Rev. Lett. 95, 137404 (2005).

[4] S. Zhang, W. Fan, B. K. Minhas, A. Frauenglass, K. J. Malloy, and S. R. J. Brueck, "Midinfrared Resonant Magnetic Nanostructures Exhibiting a Negative Permeability", Phys. Rev. Lett. 94, 037402 (2005).

[5] M. A. Ordal, L. L. Long, R. J. Bell, S. E. Bell, R. R. Bell, R.W. Alexander, and C. A. Ward, "Optical properties of the metals Al, Co, Cu, Au, Fe, Pb, Ni, Pd, Pt, Ag, Ti, and W in the infrared and far infrared", Appl. Opt. 22, 1099 (1983).

[6] Vladimir M. Shalaev, Wenshan Cai, Uday Chettiar, Hsiao-Kuan Yuan, Andrey K. Sarychev, Vladimir P. Drachev, Alexander V. Kildishev, "Negative Index of Refraction in Optical Metamaterials," Opt. Lett. 30, 3356-3358 (2005).

[7] V. A. Podolskiy, A. K. Sarychev, and V. M. Shalaev, "Plasmon modes in metal nanowires and left-handed materials", J. Nonlinear Opt. Phys. Mater. 11, 65 (2002).

[8] V. A. Podolskiy, A. K. Sarychev, and V. M. Shalaev, "Plasmon modes and negative refraction in metal nanowire composites", Opt. Express 11, 735 (2003).

[9] Y. Svirko, N. Zheludev, and M. Osipov, "Layered chiral metallic microstructures with inductive coupling", Appl. Phys. Lett. 78, 498 (2001).

[10] D. L. Mills and E. Burstein, "Polaritons: the electromagnetic modes of media", Rep. Prog. Phys. 37, 817 (1974).

[11] R. E. Camley and D. L. Mills, "Surface polaritons on uniaxial antiferromagnets", Phys. Rev. B 26, 1280 (1982).

[12] C. Enkrich, M. Wegener, S. Linden, S. Burger, L. Zschiedrich, F. Schmidt, J. F. Zhou, Th. Koschny, and C. M. Soukoulis, "Magnetic Metamaterials at Telecommunication and Visible Frequencies", Phys. Rev. Lett. 95, 203901 (2005).

[13] A. N. Grigorenko, A. K. Geim1, H. F. Gleeson1, Y. Zhang, A. A. Firsov, I. Y. Khrushchev & J. Petrovic, "Nanofabricated media with negative permeability at visible frequencies", Nature 438, 335-338 (2005).

[14] S. Anantha Ramakrishna, Tomasz M. Grzegorczyk, Edt.: *Physics and Applications of Negative refractive index materials*. CRC Press, 2009.

[15] Richard W. Ziolkowski, "Design, Fabrication, and Testing of Double Negative Metamaterials", IEEE Transactions on Antennas And Propagation 51(7), 1516-1529 (2003).

[16] Shuang Zhang, Wenjun Fan, K. J. Malloy and S. R. J. Brueck, N. C. Panoiu and R. M. Osgood, "Near-infrared double negative metamaterials", Opt. Express 13, 4922-4930 (2005).

[17] E. P. Wigner. "Lower limit for the energy derivative of the scattering phase shift". Phys. Rev., 98:145–147 (1955).

6 TOP-DOWN APPROACHES

Abstract: Many nanofabrication techniques were reported in books and journals at present. In this chapter, we targeted two commonly used approaches for fabrication of subwavelength metallic structures: focused ion beam (FIB) technology and laser interference photolithography. The former can realized fine nanofabrication over a local tiny area in one-step only. But it is a technique with high expenditure and small localized fabrication area. The latter can realize large area fabrication and cost effective. But it needs pattern transformation from photoresist into substrate. Some fabrication examples were presented. Problems existing in the fabrication processes were addressed as well.

6.1 Introduction

The nanofabrication processes can be divided into two well defined approaches: 1) 'top-down' and 2) 'bottom-up'. The 'top-down' approach uses traditional methods to guide the synthesis of nanoscale materials. The paradigm proper of its definition generally dictates that in the 'top-down' approach it all begins from a bulk piece of material, which is then gradually or step-by-step removed to form objects in the regime of nanometer-size scale. Well known techniques such as photo lithography, electron beam lithography, anodization, and ion- and plasma-etching, that will be later described, all belong to this type of approach. The top-down approach for nanofabrication is the method firstly suggested by Feynman in his famous American Physical Society lecture in 1959.

Top down fabrication can be likened to sculpting from a block of stone. A piece of the base material is gradually eroded until the desired shape is achieved. That is, you start at the top of the blank piece and work your way down removing material from where it is not required. Nanotechnology techniques for top down fabrication vary but can be split into mechanical and chemical fabrication techniques. The most top down fabrication technique is nanolithography. In this process, required material is protected by a mask and the exposed material is etched away. Depending upon the level of resolution required for features in the final product, etching of the base material can be done chemically using acids or mechanically using ultraviolet light, and x-rays or electron beams. This is the technique applied to the manufacture of computer chips.

Top down approach mainly includes the following techniques:
• Lithographic techniques
  – Electron beam lithography
  – Focused ion beam milling (FIBM)
  – Holographic 3D lithography / laser interference lithography
  – Nano-lithography
• Etching techniques
• Deposition techniques
  – Sputtering, evaporation
  – Chemical vapor deposition (CVD)
  – Molecular beam epitaxy (MBE)
  – Atomic layer deposition (ALD)
• Replication technologies
  – Nanoimprinting
  – Chemical inversion processes for 3D replication in different classes of materials

For the fabrication of plasmonic nanostructures, currently, the most commonly used top-down techniques are FIBM and laser interference lithography. The former is used for fabrication of localized nanostructures, and the latter for one-dimensional grating structures, and two-dimensional (2D) metallic nanoparticles etc.

6.2 FIB technology

The FIB technique was mainly developed during the late 1970s and the early 1980s, and the first commercial instruments were introduced more than a decade ago [1]. Modern FIB systems are becoming widely available in semiconductor, microelectronics, and processing environments, as well as in failure analysis and chip-design centers. The technology enables localized milling and deposition of conductors and insulators with high precision. Hence it obtains a significant success in applications of device modification, mask repair, process control, and failure analysis [2–6]. Also, the preparation of specimens for a transmission electron microscopy (TEM) and the trimming of thin-film magneto-resistive heads (for magnetic storage disks) are important applications of the FIB, which were discussed in Ref. [7] and [8]. Only recently, a number of authors reported the use of FIB in micromachining applications for MEMS devices [9, 10].

The ion beam has many advantages over other high-energy particle beams. For example, as compared to photons or electrons, ions are much heavier and can strike with much greater energy density on the target to directly write or mill patterns on hard materials, such as semiconductors, metals or ceramic substrates, and even ultra-hard materials, *e.g.*, diamond. On the other hand, photons and electrons can only effectively write or expose on soft materials, such as photoresists (SU8, AZ1350, AZ-P4400 etc.) or e-beam resists (PMMA). The resists are normally used as the media to transfer the patterns from mask to silicon wafer or other substrates. Commonly used techniques are photolithography, laser direct writing, and e-beam lithography (also called e-beam writing). Therefore, the direct writing capability of the FIB can reduce various hassles and defects caused by the masks and resists during the process of pattern transformation. Also, as compared to e-beams, the FIB does not generate high-energy backscattered electrons, which often limits the minimum linewidth attainable by the e-beam lithographic or lift-off process. On the other hand, because ions are much heavier, the lateral scattering of the FIB is relatively low, resulting in striking only the intended regions. Thus a fine FIB can directly write a very narrow line.

6.2.1 Introduction of FIB machine

6.2.1.1 Ion sources

With the recent advent of powerful sources such as the liquid metal ion source (LMIS) in the late 1970s as well as the advances in ion optics since the late 1980s, the FIB technology has been developed rapidly and many FIB-related applications appeared. The range of materials being used in FIB systems is also expanding to further increase the extent of their applications. The ion sources that are currently available include Al, As, Au, B, Be, Bi, Cs, Cu, Ga, Ge, Er, Fe, H, In, Li, Ni, P, Pb, Pd, Pr, Pt, Si, Sn, U and Zn. Many of these ion species are produced from liquid-metal alloy sources because of the high melting temperature and the reactivity or volatility associated with the pure metal species. Ar, B and P ions are particularly interesting because of their use in implantation of semiconductor materials. The popular ion species for microfabrication are As, Be, Ga and Si.

6.2.1.2 Ion column

A schematic diagram of a FIB ion column is shown in Fig. 6-1. The structure of the column is similar to that of a scanning electron microscope (SEM). The major difference between FIB and SEM is the usage of a gallium ion (Ga^+) beam instead of an electron beam. Therefore, the ion energy for bombardment/collision is much higher than that of the electron energy. Direct material removal can be realized accordingly. A vacuum of about 1×10^{-7} mbar is maintained inside the column. The ion beam is generated from a liquid-metal ion source (LMIS) under the electric field with high voltage. This electric field causes the emission of positively charged ions from a liquid gallium cone, which is formed on apex of a tungsten needle. A typical extraction voltage is 7000 V. The extraction current under normal operating conditions is 2 µA. [11] After a first refinement through the spray aperture, the ion beam is condensed in the first electrostatic lens. The upper octopole then adjusts the beam stigmatism. The ion beam energy is typically between 10 keV and 50 keV (most commonly used energy is 30 keV for FEI FIB machines) with beam currents varying from 1 pA to 10 nA. Using the variable aperture mechanism, the beam current can be varied over four decades, allowing a fine beam for both high-resolution imaging on sensitive samples and a heavy beam for fast and rough milling. Typically, total seven values of the beam current can be selected for users in the commercial FIB machines.

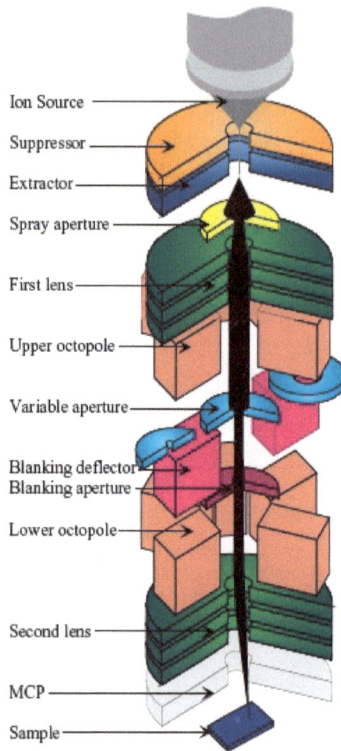

Ion Source
Suppressor
Extractor
Spray aperture
First lens
Upper octopole
Variable aperture
Blanking deflector
Blanking aperture
Lower octopole
Second lens
MCP
Sample

Fig.6-1 Schematic diagram of a two-lens FIB system. Reprinted with permission from "Robert Puers, J. Micromech. Microeng. 11, 287, 2001" with Copyright ©2001 of Institute of Physics Publishing.

Blanking of the beam is accomplished by the blanking deflector and aperture, while the lower octopole is used for raster scanning the beam over the sample in a user-defined pattern. In the second electrostatic lens, the beam is focused to a fine spot, enabling a best resolution in the scale of sub 10 nm. The multi-channel plate (MCP) is used to collect secondary particles for imaging. The imaging principle is discussed in the following sections.

6.2.1.3 Work chamber

The samples that are treated by FIB machines are mounted on a motorized five-axis stage, inside the work chamber. Under normal operating conditions, the vacuum as low as 10^{-7} mbar is maintained inside this stainless-steel chamber. Loading and unloading of the samples is usually done through a loadlock subsystem, in order to preserve the vacuum inside the work chamber as low as possible. It typically takes tens minutes to load or unload a sample.

6.2.1.4 Vacuum system and gas delivery system

A subsystem of vacuum pumps is needed to maintain the vacuum inside the column and the work chamber. Two or three pumping steps are used in combination with a mechanical pump, a turbo pump, and ion pumps for pumping the work chamber. The ion column is additionally provided with one or two ion pumps. For majority FIB machines, the system is available for delivering a variety of gases to the sample surface. To this end, a gas cabinet containing all applicable gases is equipped outside the vacuum chamber. The gas containers are connected to a so-called nozzle assembly inside the vacuum chamber through an appropriate piping system. The chemical gases are used for faster and more selective etching, as well as for the deposition of materials (see below).

6.2.1.5 User interface

All operations such as loading and unloading of samples (partly), manipulating the stage, controlling valves for gas delivery, turning on and off pumps and manipulating the ion beam are carried out via operation software. Indeed, the complete user interface is realized by means of a computer workstation. Currently, there are two operation systems have been used in the commercial FIB-equipped computer workstation, UNIX (e.g. Micrion 9500EX) and Windows (*e.g.*, FEI Quanta 200 3D, Strata, and Nova series).

6.2.2 Principle of FIB imaging, milling, and deposition

When energetic ions hit the surface of a solid sample, they lose energy to the electrons of the solid as well as to its atoms. The most important physical effects of the incident ions on the substrate are: sputtering of neutral and ionized substrate atoms (this effect enables substrate milling), electron emission (this effect enables imaging, but may cause charging of the sample), displacement of atoms in the solid (induced damage) and emission of phonons (heating). Chemical interactions include the breaking of chemical bonds, thereby dissociating molecules (this effect is exploited during deposition).

6.2.2.1 Imaging

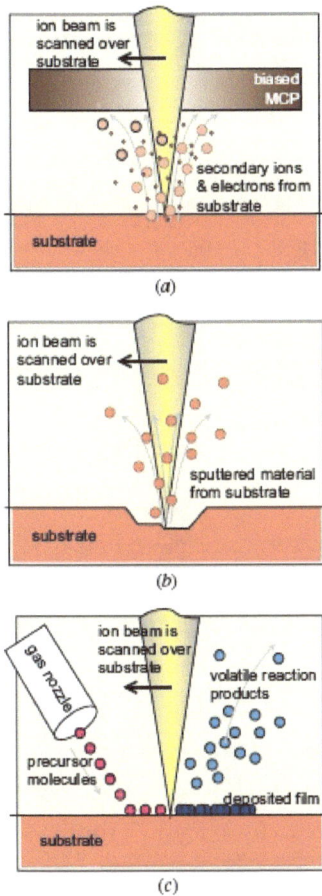

Fig.6-2 Principle of FIB (a) imaging, (b) milling and (c) deposition. Reprinted with permission from "Robert Puers, J. Micromech. Microeng. 11, 287, 2001" with Copyright © 2001 of Institute of Physics Publishing.

As illustrated in Fig. 6-2 (a), during FIB imaging the finely focused ion beam is raster scanned over a substrate, and secondary particles (neutral atoms, ions and electrons) are generated in the sample. As they leave the sample, the electrons or ions are collected on a biased detector (MCP). The detector bias is a positive or a negative voltage, respectively, for collecting secondary electrons or secondary ions. The secondary ions that are emitted can be used for secondary ion mass spectroscopy (SIMS) of the target material in a mass spectrometer attached to the system. Inevitably, during FIB operations, a small amount of Ga^+ ions are implanted in the sample, and large numbers of secondary electrons leave the sample. To prevent positive surface charges from building up, the substrate can be flooded with electrons from a separate electron source (only when collecting secondary ions for imaging). The system thus prevents damage due to electrostatic discharge and it enables the reliable imaging of non-conducting materials such as glass (which is often used in microsystems).

The best resolution of FIB images equals the minimum ion beam spot size, *i.e.*, below 10 nm. In crystalline materials such as aluminum and copper, the ion penetration depth varies due to channeling along open columns in the lattice structure. Since the secondary electron emission rate depends on the penetration depth, FIB can be used to image crystal grains, revealing different crystal orientations. Changing the angle between the ion beam and the sample surface (equivalent to changing the sample tilt angle) causes the image to be oriented differently with respect to the beam. This manifests itself in the changing brightness of the grains under different tilting angles. It should be mentioned that imaging with FIB inevitably induces some damage to the samples. Most of the Ga^+ ions that arrive at the sample surface enter the sample; thus ion implantation occurs. The depth of this implanted region is related to the ion energy and the angle of incidence. Besides implantation, some milling always occurs when the ion beam is scanned across the sample surface. Of course this milling effect can be drastically reduced when using a fine ion beam (fine spot and low ion current).

6.2.2.2 Milling

An energetic ion can interact with a target surface in various ways. Depending on the ion energy, the interaction can be swelling, deposition, sputtering, redeposition, implantation, backscattering or nuclear reaction. However, some of the interactions are not completely separable and may lead to unwanted side effects that need to be understood and avoided for a specific application. For milling applications, it is desirable that the incoming ions interact only with the atoms at the surface or near the surface layer of the target substrate and also lead to a collision cascade on the atoms. If the ion energy (or momentum) is adequate, the collision can transfer sufficient energy to the surface atom to overcome its surface binding energy (3.8 eV for Au and 4.7 for Si), and the atom is ejected as a result. This interaction is called sputtering and is the governing effect in the FIB milling. Because the interaction depends solely on momentum transfer to remove the atoms, sputtering is a purely physical process. The sputtering yield, defined as the number of atoms ejected per incident ion, is a measurement of the efficiency of material removal. The yield is normally ranging from 1 to 50 atoms per ion and is a function of many variables, including masses of ions and target atoms, ion energy, direction of incidence to the surface of the target, target temperature and ion flux. Initially, the sputtering yield increases as the ion energy increases, but the yield starts to decrease as the energy is increased past the level where the ions can penetrate deep into the substrate. At this stage of interactions, implantation or doping can take place in which the ions become trapped in the substrate as their energy is expended. As a result, the proper energy for sputtering is between 10 and 100 keV for most of the ion species used for milling. During sputtering, a portion of the ejected atoms or molecules is frequently redeposited into the sputtered region and this redeposition makes it difficult to control the amount of material removed by sputtering. In fact, the essence of FIB milling is to carefully control both the material sputtering and the redeposition, so that a precise amount of material can be removed. It should also be noted that since the FIB implantation is mainly material property alternation in nature instead of material removal, only sputtering and redeposition will be examined in this review. The removal of sample material is achieved using a high ion current beam. The result is a physical sputtering of sample material, as illustrated schematically in Fig. 6-2 (b). By scanning the beam over the substrate, an arbitrary shape can be etched. However, these numbers cannot be used directly to calculate the etch rate, because, depending on the scanning style, redeposition occurs, which drastically reduces the effective etch rate. Furthermore, the sputtering yield is dependent of the angle of incidence: it roughly increases with $1/\cos(\theta)$, whereas θ is the angle between the surface normal and the ion beam direction.

6.2.2.3 Deposition

FIB enables the localized maskless deposition of both metal and insulator materials. The principle is chemical vapour deposition (CVD) and the occurring reactions are comparable to, for example, laser induced CVD [12, 13]. The main difference is the better resolution but lower deposition rate of FIB. The metals that can be deposited on commercially available machines are platinum (Pt) and tungsten (W). In the case of W, the organo-metallic precursor gas is $W(CO)_6$. The deposited insulator material is SiO_2, with 1, 3, 5, 7- tetramethylcyclotetrasiloxane (TMCTS) and oxygen (O_2) or alternatively water vapor (H_2O) as precursors. The deposition process is illustrated in Fig. 2(c). The precursor gases are sprayed on the sample surface by a fine needle (nozzle), where they adsorb the gases. In a second step, the incoming ion beam decomposes the adsorbed precursor gases. Then the volatile reaction products are desorbed from the surface and removed through the vacuum pumping system, while the desired reaction products (W or SiO_2) remain on the localized surface area as a thin film. The deposited material is not fully pure however, because organic contaminants as well as Ga^+ ions (from the ion beam) are inevitably included.

6.2.3 Fabrication of plasmonic structures

As some fabrication examples here, the milling experiments were carried out using a FIB machine (Micrion 9500EX) with a liquid gallium ion source. This FIB machine is integrated with a scanning electron microscope (SEM), energy dispersion X-ray spectrometer (EDX) facilities, and a gas-assistant etching (GAE) functions. This machine has a focused Ga^+ ion beam with maximum energy of 50 keV, a probe current ranging from 4 pA to 19.7 nA, and beam limiting aperture size ranging from 25 μm to 350 μm. For the smallest beam currents, the beam was focused down to as small as 7 nm in diameter at site of full width and half maximum (FWHM). Using a computer program, the milling process is carried out by means of varying the ion dose for different relief depths of the milled patterns. The defined area for the FIB milling with bitmap function and zero-overlap scanning was 15×15 μm^2 and 5×5 μm^2, respectively.

A bitmap function of our FIB machine was used firstly in our experiments. The bitmap file is called in the operation window before the FIB milling that acted as a virtual mask. Data-format transformation is important for both two-dimensional (2D) and three-dimensional (3D) nanofabrication using FIB technology. The reason is that there is only one bitmap file format which has the extension name *.XBM, in operation software of the FIB workstation. This XBM format editing has limited drawing functions (it can only draw points, lines, rectangles, and circles). It is very difficult to create a complicated map file *e.g.*, a spiral, gear, and so on. Therefore, it is necessary to make use of a specialized program which can transfer other bitmap files created by other drawing software (such as AutoCAD and CorelDRAW) into the FIB-required XBM format file. In other words, the data transfer function, which acquires inspection data generated by any of several kinds of wafer inspection systems and then converts it to the internal format (XBM). As can be seen in Fig. 6-3, XBM is the data format recognized for all the Micrion FIB machines-based applications. It may be different for the different models, *e.g.*, JPG bitmap format is accepted only for the machines from FEI Corporation. The commonly used map formats (*e.g.*,TIF, DXF etc.) are selected for the transformation. Firstly, a bitmap file created by AutoCAD or CorelDRAW is transferred into the TIF format. This transformation is easily realized in the Windows95/98 environment. Next, the specialized transferring software transferred the TIF format file into the required XBM format.[14] Considering this, suitable transformation software is necessary for the FIB microfabrication requirement. For practical 2D and 3D micromanufacturing using FIB, this software is required. During the milling process, the beam current of 209 pA and 99 pA were used respectively. A series results were

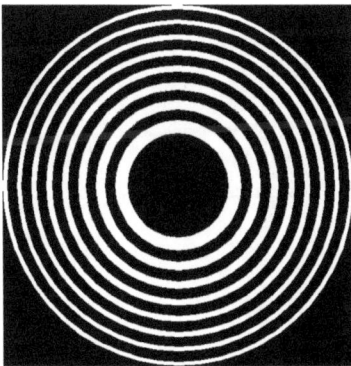

Fig. 6-3 Designed pattern with bitmap format of *.XBM (JPG format is available for FEI FIB machines) which is the only format accepted by our FIB. The computer workstation of the FIB reads the bitmap file and scans the green area by controlling beam deflection ("0" is undo and "1" is do). The pattern is named the defined area before milling.

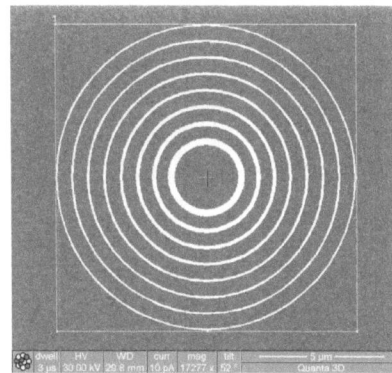

Fig. 6-4 Designed pattern with bitmap format of *.JPG which is the only format accepted by the FEI Quanta 200 3D machine. The designed pattern can be called in the defined box with size the same as the outer diameter of the pattern. This figure shows the operation window.

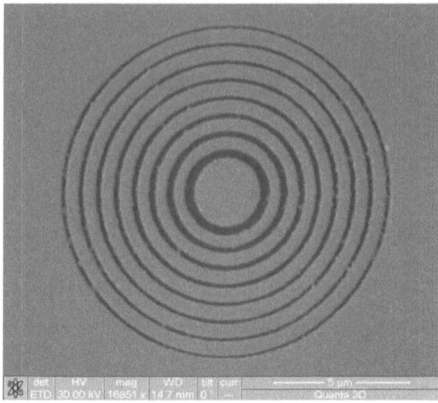

Fig. 6-5 SEM micrograph of the FIB fabricated plasmonic structure with variant period. Design outer diameter and focal length is 13.5 μm and 1 μm. The structure performance was reported in Ref. [15].

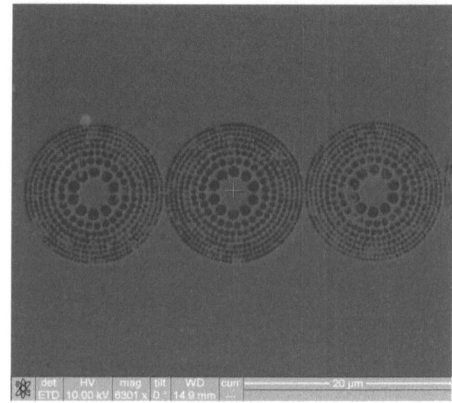

Fig. 6-6 Anther plasmonic structure with variant pinholes for superfocusing. Working performance of the structure was reported in Ref. [16].

obtained using the different process parameters: ion doses ranging from 0.25 nC/μm^2 to 2 nC/μm^2, the stage-tilted angle ranging from 10° to 45°, and ion energy of 50 KeV. Working distance from sample surface to facet of ion column was 20 mm. Electron mode of the FIB milling was selected for the substrate material of Si (100).

Figure 6-3 and Fig. 6-4 is the designed pattern with XBM and JPG format, respectively. Before the FIB milling, the pattern is called in a defined frame with size the same as outer diameter of the designed pattern. Figure 6-5 is a SEM micrograph of the FIB fabricated structure using process parameters of 10 pA beam current, 30 keV ion energy, and 10 min. milling time. Figure 6-6 is an Ag film coated plasmonic structure with circular nanopinholes for superfocusing. In comparison to the structure shown in Fig. 6-5, it can suppress sidelobes of the transmission intensity. Figure 6-7 is another FIB fabrication example, an Ag nanoparticles array with five-star shape and 170 nm period. It can be used for biosensing or nanophotonic devices for field excitation. Another typical FIB fabrication example is fiber probe of near-field scanning optical microscope, see pictures of the modified probe in Chapter 4. The central aperture is penetrated through the coated Al thin film with 100~200 nm in thickness so as to enable light beam reaching the inside fiber core for further propagation. Size of the aperture is ranging from 50 nm to 250 nm. The fiber probe with different aperture size can be selected before use according to concrete usages and sample requirements.

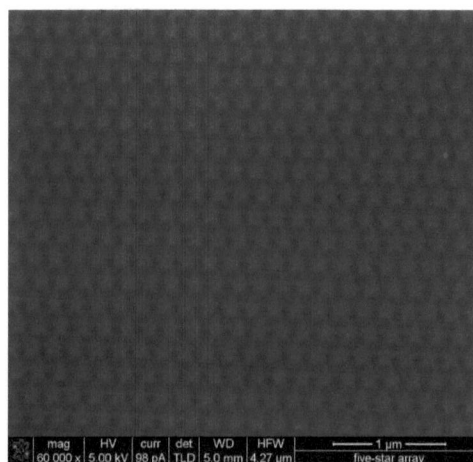

Fig. 6-7 Ag nanopaticles with five-star shape fabricated using FIB machine (FEI Nova). Beam current, ion energy, and milling time is 10 pA, 30 keV, and 10 min., respectively.

6.3 Laser interference photolithography

In last decades, many research papers reported the applications of noble metallic nanoparticles for biosensing due to an effect of localized surface plasmon resonance (LSPR) which enhances sensitivity of the detection greatly [17-21].

Currently, the commonly used fabrication technique for the nanoparticles is nanosphere lithography (NSL), also named self-assembly monolayer (SAM) technique [22-24]. However, this method has shortages of worse uniformity in both size and shape, unstable process, and low repeatability due to inherent characteristics originating from the chemical process itself. Considering this, we put forth another approach, laser interference lithography (LIL) technique for the purpose of fabrication of the Ag dots array in this chapter. Our experimental results demonstrated that this technique is applicable to gain the $\lambda/4$ feature size over large surface area of the substrate. This fabrication process is maskless, cost effective, and better controllability in comparison to the NSL method.

A Lloyd's mirror interferometer system was built, as shown in Fig. 6-8.[25] Light source is He-Cd laser at 442 nm working wavelength. The laser beam is filtered and expanded by a spatial filter which is composed of Lens 1 and a pinhole. The expanded beam is collimated by Lens 2. A part of the incident beam is reflected back by the mirror which is positioned in normal to the substrate and interferes with the other non-reflection beam to form the interference patterns. Since the beam is only split with a short path length near the substrate, this setup is very insensitive to the mechanical vibration caused instabilities. Hence no extra feedback control system is required to stabilize the interference fringe patterns [26]. Exposure time is controlled by a shutter. If the light intensity of the collimated beam is I_0, the radiation on the surface is given by

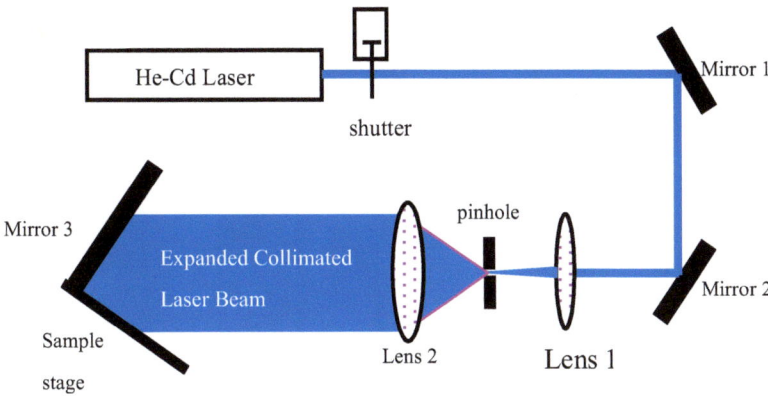

Fig.6-8 Experimental setup of the Lloyd's mirror laser interference lithography system.

$$I_{2-beam} = 2I_0[1+\cos(2kx\sin\theta)] \qquad (6-1)$$

where $k = 2\pi/\lambda$, here θ is the half angle between the two beams. The grating period d is given by

$$d = \lambda/(2\sin\theta) \qquad (6-2)$$

After the first time exposure, the substrate is rotated over 90° and exposed again with the same exposure time. The distributed intensity on the surface is written as

$$I_{double-2b} = 2I_0[2+\cos(2kx\sin\theta)+\cos(2ky\sin\theta)] \qquad (6-3)$$

The period of the interference pattern generated by this method can be easily tuned by means of changing the incident angle on the mirror as well as wavelength of the light source.

Figure 6-9 is a schematic diagram of the fabrication process. Firstly, the quartz substrate was dipped in nitric acid solution for ~6 h. Then it was cleaned by ultrasonic vibrations and acetone to remove the dust attached on surface of the substrate. We used an oven baking the substrate for half an hour at 150 °C for the purpose of removing the solvent absorbed on the surface. After that, an etched mask layer of Cr with ~10 nm in thickness was deposited in the front side of the cleaned quartz. The Cr thin layer coated on the substrate is for the purpose of enhancing adhesion between photoresist and substrate, and can also be used as a protection layer for the dry etching in next step. On top surface of this Cr layer, a layer of positive resist (AR-P3170, Allresist Co.) in 100 nm thickness of was spin-coated, and followed by pre-baking time of 20 min. at 95 °C. This fabrication process is shown in Fig. 6-9 (a). Figure 6-9 (b) is the schematic diagram of this interference lithographic process.

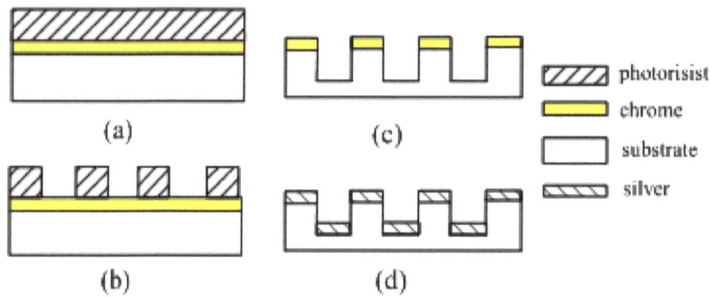

Fig. 6-9 Schematic diagram of the fabrication process: (a) 10 nm Cr and 100 nm photoresist on a SiO$_2$ substrate, (b) laser interference exposure and photoresist development, (c) wet etching and reactive ion etching for pattern transferred, (d) lift-off Cr layer and deposit Ag structure of 25 nm in thickness.

The dimension of the exposed area strongly depends on the exposure ion dose. Because of the cosine instead of rectangular distribution of the exposure intensity, energy at wings or tails of the beam profile still has energy contribution to a certain extent on the exposure process. The wing energy causes line broadening at edge of the dots, and thus makes the dots dimension enlarged slightly. Generation of the structures with high aspect ratio will be limited due to this broadening effect accordingly. For compensation, the corrected exposure time should be slightly shorter than the normal value.

The photoresist was exposed by the collimated beam from the Lloyd's mirror interferometer system. A 1D grating pattern of the photoresist layer was formed after the first time exposure and development. Exposure dose was measured to be 1.5 mW/cm^2 in normal incidence. The developer adopted in our experiments is AR 300-35 (Allresist Co.). Figure 6-10 is AFM measurement results of the Cr patterns.

The Cr patterns were obtained by the wet etching in a mixture of ceric ammonium nitrate, perchloric acid and distilled water, as shown in Fig. 6-10 (a) and Fig. 6-10 (b). Subsequently, the Cr pattern was transferred to the quartz by means of CHF$_3$/O$_2$ reactive ion etching at a flow rate of 35 sccm/1sccm, and 8 min. etching time. After the remained Cr layer was removed, the AFM was employed to view the samples (see Fig. 6-10(c)). The schematic diagram of this process is shown in Fig. 6-9.

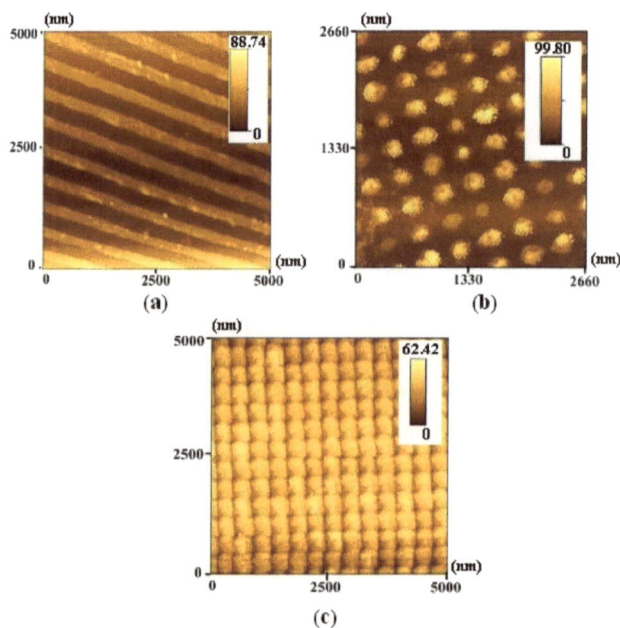

Fig.6-10 (a) Cr gratings with wet etching method, (b) Cr dots array with wet etching method, and (c) pattern of the quartz substrate with reactive ion etching technique.

A 2D photonic crystals (PCs) fabricated using interference lithography and CdSe electro-deposition was reported.[27] Characterization of CdSe 2D PCs with a hexagonal array of air voids demonstrated a well-defined drop in the transmission spectra at 4.23 µm for incident angles of 0° and 40° relative to surface normal. The drop in transmission increased with the incident angle, reaching a maximum of approximately 2.6 dB at 40°. Moreover, transmission of the polymeric PCs with an identical structure has a less pronounced drop in transmission, indicating that the drop in transmission is strongly dependent of the index contrast of the crystal. This fabrication process offers advantages over alternative approaches because it is simple and can be integrated onto arbitrary substrates. It can also be scaled to near-IR and visible wavelengths by reducing the diameter and pitch of the air voids. This can be accomplished by generating submicron interference gratings using electron-beam lithography and by using a shorter wavelength laser to create the interference pattern. The diffraction pattern is produced with the optical arrangement in Fig. 6-11, which is comprised of a collimated Nd: YAG laser beam impinging on a mask that has three gratings oriented 120° relative to one another. Each grating has 2 µm wide, 4 mm long features separated by 2 µm.

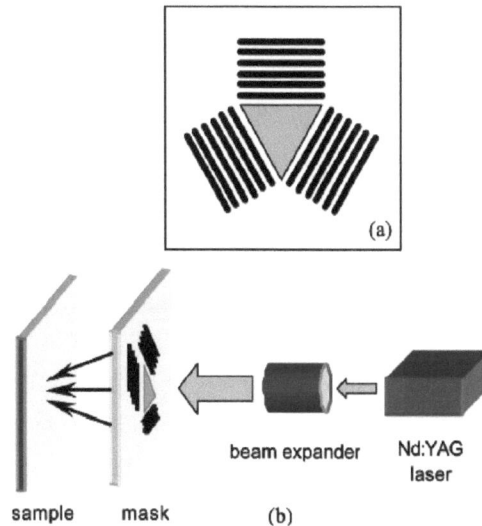

Fig. 6-11 Schematic representation of the (a) three-grating diffraction mask, and (b) optical setup for creating the hexagonal interference pattern on the ITO coated substrate. Reprinted with permission from "Ivan B. Divliansky, Atsushi Shishido, Iam-Choon Khoo, and Theresa S. Mayer, Fabrication of two-dimensional photonic crystals using interference lithography and electrodeposition of CdSe, Appl. Phys. Lett. 79, 3392 (2001)" with copyright ©2001 of American Institute of Physics.

The sample is exposed by placing it at the focal point of the diffraction pattern, yielding a hexagonal array of photoresist columns with a diameter and pitch of 1.3 µm and 2.7 µm in negative-tone photoresist. The inverse pattern containing a hexagonal array of air voids can be fabricated by the same mask using a positive-tone photoresist. This pattern creates a template that should yield a PC with a mid-infrared (IR) band gap at approximately 5 µm. The band gap can be scaled towards the visible wavelength regime by reducing the dimensions of the grating and the wavelength of the laser. The template was made from SU-8 negative photoresist (MicroChem Corp.) deposited on a glass substrate coated on one side with indium thin oxide (ITO). The ITO served as an optically transparent seed layer for electro-deposition of the CdSe. The photoresist was spun onto the substrate at 1500 rpm for 20 s and soft baked on a hotplate for 1 min at 65 °C and 1 min at 95 °C to achieve a 3 µm thick film. The sample was then placed at the focal point of the diffraction pattern, which was 50 mm behind the mask plate, and was illuminated for 2 min with l. 03 µm, 10 ns laser pulses from a Spectra Physics GCR-13 Nd: YAG laser with an intensity of 30 mWcm^{-2}. After exposing the sample, it was baked for 1 min at 50 °C and for 1 min at 95 °C, developed in PGMEA developer (MicroChem Corp.), and rinsed with isopropanol. Figure 6-12 (a) and (b) show optical and scanning electron microscope (SEM) images of the resulting hexagonal array of photoresist columns, which are 1.3 µm in diameter and have a period of 2.7 µm. Finally, a simple synthetic approach to fabricate the 2D mid-infrared CdSe PCs by electro-deposition of CdSe in a polymer template defined using interference lithography is adopted for the final formation.

Fig. 6-12 Optical and SEM images showing the (a) top view of the negative photoresist columns created by interference lithography, and (b) side view of the same sample. Reprinted with permission from "Ivan B. Divliansky, Atsushi Shishido, Iam-Choon Khoo, and Theresa S. Mayer, Fabrication of two-dimensional photonic crystals using interference lithography and electrodeposition of CdSe, Appl. Phys. Lett. 79, 3392 (2001)" with copyright ©2001 of American Institute of Physics.

6.4 Summary

As an important top-down approach for nanofabrication of plasmonic structures, focused ion beam technology and laser interference lithography were highlighted and introduced in this chapter. Working principle, system configuration and fabrication examples of the FIB technology were presented firstly. It has advantages of one-step fabrication, no materials selectivity, high scanning resolution, maskless, writing of arbitrary 2D patterns, and localized fabrication etc. However, it also has shortages of low fabrication speed, high expenditure, and small fabrication area. Considering this, it is necessary to employ other techniques for the purpose of large area fabrication.

In contrast, laser interference lithography techniques can realize cost effective fabrication over large area. However, shape of the exposing patterns is limited in comparison to the FIB writing technique. It is an effective and practical technique for fabricating metallic nanoparticle arrays which can be used as a type of biosensor for LSPR-based immuonoassay and photothermal effect-based detection of cancer cells.

Reference

[1]   John Melngailis, "Focused ion beam technology and applications", J. Vac. Sci. Technol. B 5, 469 (1987).

[2]   Stewart D K, Doyle A F and Casey J D Jr 1995, "Focused ion beam deposition of new materials: dielectric films for device modification and mask repair, and Ta films for x-ray mask repair". Proc. SPIE 2437, 276

[3]   Reyntjens S, De Bruyker D and Puers R 1998, "Focused ion beam as an inspection tool for microsystem technology", Proc. 1998 Microsystem Symp. (Delft, the Netherlands) p 125.

[4]   Ward B W, Economou N P, Shaver D C, Ivory J E, Ward M L and Stern L A 1988, "Microcircuit modification using focused ion beams", Proc. SPIE 923, 92.

[5]   Glanville J, "Focused ion beam technology for integrated circuit modification", Solid State Technol. 32, 270 (1989).

[6]   Stewart D K, Stern L A, Foss G, Hughes G and Govil P 1990, "Focused ion beam induced tungsten deposition for repair of clear defects on x-ray masks", Proc. SPIE 1263, 21.

[7]   Walker J F, Reiner J C and Solenthaler C 1995, Proc. Microsc. Semicond. Mater. Conf. (Oxford, 20–23 March 1995) p 629.

[8]   Athas G J, Noll K E, Mello R, Hill R, Yansen D, Wenners F F, Nadeau J P, Ngo T and Siebers M 1997, "Focused ion beam system for automated MEMS prototyping and processing", Proc. SPIE 3223, 198.

[9]   Daniel J H and Moore D. F, "A microaccelerometer structure fabricated in silicon-on-insulator using a focused ion beam process", Sensors Actuators A 73, 201(1999).

[10] Brugger J, Beljakovic G,Despont M, de Rooij N.F and Vettiger P, "Nanomechanical device fabrication based on focused ion

beam in microsystems fabrication", Microelectron. Eng. 35, 401(1997).

[11] Steve Reyntjens and Robert Puers, "A review of focused ion beam applications in microsystem technology", J. Micromech. Microeng. 11, 287 (2001).

[12] Thornell G and Johansson S, "Micromachining at the fingertips", J. Micromech. Microeng. 8, 251 (1998).

[13] Johansson S, Schweitz J-A, Westberg H and Boman M, Microfabrication of three-dimensional boron structures by laser chemical processing, J. Appl. Phys. 72, 5956 (1992).

[14] Y. Fu, and N. K. A. Bryan, "Influence of redeposition effect for focused ion beam three-dimensional micromachining." International Journal of Advanced Manufacturing Technology. 16 (8), 600-602 (2000).

[15] Yongqi Fu, Wei Zhou, Lim Enk Ng Lennie, Chunlei Du, Xiangang Luo, "Plasmonic microzone plate: superfocusing at visible regime", Applied Physics Letter 91(6), 061124 (2007).

[16] Yongqi Fu, Wei Zhou, Lim Enk Ng Lennie, "Nano-pinhole-based optical superlens", Research Letter in Physics, Vol. 2008, 148505 (2008).

[17] Malinsky MD, Kelly KL, Schatz GC, Van Duyne RP, "Chain length dependence and sensing capabilities of the localized surface plasmon resonance of silver nanoparticles chemically modified with alkanethiol self-assembled monolayers," J. Am. Chem. Soc. 123(7), 1471–1482 (2001).

[18] Haes AJ, Van Duyne RP, "A nanoscale optical biosensor: sensitivity and selectivity of an approach based on the localized surface plasmon resonance spectroscopy of triangular silver nanoparticles," J. Am. Chem.Soc. 124(35), 10596–10604 (2002).

[19] Haes AJ, Van Duyne RP, "Nanoscale optical biosensors based on localized surface plasmon resonance spectroscopy." Proc of Int. Soc. Opt. Eng. (SPIE) 5221, 47–58 (2003).

[20] Haes AJ, Van Duyne RP, "Nanosensors enable portable detectors for environmental and medical applications," Laser Focus World 39, 153–156 (2003).

[21] Haes AJ, Zou S, Schatz GC, Van Duyne RP, " A nanoscale optical biosensor: the long range distance dependence of the localized surface plasmon resonance of noble metal nanoparticles," J. Phys. Chem. B 108(1), 109–116 (2004).

[22] Hulteen JC, Van Duyne RP, "Nanosphere lithography: a materials general fabrication process for periodic particle array surfaces," J. Vac. Sci. Technol. A 13, 1553–1558 (1995).

[23] Shaoli Zhu, Xiangang Luo, Chunlei Du, Fei Li, Shaoyun Yin, Qiling Deng, and Yongqi Fu, "Hybrid metallic nanoparticles for excitation of surface plasmon resonance", J. Appl. Phys. 101, 064701 (2007).

[24] Shao-li ZHU, LUO Xian-gang, DU Chun-lei, "Discrete Dipole Approximation Aided Design Method for Nanostructure Arrays", Chin. Phys. Lett. 24(10), 2902-2905 (2007).

[25] Haiying Li, Xiangang Luo, Chunlei Du, Xunan Chen, Yongqi Fu, Ag particles array fabricated using laser interference technique for biosensing，Sensors and Actuators B Chemical 134, 940–944 (2008).

[26] Cees J. M. van Rijn, "Laser interference as a lithographic nanopatterning tool," J. Microlith., Microfab., Microsyst. 5(1), 0110121-0110126 (2006).

[27] Ivan B. Divliansky, Atsushi Shishido, Iam-Choon Khoo, and Theresa S. Mayer, "Fabrication of two-dimensional photonic crystals using interference lithography and electrodeposition of CdSe", Appl. Phys. Lett. 79, 3392 (2001).

7 BOTTOM-UP APPROACHES

Abstract: Two commonly used approaches for fabrication of plasmonic nanostructures: self assembled monolayer and electrochemistry techniques were introduced. The bottom-up approach has advantages of cost effective, large area fabrication, and simple process. But it is chemistry-based process, and thus has shortages of low repeatability, unstable, and low uniformity. It is suitable for fabrication of micro-/nanostructures which can be used as biosensors in life science.

7.1 Introduction

The "bottom-up" approach on the other hand takes the idea of "top down" approach and flips it right over. In this case, instead of starting with large materials and chipping it away to reveal small bits of it, it all begins from atoms and molecules that get rearranged and assembled to the nanostructures over large area. It is the new paradigm for synthesis in nanotechnology world as the 'bottom-up' approach allows a creation of diverse types of nanomaterials, and it is likely to revolutionize the way of making materials. It requires a thoroughly understanding of the short range forces of attraction such as Van der Waals forces, electrostatic forces, and a variety of inter-atomic or intermolecular forces. Since it is not possible to have various minute things come together without some attractive force or active field of force in the region, having the fundamental forces "doing all the work" for you is the key principle underlying this approach.

Bottom up fabrication can be described as building a brick house. Instead of placing bricks one-by-one at a time to produce a house from bottom, bottom up fabrication technology places atoms or molecules one-by-on at a time to build the desired nanostructure. Such processes are time consuming and so self assembly techniques appeared where the atoms arrange themselves as required. Self assembling nanomachines are regularly mentioned by science fiction writers but significant obstacles including the laws of physics will need to be overcome or circumvented before this becomes a reality. Other areas involving bottom up fabrication are already quite successful. Manufacturing quantum dots by self-assembly quantum dots has rendered the top down lithographic approach to semiconductor quantum dot fabrication virtually obsolete.

Typical bottom-up approaches for nanostructures mainly include the following techniques:

1) Chemical or electrochemical reactions for precipitation of nanostructures,

2) Self-assembly of nanoparticles or monomer/polymer molecules,

3) Sol-gel processing,

4) Laser pyrolysis,

5) Chemical vapor deposition, physical vapor deposition,

6) Plasma or flame spraying synthesis,

7) Atomic or molecular condensation,

8) Sputtering and thermal evaporation,

9) Bio-assisted synthesis of nanomaterials.

One of the basic 'bottom-up' techniques is chemical precipitation by which nanoparticles of metals, alloys, and oxides etc. are prepared in aqueous or organic solutions. There are several ways to obtain nanoscale precipitates. They can be derived by

1) A controlled phase transformation (*i.e.*, liquid state diffusion) guided by the free energy diagrams or

2) By controlling the solid state diffusion: following a composite route approach mixing, *e.g.*, two different materials and stirring them mechanically.

3) Other approaches can be found in exploiting internal oxidation of materials or

4) Thin film deposition of coatings or sputtering.

For fabrication of plasmonic nanostructures, the most commonly used methods are self-assembly of nanoparticles or monomer/polymer molecules and electrochemical deposition (includes electroplating and electrodeposition).

It is worthy to point out that in spit of being so promising and inviting, the ability to build things in the order of from bottom to up is fairly limited in scope. Through we can assemble relatively simple structures, we cannot produce complex integrated devices using the bottom up approach. Any kind of overall ordered arrangement aside from repeating regular patterns cannot be done without some sort of top-down influence like lithographic patterning.

7.2 Self-assembly monolayer

A self assembled monolayer (SAM) also named nanosphere lithography, is an organized layer of amphiphilic molecules in which one end of the molecule, the "head group" shows a special affinity for a substrate. SAM also consists of a tail with a functional group at the terminal end, as shown in Fig. 7-1. SAM are created by the chemisorption of hydrophilic "head groups" onto a substrate from either the vapor or liquid phase [1] followed by a slow two-dimensional organization of hydrophobic "tail groups"[2]. Initially, adsorbate molecules form either a disordered mass of molecules or a "lying down phase", and over a period of hours, begin to form crystalline or semicrystalline structures on the substrate surface [3, 4]. The hydrophilic "head groups" assemble together on the substrate, while the hydrophobic tail groups assemble far from the substrate. Areas of close-packed molecules nucleate and grow until the surface of the substrate is covered in a single monolayer.

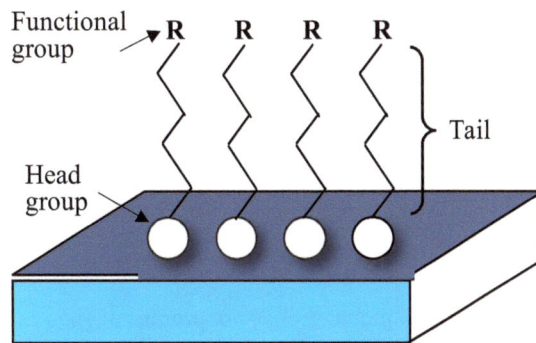

Fig. 7-1 Schematic diagram of bio-molecular binding with particles on substrate.

The forces and interactions in the self-assembly process are briefly introduced as follows:

7.2.1 Forces and Interactions of Self-Assembly [5]

Relevant bonds affecting the bottom up fabrication can be categorized as covalent and noncovalent: covalent bonds, as is well known, are strong; a single C–C bond has energy of ~90 kcal/mol and is responsible for the bonding between the adjacent subunits of the macromolecules. In contrast, normally, the noncovalent bonds are weak due to the following factors:

(1) Van der Waals interactions (~0.1 kcal/mole per atom),

(2) Hydrogen bonding (~1 kcal/mol),

(3) Ionic bonding (~3 kcal/mol in water, 80 kcal/mole in vacuum), and

(4) Hydrophobic interactions.

These interactions (or a combination thereof) determine the specific recognition and binding between specific molecules and macromolecules that are used in the self-assembly processes

7.2.2 Van der Waals–London Interactions

The van der Waals–London interactions are generated between two identical inert atoms that are separated from each other by a distance that is large in comparison to the radii of the atoms. Because the charge distributions are not rigid, each atom causes the other to slightly polarize and induce a dipole moment. These results in dipole moments and causes an attractive interaction between the two adjacent atoms. This attractive interaction varies as the minus sixth power of the separation of the two atoms, $\Delta U = -(A/R^6)$, where A is a constant. The van der Waals–London interaction is a quantum effect and will be resulted in between any two charged bodies whereas the charge distributions are not rigid, so that they can be perturbed in space to induce a dipole moment. Two atoms will be attracted to each other until the distance between them equals the sum of their van der Waals radii. Brought closer than that, the two adjacent atoms will repel each other. Individually, these forces are very weak, but they can play an important role in determining the binding of two macromolecular surfaces.

7.2.3 Hydrogen Bonding

Another important interaction that binds different molecules together is hydrogen bonding. This is how the nucleotides of DNA (described below) complementarily bind to each other. Hydrogen bonding is largely ionic in nature and can be generated when a hydrogen atom, which is covalently bonded to a small electronegative atom (e.g., F, N, O), develops a positive-induced dipole charge. This positive charge can then interact with the negative end of a neighboring dipole, resulting in a hydrogen bond. This interaction typically varies as $1/R^3$, where R is the distance between the two neighboring dipoles. Hydrogen bonding is an important part of the interaction between water molecules.

7.2.4 Ionic Bonding

Ionic interaction can also take place in partially charged groups or fully charged groups (ionic bonds). The force of attraction is given by Coulomb's law as $F = K((q^+q^-)/R^2)$, where R is the distance of separation. When present among water and counterions in biological mediums, ionic bonds become weak because the charges are partially shielded by the presence of counterions. Still, these ionic interactions between groups are very important in determining the recognition between different macromolecules.

7.2.5 Hydrophobic Interactions

Hydrophobic interactions between two hydrophobic groups are produced when these groups are placed in water. The water molecules tend to move these groups close in such a way as to keep these hydrophobic regions close to each other. Hence, this coming together of hydrophobic entities in water can be termed as a hydrophobic bond. Again, these interactions tend to be weak but are important, for example, two hydrophobic groups on the surface of two macromolecules come together when placed in water and bind the two surfaces. The surface energy at the interface of a hydrophobic surface and water is high because water tends to be moved away from the surface. Therefore, two hydrophobic surfaces will join together to minimize the total exposed surface area and the total energy. The phenomenon plays a very important role in the protein folding and also to assemble objects in fluids using capillary forces.

7.2.6 Classical nucleation theory for clusters formation [6]

Nucleation of new particles from a continuous phase can occur heterogeneously or homogeneously: phase heterogeneous nucleation from the vapor occurs on foreign nuclei, dust particles, ions or surfaces. Homogeneous nucleation occurs in the absence of any foreign particles or ions when the vapor molecules condense to form embryonic droplets of nuclei. For more information regarding the nucleation theory, please see Ref. 6.

As an example, we gave some fabrication results of metallic nanoparticles array fabricated using the SAM process.[7, 8]

Localized surface plasmon resonance (LSPR)-based bio-nanochips are of great interest in various applications such as environmental protection, [9, 10] biotechnology, [11] and food safety [12]. The LSPR-based nanosensor, is a device sensing variation of effective refractive index of bio-samples, which relies on the extraordinary optical properties of the noble metal (*e.g.*, Ag, Au and Cu) nanoparticles [9-12]. The sensing capability of the LSPR-based sensor can be altered by tuning shape, size, and material composition of the metallic nanoparticles.

In this research domain, representative works were performed by a research group in Northwestern University. One of their research subjects was focused on the measurement of binding signals between the antigen and antibody with the triangular Ag nanoparticles [12-14]. But it was found that the arm length between carbon chain of biotin and the triangular metallic nanoparticles is short. A few hot spots originating from the particles can be used in their works, and thus limited its sensitivity. In contrast, the original rhombic Ag nanoparticles having much more hot spots compared to that of the triangular Ag nanoparticles and a modified biotin with longer carbon chain were proposed for the purpose of improving detection sensitivity. It is well known that the more binding hot spots, the stronger the localized surface plasmon resonance can be excited from the Ag nanoparticles. Here in the presented work, each rhombic Ag nanoparticle binds the number of hot spots that are twice and even more than that of the triangular Ag nanoparticles. It apparently gives rise to millions of binding hot spots throughout the whole metal array. Therefore, the rhombic Ag nanoparticles can greatly enhance signal intensity of the LSPR-based spectrum. Moreover, the elongated arm length of the carbon chain of the biotin can make the binding reaction to produce streptavidin easily.

The fabrication procedures are described as follows:

(1) Materials sources

11-Mercaptoundecanoic acid (11-MUA), 1-octanethiol (1-OT), 1-ethyl-3-[3-dimethylaminopropyl] carbodiimide-hydrochloride (EDC) were acquired from Sigma-Aldrich Corp. Biotin and streptavidin were purchased from Pierce. Absolute ethanol and 10 mM phosphate-buffered saline (PBS) with pH = 7.4, were purchased from Jinshan company. Ag wire (99.99%, 1 mm diameter) was obtained from Jubo company. Quartz glass substrates with diameters of 25 mm were bought from Juke. Polystyrene nanospheres with diameters of 500 ± 20 nm and glass nanospheres with diameters of 200 ± 8 nm were received as a suspension in water (Duke, LTD)) and were prepared for further treatment.

(2) Substrate Preparation

Firstly, the glass substrate was cleaned in a piranha solution (1:3 30% H_2O_2/H_2SO_4) at 80 □ for 30 min., and then cooled by high-pressure N_2. Once cooled, the glass substrates were rinsed with copious amounts of second distilled water and then sonicated 60 min in 5:1:1 $H_2O/NH_4OH/30\%$ H_2O_2. Next, the glass was rinsed repeatedly with water and stored in water until used.

(3) Nanoparticle Preparation

The extended NSL is employed here to create the surface-confined rhombic Ag nanoparticles supported on a glass substrate. This method is developed on the basis of the NSL [12]. For these experiments, the single-layer of size-monodisperse polystyrene nanospheres and glass nanospheres solution ~10□liter were spin-coated onto the glass substrate to form a deposition mask, followed by a process of etching off the nanospheres from the glass using hydrofluoric acid. After that, Ag thin film was deposited on the nanosphere masks using thermal or electron beam evaporation. After removal of polystyrene nanospheres by sonication in absolute ethanol for 3 min., well ordered two-dimensional (2D) rhombic nanoparticle arrays were finally achieved on the substrate. By altering the nanosphere diameter and thickness of the deposited metallic film, the nanoparticles with different in-plane width, out-of-plane height and interstructure space can be derived.

Extended nanosphere lithography (NSL) technique was employed to create the surface-confined rhombic Ag nanoparticles supported on a glass substrate, as shown in Fig. 7-2. Firstly, the glass substrate was cleaned. Then the self-assembly of size-monodisperse polystyrene nanospheres and glass nanospheres were applied to form a deposition mask, and followed by hydrofluoric acid to etch off the glass nanospheres. After that, the Ag particles were deposited through the nanosphere masks using thermal or electron beam evaporation technique. After removal of the polystyrene nanospheres by sonication in absolute ethanol for 3 min., well ordered two-dimensional rhombic Ag nanoparticle arrays were finally obtained on the substrates. By changing the nanosphere diameter and thickness of the deposited metal film, the nanoparticles with different in-plane width, out-of-plane height, and inter-particle space can be derived. Figure 7-2 shows flow-chart of fabrication process of the metallic nanoparticles.

Imaging of the surface topography of the samples was carried out using JSM-5900LV Scanning Electron Microscope (SEM). The Ag nanorhombus have in-plane widths of ~140 nm and out-of-plane heights of ~40 nm as measured by the SEM and sidestep apparatus, respectively. We can see that the size of the rhombic Ag nanoparticles fabricated is not identical as limitation of the SAM method, which accounts for the errors in our following experiment. Figure 7-3 is another fabrication result for a triangular Ag nanoparticles array by means of SAM technique.

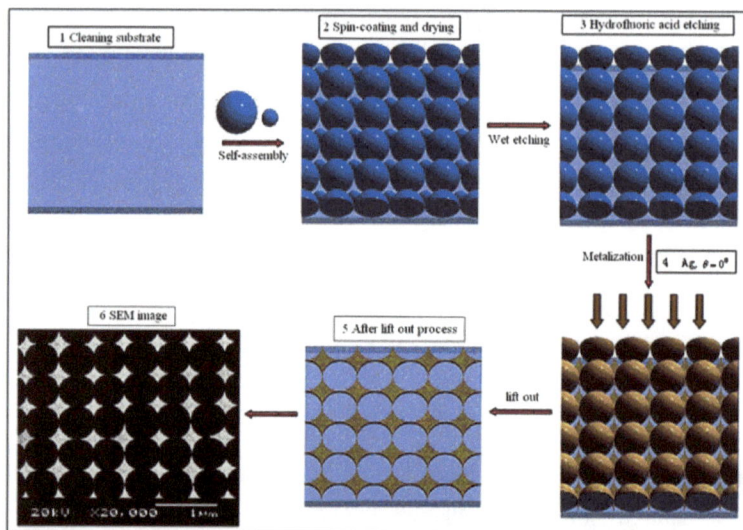

Fig.7-2 Flow chart of procedures of self-assembly monolayer method: 1. cleaning substrate, 2.spin-coating and drying, 3.hydrofluoric acid etching, 4. metalization, 5.lift-off process, and 6. finished prototype.

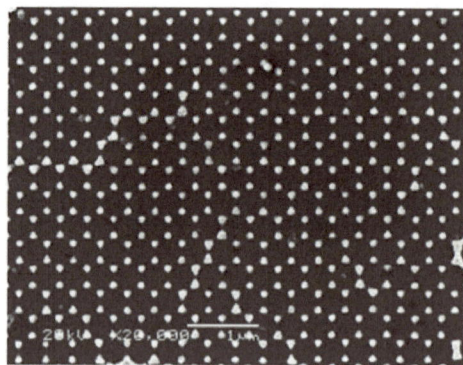

Fig.7-3 SEM micrograph of the fabricated metallic triangular nanoparticles using SAM technique.

7.3 Electroplating and electrodeposition

Electrochemistry (also named as "electroplating") represents one of the most powerful techniques for high density, high aspect ratio, and high fidelity designs. Similarly to the lithographic techniques, it requires a polymer mask through which metal is deposited with extremely high resolution. Differently to other mask-based techniques like isotropic etching, ion milling, and reactive ion etching (RIE), it avoids the problem of shadowing and builds structures atom-by-atom. The presence of an electric field in the solution allows that metal ions discharge and conform the smallest features of the mold. Metal deposition can be performed under two major techniques: with current (pulsed or DC) or electroless (via catalytic, exchange or electrophoretic reactions), as shown in Fig. 7-4. The first method takes place in an electrolytic cell and involves a reaction under an imposed bias and current flow. In this case important parameters for the process control are pH, current density, temperature, agitation, and solution composition. The electroless deposition is performed on the basis of a substantial oxidation reaction that replaces the dissolution of a substrate as it is the case for immersion plating.

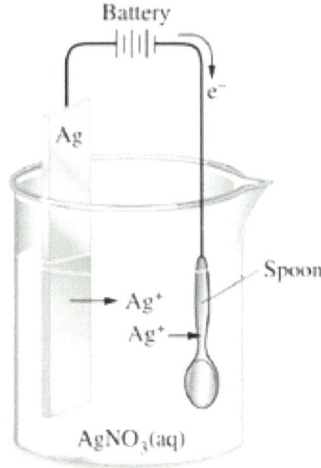

Fig. 7-4 Schematic diagram of the electroplating principle.

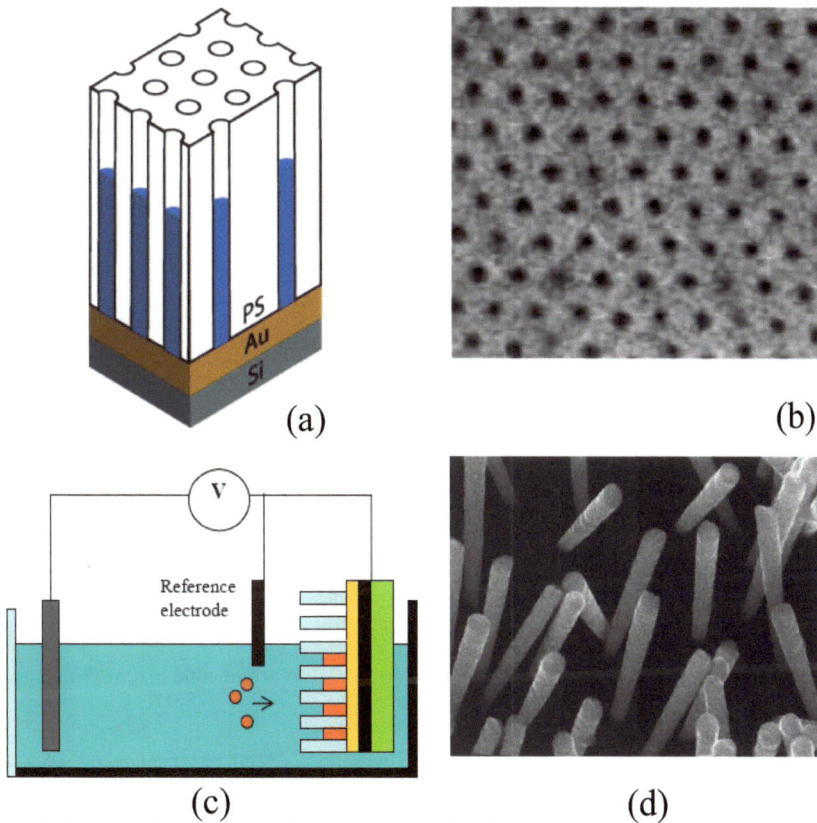

(a)   (b)

(c)   (d)

Fig. 7-5 (a) Schematic diagram that shows a Si substrate coated with Au thin film as conductive layer. (b) Polycarbonate material is perforated as a template on the Au layer. (c) An Au nanowires array is completed. (d) SEM picture of the finished Au nanowires.

Electro-deposition is a technique used for manufacturing ordered arrays of nanomaterials like quantum dots on a flat surface, when the deposition is done in the presence of a shadow mask. Figure 7-5 (a) is a schematic diagram that shows a Si substrate coated with an Au thin film as a conductive layer. Film thickness can be ranging from ~300 nm to 600 nm. Polycarbonate material is perforated as a template (see Fig. 7-5 (b)) on the Au layer. After the electro-deposition process and removal of the polycarbonate material, fabrication of an Au nanowires array is completed, as shown in Fig. 7-5 (c). The template can be fabricated using heavy ions direct bombing [15], and top-down techniques. Nanowire size depends on acid and applied voltage ranging from 10 nm to ~150 nm. The Au wires grow from thiosulphate solution. This process has a capability of large area formation (several square centimeters).

7.4 Summary

Bottom-up approach was described in this chapter. Among the reported techniques at present, the SAM and electrodeposition techniques were targeted and illustrated briefly. In comparison to the top-down approach described in previous chapter, the bottom-up approach has advantages of allowing a creation of diverse types of nanomaterials, cost effective, simplified process procedure, and large area formation. However, it is a chemistry-based process. Therefore, some inherent shortages originated from chemistry process still exist here, *e.g.*, irregular shape, non-uniform distribution, and low repeatability. If combining with the top-down approach together, nanostructures over a large area with localized fine and tiny particles, holes, and cavities in complicated shapes can be realized. Undoubtedly, design freedom of the plasmonic structures and nanophotonic structures will be greatly improved by the combination approach.

Reference

[1] Schwartz, D.K., "Mechanisms and Kinetics of Self-Assembled Monolayer Formation". Annu. Rev. Phys. Chem. 52, 107 (2001).

[2] Wnek, Gary, Gary L. Bowlin. Edt.: *Encyclopedia of Biomaterials and Biomedical Engineering*. Informa. Healthcare 1331-1333 (2004).

[3] Love, "Self-Assembled Monolayers of Thiolates on Metals as a Form of Nanotechnology". Chem. Rev. 2005, 105, 1103-1169.

[4] Vos, Johannes G., Robert J. Forster, Tia E. Keyes. *Interfacial Supramolecular Assemblies*. Wiley, John & Sons, Incorporated, 2003, 88-94.

[5] W.A.Goddard III, D.W. Brenner, S. E.Lyshevski, G.J. Iafrate, Edt.: *Handbook of Nanoscience, Engineering and Technology*, CRC Press, 2007.

[6] A. S. Edelstein and R. C. Cammarata, Edt.: *Nanomaterials: Synthesis, Properties and Applications*, Institute of Physics Press (UK), 1996.

[7] Shaoli Zhu, Fei Li, Chunlei Du, Yongqi Fu, "A novel bio-nanochip based on localized surface plasmon resonance spectroscopy of rhombic nanoparticles", *Nanomedicine* Vol. 3, No. 5, 669-677 (October 2008).

[8] Shaoli Zhu, Fei Li, Xiangang Luo, Chunlei Du, Yongqi Fu, "A Localized Surface Plasmon Resonance Nanosensor Based on Rhombic Ag Nanoparticle Array", *Sensors and Actuators B Chemical* 134, 193–198 (2008).

[9] Malinsky MD, Kelly KL, Schatz GC,Van Duyne RP., "Chain length dependence and sensing capabilities of the localized surface plasmon resonance of silver nanoparticles chemically modified with alkanethiol self-assembled monolayers." *J. Am. Chem. Soc.* 123(7), 1471–1482 (2001).

[10] Haes AJ, Van Duyne RP., "A nanoscale optical biosensor: sensitivity and selectivity of an approach based on the localized surface plasmon resonance spectroscopy of triangular silver nanoparticles". *J. Am. Chem. Soc.* 124(35), 10596–10604 (2002).

[11] Haes AJ, Van Duyne RP., "Nanosensors enable portable detectors for environmental and medical applications". *Laser Focus World* 39, 153–156 (2003).

[12] Yonzon CR, Jeoung E, Zou S, Schatz GC, Mrksich M, Van Duyne RP, "A comparative analysis of localized and propagating surface plasmon resonance sensors: the binding of concanavalin A to a monosaccharide functionalized self-assembled monolayer". J. Am. Chem. Soc.126 12669-12676 (2004).

[13] Riboh JC, Haes AJ, McFarland AD, Yonzon CR, Van Duyne RP., "A nanoscale optical biosensor: real-time immunoassay in physiological buffer enabled by improved nanoparticle adhesion". *J. Phys. Chem. B* 107, 1772–1780 (2003).

[14] Haes AJ, Zou S, Schatz GC, Van Duyne RP., "A nanoscale optical biosensor: the long range distance dependence of the localized surface plasmon resonance of noble metal nanoparticles". *J. Phys. Chem. B* 108(1), 109–116 (2004).

[15] E. Ferain and R. Legras, "Track-etch templates designed for micro- and nanofabrication", *Nuclear Instruments and Methods in Physics Research B* 208, 115–122 (2003).

8 CHARACTERIZATIONS

Abstract: Approaches for characterization of plasmonic structures were presented in this chapter. Firstly, scanning probing microscopy-based geometrical method was introduced. Probing problems for the commercial probes were addressed. Then we gave a major part for illustration of optical characterization of the plasmonic structures, including near-field scanning optical microscope, Raman spectroscopy, confocal microscopy, and multiphoton microscopy.

8.1 Introduction

Normally, characterization of plasmonic nanostructures and nanophotonic devices includes geometrical and optical characterizations. Currently, scanning probe microscope (SPM) is a key approach for the geometrical characterization. SPM is a general name for the probe scanning technology-based point-to-point measurement method. It involves atomic force microscope (AFM), scanning tunneling microscope (STM), and near-field scanning optical microscope (NSOM) or scanning near-field optical microscope (SNOM). Amongst, AFM is a most commonly used tool for the geometrical characterization. Optical characterization technology includes NSOM, surface enhanced Raman spectroscopy (SERS), confocal microscopy, and multiphoton microscopy. NSOM and SERS are the most popular tools for the optical characterization. The following sections will describe them one-by-one in detail combining with corresponding experimental results of the relevant projects.

8.2 Geometrical characterization

AFM as an important characterization tool was invented in 1986 by Binnig, Quate and Gerber. Like all other scanning probe microscopes, the AFM utilizes a sharp probe moving over the surface of a sample in a raster scan. In the AFM systems, the probe is a tip on the end of a cantilever which bends in response to the force between the tip and the sample. There are numerous technical papers reported and illustrated this technique. Thus we did not waste our pages to introduce its working principle and characteristics here. Interested readers can read Refs. [1-4] and other relevant papers. In this chapter, we intended to highlight a technical problem firstly which was found during characterization of our fabricated plasmonic structures. [5]

Enhanced surface plasmon polaritons (SPPs) is appealing in recent years due to its advantages of long propagation depth, extraordinary transmission, and intensified intensity [6-11]. It can be locally excited and has a potential market in applications of data storage, nanophotolithography, biomedical, and photonics etc.[6] It is promising to develop a nano-optical lens on the basis of the enhanced SPPs to realize beam shaping, *i.e.*, collimation, deflection, or focusing. Reported plasmonic nanostructures have no depth-tuning (depth is the same for all grooves flanked in one or both sides of metal films), and width-tuning of the grooves [12-15]. In contrast, a novel metallic nanostructures design, depth-tuning-based plasmonic structure was presented in this section. Higher spatial optical resolution is expected for the future applications because it overcomes the diffraction limitation existed in the conventional optical systems. However, all the enhanced SPPs-based nanostructures are designed with feature size at nanolevel. Currently, the AFM is commonly used to measure depth and width of the nanostructures. A problem exists for the geometrical characterization using the AFM after fabrication of the nanostructures. Our experiments demonstrated that it is difficult to accurately measure the depth of the nanostructures if aspect ratio (depth to width) of the nanostructure is ~1:1 and larger using the AFM with commercial probe working in the tapping mode due to the apex wearing and large cone angle of the pyramid probe.

In this section, we put forth a new approach to solve this problem. A modified AFM probe was proposed to reduce the measurement error. After characterization of the fabricated nanostructures using the modified probe, measurement of optical transmission is carried out to evaluate performance of the fabricated nanostructure and verify the geometrical

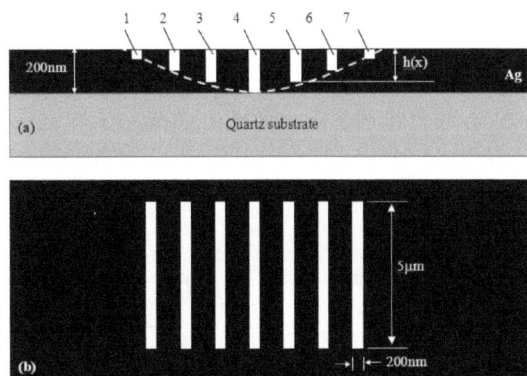

Fig.8-1 Design configuration of the enhanced SPPs based nanostructure. (a) Side view of the structure. A parabolic profile is designed to focus the enhanced SPPs. (b) Top view of the structure. Black color means the Ag thin film, and white color means the air.

characterization results indirectly. In addition, design of the novel depth-tuned metallic nanostructure was addressed. Design issue of the plasmonic structures can be seen in Ref. 10. Our target is that design a nanostructure which can act as a "nano-lens" to realize beam shaping such as collimation or focusing for the future possible usages such as detection and inspection. A higher spatial resolution can be expected through this type of lens. Figure 8-1 is our designed enhanced SPPs-based nanostructure, which consists of 7 slits. The central slit is thoroughly penetrated through the Ag thin film. Depth distribution of the other 6 slits is symmetrical to the central slit, and gradually decreases. The outline of the groove bottom was designed being a curve of parabolic. Beam shaping can be realized theoretically by a way of controlling phase distribution along one direction.

The milling experiments were carried out using our FIB machine (FEI Quanta 200 3D) with ion source of liquid gallium, integrated with scanning electron microscope (SEM). This machine uses a focused Ga^+ ion beam with energy ranging from 5 keV to 30 keV, a probe current ranging from 1 pA to 20 nA, and beam limiting aperture size ranging from 15μm to 350 μm. For the smallest beam currents, the beam can be focused down to nominal value of 10 nm in diameter at site of full width and half maximum (FWHM). The milling process is performed under programming control, by means of varying the ion dose for different relief depth. The ion column employs two focusing lens and octo-pole deflection, which can realize scanning with resolution as small as several nanometers. The stage is commercially supplied with ±50 mm moving distance in X and Y axes. Positioning accuracy is ±1 μm. The stage can be rotated to 360° in horizontal, and tilted from -5° to maximum angle of 60°.

After calibration, the nanostructure was directly milled using the FIB, as shown in Fig. 8-2. Process parameters, ion energy, beam current, and spot overlapping, are 30 keV, 10 pA, and 50%, respectively. In order to avoid charging problem during the milling, protective method of wrapping the insulate sample of quartz with copper tape was adopted to form a conductive path.

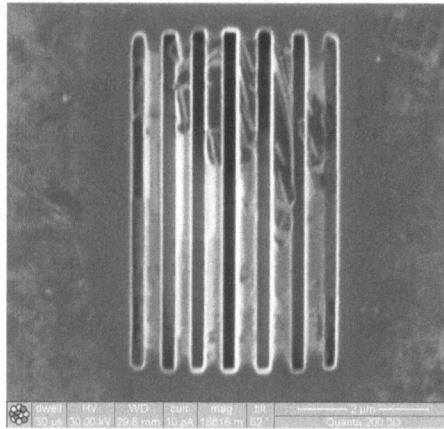

Fig.8-2 FIB image of the fabricated nanostructure on the
Ag thin film with thickness of 200 nm.

The geometrical characterization was performed using a AFM. The scanner was calibrated using a standard sample with 160 nm in height. The AFM (DI NanoScope IIIa) image typically had a scan range of 1~5 μm in vertical z direction. The silicon probe has a pyramid shape. Base of the pyramid is 6 μm in size, the height of the pyramid probe is 20 μm, and the height-to-base ratio is ~3. The AFM scan was performed with the tapping mode, where the change in the oscillation amplitude of the probe was sensed by the instrument. A large measurement error was found during the calibration using the AFM. The measured depth of the nanostructure is much smaller than that of the actual value. It attributes to the apex wearing of the commercial probe during scanning. Radius of the apex will gradually broaden due to wearing while the probe scans continuously on the sample surface. Material of the probe with tapping mode is Si_3N_4. Silicon nitride cantilevers are less expensive than those made of other materials. They are rugged and well suited to be used for image scan in almost all environments. They are especially compatible to organic and biological materials. The pyramidal tips are highly symmetric with its end having a radius of ~ 20 nm. But for characterization of the nanostructures, the problem of apex wearing of this type of probe is a sensitive issue even if being used several hours only.

The other reason is large cone angle (full angle is 34°) of the pyramid commercial probe which blocks the probe to detect the deeper area of the nanostructures. Therefore, measurement error will be generated in the scanned results. To solve this problem, we designed a new AFM probe with high aspect ratio (~24:1) and triangle shape in cross section for the usage of geometrical characterization of the depth-tuned structures. Optimization and simulation was carried out before using professional software (Pro-E and Ansys). The shape, dimension, and aspect ratio of the tips are determined in terms of the acceptable deformation and strengthen of the tips while design the modified probe [16]. The new probe was modified using the FIB milling, as shown in Fig. 8-3 (a) and (b). The FIB milled tips allow high resolution probing while scans the plasmonic nanostructure without sacrificing rigidity.

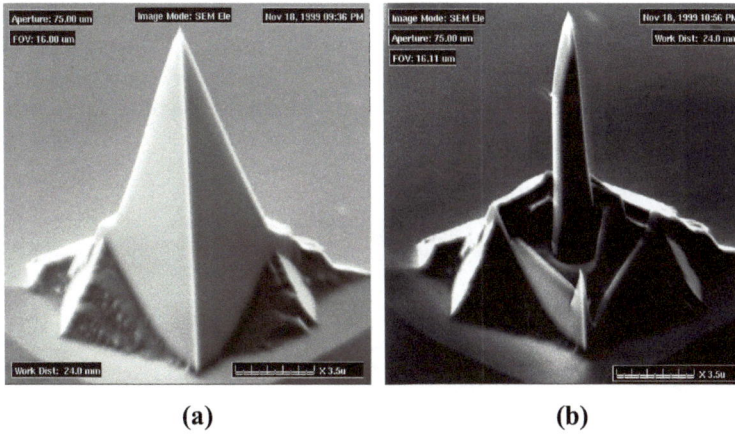

Fig. 8-3 AFM probe for tapping mode. (a) the commercial probe with half cone angle of 17° and material of Si_3N_4. (b) FIB trimmed probe with high aspect ratio.

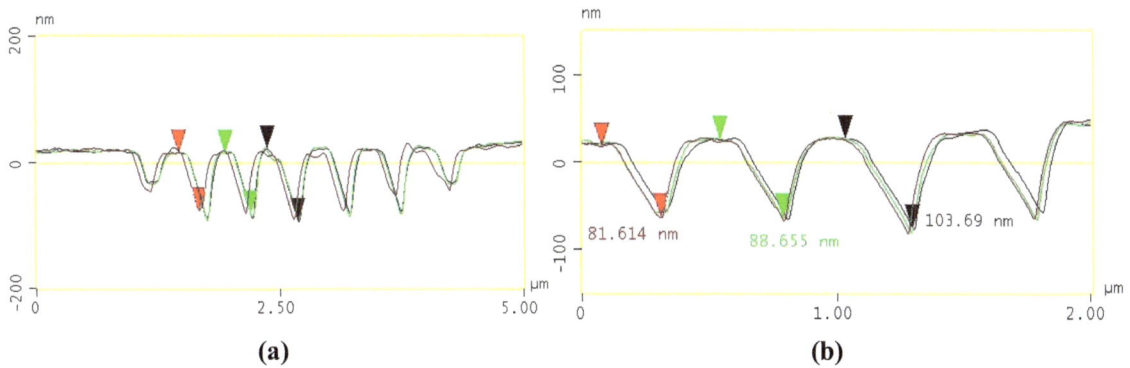

Fig. 8-4 Measurement results using the commercial AFM probe with tapping mode which had been used 1 day before this characterization. (a) overall profile of the structure. (b) zoom in AFM image of the partial profile.

In contrast, we show the characterized results of the AFM using both the commercial probe and the FIB trimmed probe, as shown in Fig. 8-4 and Fig. 8-5. It can be seen that the measurement difference is as large as 123 nm for the central slit of the same structure. It is shown from the zoom-in results of Fig. 8-4 (b) and Fig. 8-5 (b) that the grooves are still "V" shape even the scan range was reduced to be 600 nm. However, using the new FIB trimmed tip, even the

Table1 The nanostructure measured with the FIB trimmed AFM probe. The statistical data was calculated through 14 randomly selected sites on the structure.

Ag nanostructure	Design width=200nm					
Unit: nm	slit1	slit2	slit3	slit4	slit5	slit6	slit7
MAX	50	92	139	200	122	91	39
MIN	32	45	81	171	77	56	26
AVERAGE	40.72	76.36	118.54	182.63	103.09	72.82	33
RMS	5.04	11.03	17.54	6.84	13.82	10.19	3.46

Table2 The nanostructure measured with the commercial AFM probe which had been used 1 day. The statistical data was calculated through 14 randomly selected sites on the structure.

Ag nanostructure	Design width=200nm					
Unit: nm	slit1	slit2	slit3	slit4	slit5	slit6	slit7
MAX	67	89	94	108	93	84	56
MIN	52	69	81	101	83	76	50
AVERAGE	53.42	80.54	89.65	104.52	88.26	81.02	53.12
RMS	7.84	13.01	19.04	9.54	14.36	13.24	6.84

part of flat bottom of some grooves can be can be detected when the scan range was reduced to be 680 nm. Table 1 and table 2 show the measurement difference for each slit of the same nanostructure. It can be seen that the deeper the measured slits, the larger difference of the characterization data between the two types probe will be.

The difference between the theoretical and practical values is caused by the FIB fabrication errors. It can be calculated from the table 1 that the average value of *rms* for the measured depth of the 7 slits is ±5%. In addition, the Ga^+ implantation is another reason that causes the refractive index and extinction coefficiency of the Ag film changed

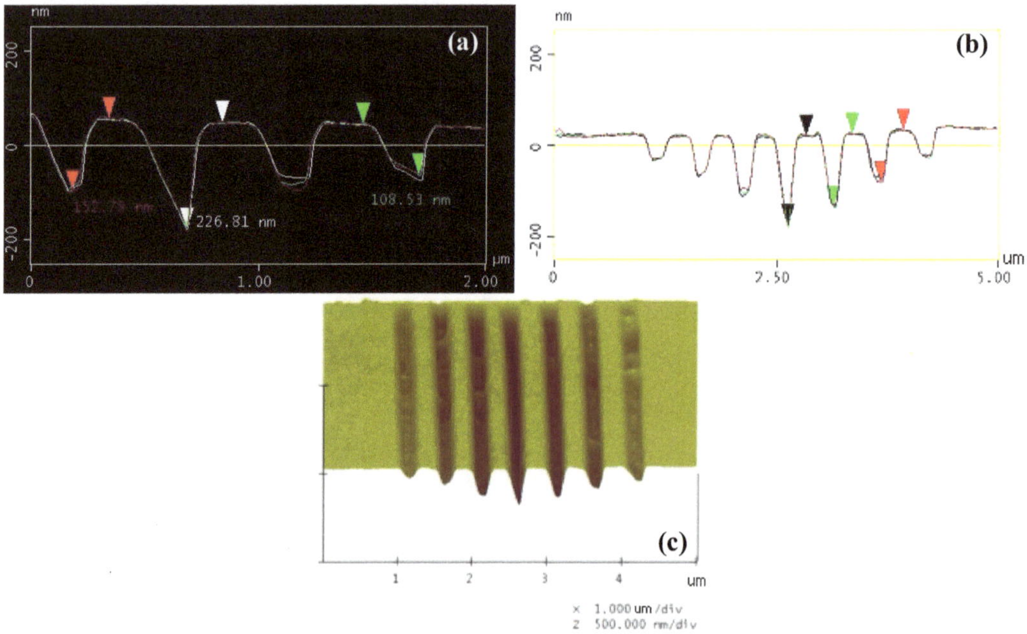

Fig.8-5 Measurement results using the FIB trimmed AFM probe with tapping mode for this characterization. (a) overall profile of the structure. (b) zoom in AFM image of the central slits profile. (c) 3D image of the AFM.

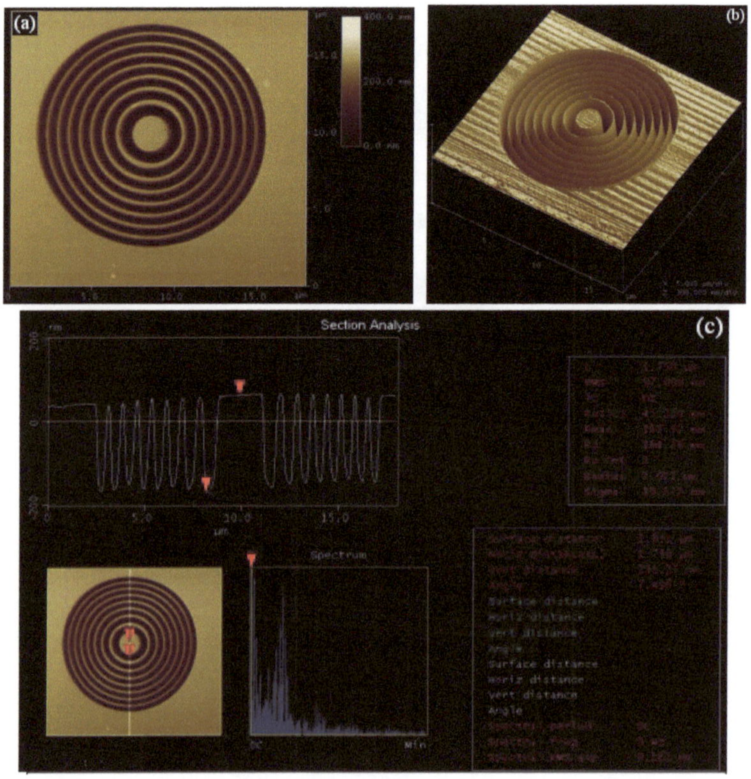

Fig. 8-6 AFM measurement results of a micro-Fresnel zone plate directly etched on an ITO glass. Designed depth of the slits is 200 nm. (a) 2D top-view of the device; (b) 3D side-view; and (c) 2D profile along central vertical direction as indicated by the white line. The AFM probe is a newly installed tip working for tipping mode.

[17]. For more deep slits the effect is stronger because the amount of the implanted material and therefore the absorption is higher. Implantation can be considered as increase in the imaginary part of the refractive index of the Ag layer.

Figures 8-6 (a)-(c) show another FIB milling example measured using AFM. It is a micro-Fresnel zone plate with outer diameter of 13 μm and depth of 200 nm. The commercial AFM probe is a newly installed tip working for tipping mode. It can be seen that the depth measured is acceptable. But 2D profile shows a "V" shape for all the grooves because only one point scanned at bottom of the groove due to blocking of the cone angle of the commercial pyramid tip.

8.3 Optical characterization

8.3.1 NSOM/SNOM

NSOM system was introduced in detail in Chapter 4 already. In this section, we intended to highlight two technical issues during the NSOM probing: (1) Interaction between the fiber probe and nanoparticles/nanorods;[18] and (2) phase sensitivity. [19]

8.3.1.1 Interaction issue [18]

The interaction issue was found for imaging periodic metallic structures. Periodic arrays of paired and single gold nanorods were imaged in the near field using reflection and transmission modes of a NSOM at various wavelengths and polarizations of incident light in the visible regime. The paired nanorods act like nanoantenna, and an array of them was initially designed as a negative-index material for the near infrared. Reverse contrast in reflection and transmission images is observed under illumination from the small aperture of a metal-coated fiber probe. By changing the relative orientation of the rods to the polarization, the reverse contrast switches to the normal contrast of near-field imaging. Coupling between the aperture and the nanorod array makes the contrast higher. Transmission through the aperture is enhanced if the aperture probe is positioned between the nanorods. The average near-field transmission exhibits an opposite sign of anisotropy relative to the far-field case. Aperture probes with larger diameters always show normal imaging contrast. The results demonstrate that the broad angular spectra of small-aperture sources play a crucial role in near-field interactions with nanorod arrays. The results also show that angular redistributions of these spectra after transmission or reflection from the nanorod array are likely due to the excitation of localized and propagating SP wave. The problem of imaging with source–sample interactions was considered here for a metal-coated aperture tip and a nanoantenna array. The arrays under study and our NSOM technique were illustrated in Figs. 8-7 and 8-8. Several factors, including the angular spectrum of the source, the angular excitation spectra (resonant and non-resonant) of the surface-localized and propagating modes, and the angular transmission/reflection spectra (filtering) make overall system analysis becoming rather complicated. Figure 8-9 shows a comparison of the nanorod array types in the perpendicular polarization. It can be seen that the paired nanorod array (right-hand column) exhibits essentially normal transmission at both 633 nm and 785 nm, while the single nanorod array (left-hand column) showed some anomalous transmission

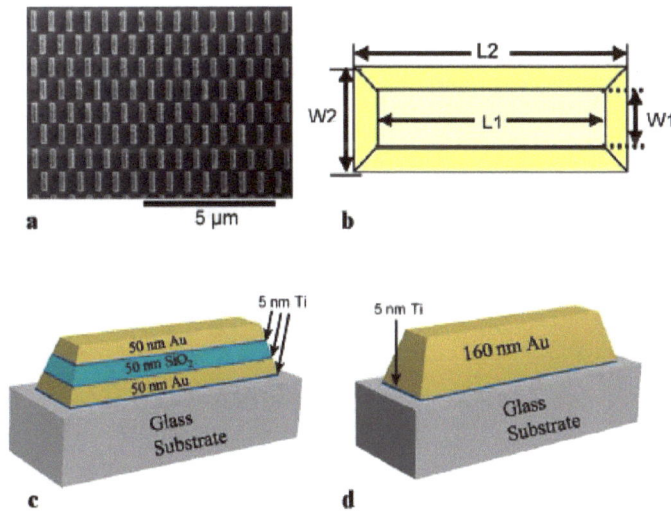

Fig. 8-7 (a) Electron microscopy image of the paired nanorod array. (b) Elementary cell dimensions: L1 and L2 are the top and bottom lengths, respectively, while W1 and W2 identify the widths (sample name: L1, L2, W1, W2, X period × Y period); (paired nanorod: 703 nm, 812 nm, 120 nm, 213 nm, 666×1852); (single nanorod: 691 nm, 781 nm, 104 nm, 205 nm, 666×1806). Vertical structures of paired (**c**) and single (**d**) nanorods. Reprinted with permission from "Ji-Young Kim, Vladimir P. Drachev, Hsiao-Kuan Yuan, Reuben M. Bakker, Vladimir M. Shalaev, Appl. Phys. B 93, 1432-0649 (2008)" with copyright © 2008 of Springer-Verlag.

images at both the wavelengths. In the near field region, the plasmonic resonance for the paired gold nanorods with such an intricate structure and geometry is complex and includes interactions between the metallic tip and the metal sample. To investigate this effect more precisely, the wavelength was changed to be 532 nm, and we again observed lower reflection from the nanorods than that of from the glass in the NSOM reflection images, as shown in Fig. 8-10 (a) at 532 nm and Fig. 8-10 (b) at 633 nm, respectively. Two basic manifestations of the near-field interaction were demonstrated. Firstly, strong changes in the anisotropy of transmission were observed which indicates a strong contribution from the large angle propagating and evanescent waves in the polarization dependence of transmission through the nanorod array. Secondly, the reversed contrast of near-field imaging in transmission and reflection is both polarized and wavelength dependent. These observations are structurally sensitive, as follows from the comparison data for both paired and single nanorod arrays of the same height profile. The contrast with large (>100 nm) aperture probes is always normal. Specific properties of the small-aperture probes affect the imaging contrast significantly in its deviation from the normal.

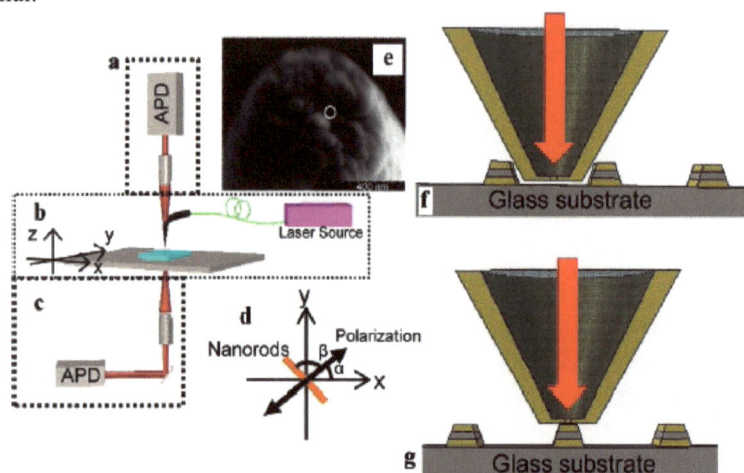

Fig. 8-8 NSOM: (a) reflection and (c) transmission detection modules; (b) NSOM stage and illumination unit; (d) orientation of the polarization axis and the nanorod; α and β denote the polarization angle and the nanorod orientation angle, respectively; (e) FESEM image of a typical metalized tip where the *circle* represents the aperture of the tip. Two key relative positions of the tip and paired gold nanorods: a gold-coated tip is in the valley between the nanorods (f) and on top of the nanorod pair (g). Reprinted with permission from "Ji-Young Kim, Vladimir P. Drachev, Hsiao-Kuan Yuan, Reuben M. Bakker, Vladimir M. Shalaev, Appl. Phys. B 93, 1432-0649 (2008)" with copyright © 2008 of Springer-Verlag.

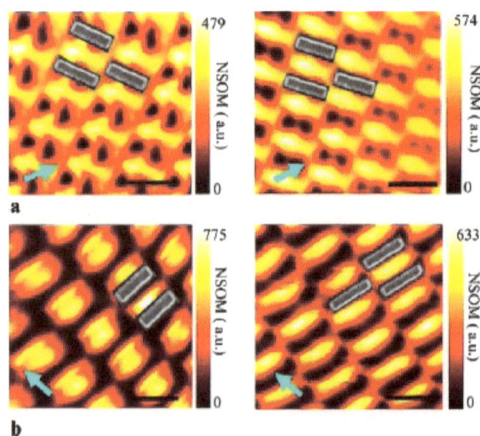

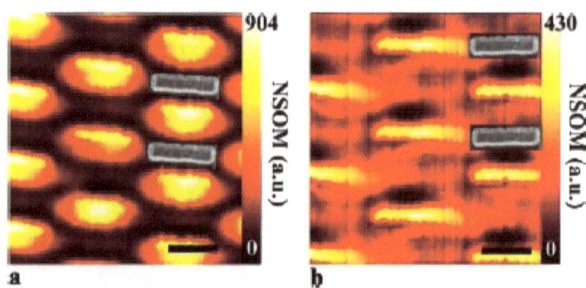

Fig. 8-9 NSOM transmission images: (a) 633 nm; (b) 785 nm. The left-hand column displays the data for the single nanorod array, and the right-hand column shows paired nanorod sample (sample A) data. All images are for the quasi-perpendicular polarization. Arrows represent the polarization axis. Scale bar is 1 μm. Reprinted with permission from "Ji-Young Kim, Vladimir P. Drachev, Hsiao-Kuan Yuan, Reuben M. Bakker, Vladimir M. Shalaev, Appl. Phys. B 93, 1432-0649 (2008)" with copyright © 2008 of Springer-Verlag.

Fig. 8-10 NSOM images in reflection mode for paired nanorod array (sample B) with 80-nm probe: (a) 532 nm, (b) 633 nm. At 532 nm, higher contrast is observed. The scanning direction is top-to-bottom. *Scale bar* is 500 nm. Reprinted with permission from "Ji-Young Kim, Vladimir P. Drachev, Hsiao-Kuan Yuan, Reuben M. Bakker, Vladimir M. Shalaev, Appl. Phys. B 93, 1432-0649 (2008)" with copyright © 2008 of Springer-Verlag.

8.3.1.2 Phase sensitivity [19]

Only recently, the phase of the induced electric fields at the surface of metal particles has been studied. Hillenbrand and Keilmann used a s-NSOM, equipped with an interferometric detection system, to simultaneously obtain intensity, phase and topography of individual metal nanoparticles.[20] Here we show how the relative phase-shift ϕ can be qualitatively visualized using an aperture NSOM. Single silver nanoparticle or small particle clusters is imaged as circular interference patterns, with the relative intensity of the central peak varying with ϕ. Although the spatial resolution is much worse than what can be obtained in a scattering-type NSOM (s-NSOM) setup, the advantages are that the present technique can be easily adapted to surface-enhanced spectroscopy measurements of individual nanoparticles or clusters and that the contrast mechanism is relatively simple.

The experiments were carried out using a commercial NSOM scanner (Nanonics NSOM-100) working in illumination mode and Cr–Al coated bent near-field probes with 100 nm apertures obtained from the same manufacturer. The sample was scanned at a constant speed of 1 μm/s in the lateral direction. The set point was kept at 50% of the free oscillation amplitude, which from a simple mechanical model corresponds to an average probe to surface distance of around 25 nm. The contrast mechanism can be qualitatively understood from a very simple model [Fig. 8-11 (a)]. In a

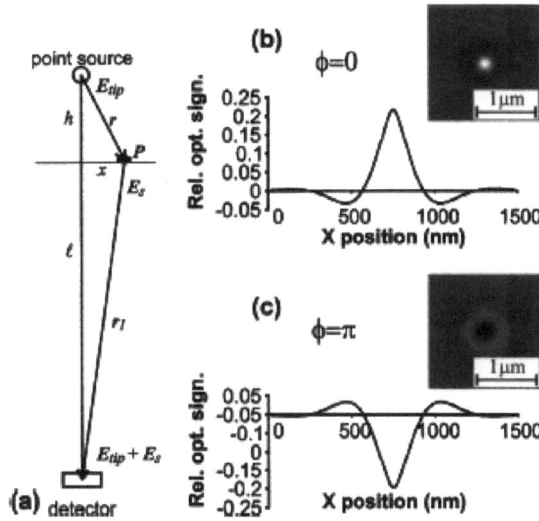

Fig. 8-11 (a) Model geometry for contrast formation (not in scale). (b,c) Simulation results from the simple dipole model with $\lambda=633$ nm, distance between the aperture and the sample $h=25$ nm, $h=\lambda/4$, $A=1$, $B=100$, $l=3$ cm, $|\alpha|=1.32\times10^{-34}$ cm^{2}V^{-1}, which corresponds to electrostatic dipole polarizability of a Ag sphere with 10 nm radius at $\lambda=633$ nm, in the two extreme cases $\phi=0$ and $\phi=\pi$. Reprinted with permission from "J. Prikulis, H. Xu, L. Gunnarsson, H. Olin, and M. Ka¨ ll, Phase-sensitive near-field imaging of metal nanoparticles, J. Appl. Phys. 92, 6211-6214 (2002)" with copyright © 2002 of American Institute of Physics.

first approximation, light emitted from the tapered fiber can be divided into a propagating wave and an exponentially decaying near field, which causes dipole excitations in particles in the vicinity of the aperture. We approximated the NSOM probe with a point source

$$E_{tip}(r) \propto \exp[i(kr-\omega t)]\left[\frac{A}{kr}+B\exp(-r/\eta)\right]$$

$$(8-1)$$

The constants A and B define the amplitudes of the far-field and near-field contributions, respectively, r is the distance from the light source, k is the wave number, v is the angular frequency, and h is the near-field decay length. Choosing the ratio $B/A=100$, and ignoring the heat dissipation of the incident intensity in the NSOM probe, corresponds to a taper throughput of approximately $T=10^{-4}$. A dipole moment $P=\alpha E_{tip}(r)$ is induced in the particle. The oscillating dipole emits scattered waves $E_s(r_1)=Z_0 k^2/(4pikr_1)dP/dt \exp(2ikr_1)$ which superimpose with the far-field emission from the fiber at the detector. Here Z_0 is the free space impedance. The measured intensity is thus approximately given by

$$I \propto \int_0^{2\pi/\omega} R[E_{tip}(\ell)+E_s(r_1)]^2 dt$$

$$(8-2)$$

where R denotes the real part and, and r is the distance between the light source and the detector. The phase of the complex nanoparticle polarizability $\alpha=|\alpha|\exp(i\phi)$ determines whether the particle appears with a dark or a bright central spot in the NSOM image, as shown in Figs. 8-11 (b) and (c).

Figure 8-12 (a) demonstrates that the ensemble averaged far-field extinction spectrum of the silver nanoparticles exhibits surface-plasmon peaks over a broad region, approximately centered at the He–Ne measurement wavelength. However, Mie theory calculations for small silver spheres (not shown here) produce SP resonances in the blue-green

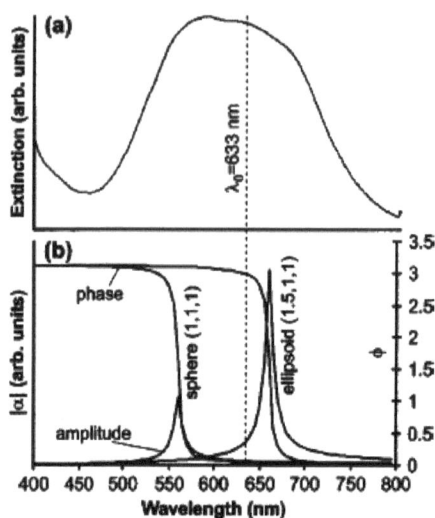

Fig. 8-12 (a) Measured ensemble-averaged extinction spectrum of silver particles deposited on cover glass. (b) Amplitude and phase of the polarizability α calculated in the electrostatic approximation for an Ag sphere and an ellipsoid with aspect ratio 1.5 in a dielectric surrounding medium with refractive index n=1.3. Reprinted with permission from "J. Prikulis, H. Xu, L. Gunnarsson, H. Olin, and M. Kä̈ll, Phase-sensitive near-field imaging of metal nanoparticles, J. Appl. Phys. 92, 6211-6214 (2002)" with copyright © 2002 of American Institute of Physics.

region. The obvious redshift of the extinction spectrum in this particular case, compared to what is expected for spherical Ag particles, indicates that the elongated particles or nanoparticle aggregates, give a dominant contribution to the far-field extinction. The huge variation of the optical signal for particles with similar height also indicates that the shape of the particles or clusters determines the phase of the near-field SP coupling at the measurement wavelength. This is also supported by polarizability calculations for ellipsoids in the electrostatic approximation, which shows that a small elongation of the particle causes a significant redshift [Fig. 8-12 (b)]. Small particles with diameter less than 60 nm always appear as bright beam spots in the NSOM images, because their resonance wavelength is much shorter than the illumination wavelength.

8.3.2 Spectrometry and SERS

Recent years, Raman spectroscopy has become an important analytical and research tool. It can be used for extensive applications ranging from pharmaceuticals, forensic science, polymers, and thin films, to semiconductors and even for the analysis of fullerene structures and carbon nano-materials. Raman spectroscopy is a light scattering technique, and can be thought of in its simplest form as a process where a photon of light interacts with a sample to produce scattered radiation at different wavelengths. It is extremely information rich (useful for chemical identification, characterization of molecular structures, effects of bonding, environment and stress on a sample). Historically, the technique of Raman spectroscopy was not that widely taught within university courses, even though the scattering process itself was established as far back as 1928 by Professor C.V Raman. FTIR, UV-VIS, and NMR etc. were generally more commonplace. In the mid 1990's, the next generation of smaller, more compact instruments started to be evolved. They utilized newer lasers, optics and detectors and began the micro Raman revolution. Raman spectroscopy provides: (1) fingerprint spectra (molecular identity); (2) information about 3D structural changes (orientation and conformation of molecular); (3) information about intermolecular interactions; and (4) dynamics. Advantages of Raman spectroscopy includes: (1) non-destructive and non-invasive detection; (2) works in-situ and in-vitro for biological samples; and (3) works under a wide range of conditions such as extreme temperature and pressure.

Briefly speaking, the population state of a molecule at room temperature is principally in its ground vibrational state. This is the larger Raman scattering effect. A small number of molecules will be in a higher vibrational level, and hence the scattered photon can actually be scattered at a higher energy (a gain in energy, and a shift to higher energy as well as a blue shifted wavelength). This is the much weaker Anti-Stokes Raman scattering. The incident photons will thus interact with the present molecule, and the amount of energy change (either lost or gained) by a photon is characteristic of the nature of each bond (vibration) presented. Not all vibrations will be observable with Raman spectroscopy (depending upon the symmetry of the molecules.) but sufficient information is usually presented to enable a very precise characterization of the molecular structure. Thus the Raman signal is very weak for the normal molecular excitation. For more information, please see the relevant papers in Ref. [21-26].

In order to improve the signal intensity, a surface enhanced Raman spectroscopy technique was reported by MIT research group. [27] Surface enhanced Raman spectroscopy (SERS) is a Raman spectroscopic (RS) technique that provides greatly enhanced Raman signal from Raman-active analyte molecules that have been adsorbed onto certain specially prepared metal surfaces. Increases in the intensity of Raman signal were regularly observed on the order of 10^4-10^6, and can be as high as 10^8 and 10^{14} for some systems.[28, 29] The importance of SERS is that it is both surface selective and highly sensitive where as RS is neither. RS is ineffective for surface studies because the photons of the incident laser light simply propagate through the bulk and the signal from the bulk overwhelms any Raman signal from the analytes at the surface. SERS selectivity of surface signal results from the presence of surface enhancement (SE) mechanisms only at the surface. Therefore, the surface signal overwhelms the bulk signal, making subtraction of the bulk signal unnecessary.

There are two primary mechanisms of enhancement described in the literature: an electromagnetic and a chemical enhancement. The electromagnetic effect is a dominant factor, and the chemical effect contributes enhancement only on the orders of one or two of magnitude.[30] The electromagnetic enhancement (EME) is dependent of the presence of the metal surface's roughness features, while the chemical enhancement (CE) involves changes to the adsorbate electronic states due to chemisorption of the analyte.[31] The structural and molecular identification power of RS can be used for numerous interfacial systems, including electrochemical, modeled and actual biological systems, catalytic, in-situ and ambient analyses and other adsorbate-surface interactions. Because of the sensitivity of SERS, detection of trace molecules can be accomplished as well. SERS is observed primarily for analytes adsorbed onto coinage (Au, Ag, Cu) or alkali (Li, Na, K) metal surfaces, with the excitation wavelength near or in the visible region. [32] Theoretically, any metal is capable of exhibiting SE, but the coinage and alkali metals satisfy calculable requirements and provide the strongest enhancement.[29] Metals such as Pd or Pt exhibit enhancements of about 10^2-10^3 for excitation in the near ultraviolet region. The importance of SERS is that the surface selectivity and sensitivity extend RS utility to a wide variety of interfacial systems which were inaccessible previously to RS because RS was not surface sensitive. These include in-situ and ambient analyses of electrochemical, catalytic, biological, and organic systems. Zou, *et. al.* [33] discussed the alternative surface techniques (sum frequency generation, infrared reflection absorption spectroscopy, and electron energy loss spectroscopy) whose limitations include a requirement for ultra high vacuum (UHV) conditions, low wave number range, low sensitivity, and bulk phase interference. SERS can be conducted under ambient conditions, has a broader wave number range, is quite sensitive and surface selective.

Like other characterization tools, SERS also has its own unique limitations. Many of these are due to the difficulty of obtaining reproducible SERS-active substrates. There may always be, however, the problem of the thermodynamics of analyte adsorption to the surface of interest. It may be difficult to obtain calibration curves for single analyte molecule, and moreso for complex mixtures.[28] Also, the SERS spectra may not be as compound-unique as scientists might like. Nevertheless, complex mixtures and trace analyses can be accomplished.

Conventional optical spectroscope can only detect a sample or target optical spectrum. It is an integral outcome of the whole illuminated area. SERS can detect the message for a certain single molecule. Therefore, its resolution is much

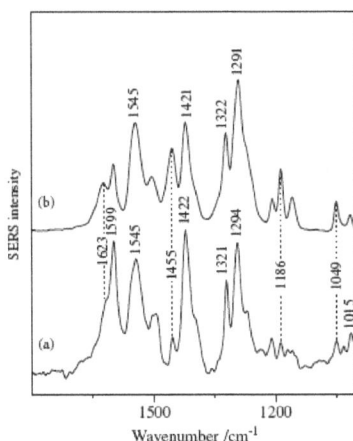

Fig. 8-13 SERS of AZ on an Ag hydroxylamine colloid at (a) 10^{-6} and (b) 10^{-3} M. Excitation at 1064 nm, pH=6.1. Reprinted with permission from "M. V. Can͂amares, J. V. Garcia-Ramos, C. Domingo and S. Sanchez-Cortes, J. Raman Spectrosc. 2004; 35: 921–927" with Copyright © 2004 of John Wiley & Sons, Ltd.

higher than that of the normal spectroscopy. Figure 8-13 is an example of Alizarin (1,2-dihydroxyanthraquinone) (AZ) measured using SERS.[34] According to the results shown above at pH =6.1, AZ is adsorbed on the metal surface in two different forms denoted temporarily as A and B here. The relationship of these two forms can be changed with the concentration, as can be seen in Fig. 8-13, whereas the spectra of AZ at the concentrations of 10^{-6} M (see Fig. 8-13 (a)) and 10^{-3} M (see Fig. 8-13 (b)) are shown, respectively. By decreasing the concentration, relative intensification of the bands corresponding to the B form is observed (*e.g.*, the bands appearing at 1599, 1500, 1422, 1294 and 1015 cm^{-1}). This means that at lower surface coverage of the AZ molecules are predominantly in B form because the first adsorption sites occupied by AZ on the metal are those implying double deprotonation of the molecule. This is probably a consequence of the stronger interaction of AZ with the primary adsorption sites, which leads to further deprotonation of the OH. A more detailed study of the concentration reveals that the changes observed with the AZ coverage were detected in the concentration ranging from 10^{-5} to 10^{-6} M owing to saturation of more active adsorption sites at 10^{-6} M (results not shown). Above the latter concentration, less active adsorption sites start to be occupied by AZ.

8.3.3 Confocal microscopy [35]

To understand confocal microscopy, it is instructive to imagine a pair of lenses that focuses light from the focal point of one lens to the focal point of the other. This is illustrated by the dark blue rays shown in Fig. 8-14. The light blue rays represent light from another point in the specimen, which is not at the focal point of the left-hand-side lens (note that the colors of the rays are purely for the purposes of distinguishing the two sets—they do not represent different wavelengths of the incident light.). Clearly, image of the light blue point is not at the same location as the image of the dark blue point. In confocal microscopy, the aim is to see only the image of the dark blue point.[36] Accordingly, if a screen with a pinhole is placed at the other side of the lens system, then all of the light from the dark point will pass through the pinhole. Note that at the location of the screen the light blue point is out of focus. Moreover, most of the light will get blocked by the screen, resulting in an image of the bright blue point that is significantly attenuated compared to the image of the dark blue point. The confocal microscope incorporates the ideas of point-by-point illumination of the specimen and rejection of out-of-focus light. One drawback with imaging a point onto the specimen is that there are fewer emitted photons to collect at any given instant. Thus to avoid building a noisy image, each point must be illuminated for a long time to collect enough light to make an accurate measurement. In turn, this increases the

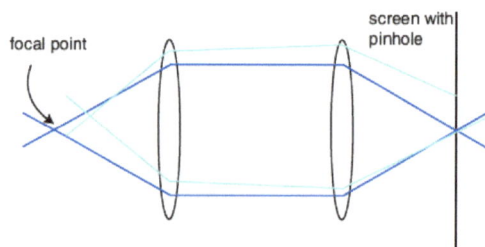

Fig. 8-14 Rejection of light not incident from the focal plane. All light from the focal point that reaches the screen is allowed through. Light away from the focal point is mostly rejected. Reprinted with permission from Denis Semwogerere, Eric R. Weeks, *Encyclopedia of Biomaterials and Biomedical Engineering* with Copyright © 2005 of Taylor & Francis.

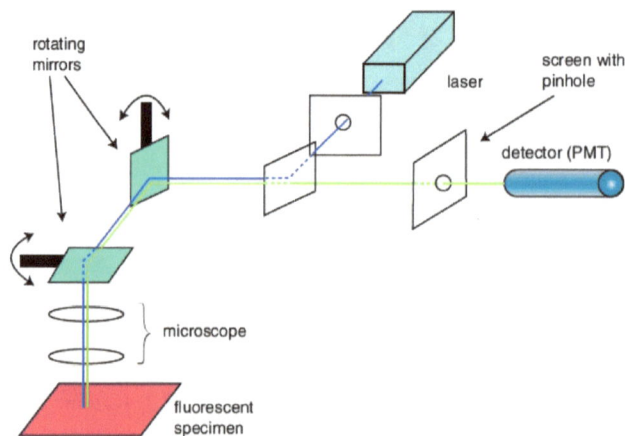

Fig. 8-15 Basic setup of a confocal microscope. Light from the laser is scanned across the specimen by the scanning mirrors. Optical sectioning occurs as the light passes through a pinhole on its way to the detector. Reprinted with permission from Denis Semwogerere, Eric R. Weeks, *Encyclopedia of Biomaterials and Biomedical Engineering* with Copyright © 2005 of Taylor & Francis.

length of time needed to create a point-by-point image. The solution is to use a light source of very high intensity, which Minsky did with a zirconium arc lamp. The modern choice is a laser light source, which has the additional benefit of being available in a wide range of wavelengths. As can be seen from Fig. 8-15, the dye in the specimen is excited by the laser light and fluoresces. The fluorescent (green) light is scanned by the same mirrors that are used to scan the excitation light (blue) from the laser and then passes through the dichroic mirror. Thereafter, it is focused onto the pinhole. The light that makes it through the pinhole is measured by a detector such as a photomultiplier tube. The image created by the confocal microscope is of a thin planar region of the specimen—an effect referred to as optical sectioning. Out-of-plane unfocused light has been rejected, resulting in a sharper, better resolved image. Figure 8-16 shows an image created with and without optical sectioning.

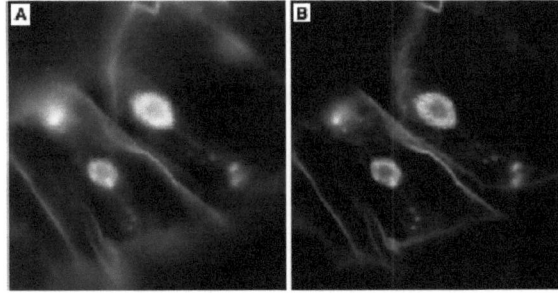

Fig. 8-16 Images of cells of spirogyra generated with and without optical sectioning. The image in (B) was created using a slit rather than a pinhole for out-of-focus light rejection. Most of the haze associated with the cell walls of the filamentous algae is absent, allowing clearer distinction of the different parts. Reprinted with permission from Denis Semwogerere, Eric R. Weeks, *Encyclopedia of Biomaterials and Biomedical Engineering* with Copyright © 2005 of Taylor & Francis.

The ability of a confocal microscope to create sharp optical sections makes it possible to build 3D renditions of the specimen. Data gathered from a series of optical sections imaged at short and regular intervals along the optical axis are used to create the 3D reconstruction. Software can combine the 2D images to create a 3D rendition. Representing 3D information in a meaningful way out of 2D data is nontrivial, and a number of different schemes have been developed. Visualization of 3D image requires relevant professional software such as SGI – VoxelView, VayTek's VoxBlast, and MAC - NIH Image etc. for further data processing. Volume rendering is a computer graphics technique whereby the object or phenomenon of interest is sampled or subdivided into many cubic building blocks, called voxels (or volume elements.) A voxel is the 3D counterpart of the 2D pixel and is a measurement of unit volume. Each voxel carries one or more values for some measured or calculated property of the volume (such as intensity values in the case of LSCM data) and is typically represented by a unit cube. The 3D voxel sets are assembled from the multiple 2D images (such as the LSCM image stack), and are displayed by projecting these images into 2D pixel space where they are stored in a frame buffer. Volumes rendered in this manner have been likened to a translucent suspension of particles in 3D space. In surface rendering, the volumetric data must firstly be converted into geometric primitives, by a process such as isosurfacing, isocontouring, and surface extraction or border following. These primitives (such as polygon meshes or contours) are then rendered for display using conventional geometric rendering techniques. Figure 8-17 shows a 3D reconstruction, from slices of a suspension of 2 mm diameter colloidal particles using ''alpha blending''—a technique that combines images by first making each of their individual pixels less or more transparent according to a computed weight called the ''alpha'' value. The result is a 3D-like structure.

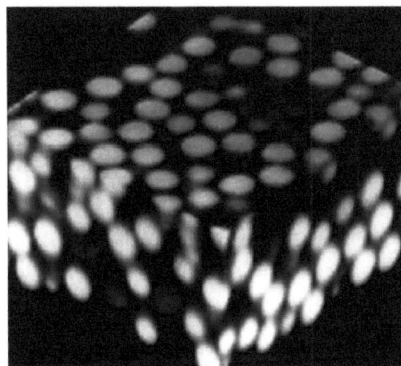

Fig. 8-17 Three-dimensional reconstruction of a series of 2D images of PMMA spheres suspended in a cyclohexylbromide and decalin solution. The image was created using "alpha blending." Reprinted with permission from Denis Semwogerere, Eric R. Weeks, *Encyclopedia of Biomaterials and Biomedical Engineering* with Copyright © 2005 of Taylor & Francis.

Applications of confocal microscopy includes cellular function, conjugated antibodies, DNA/RNA, organelle structure, cytochemical identification, and probe rationing.

8.3.4 Multiphoton microscopy

Multiple-photon excitation fluorescence microscopy is a technique that uses non-linear optical effects to achieve optical sectioning. The sample is illuminated by the light beam with a wavelength around twice the wavelength of the absorption peak of the fluorophore being used. For example, in the case of fluorescent which has an absorption peak around 500 nm, and thus 1000 nm excitation could be used. Essentially, no excitation of the fluorophore will occur at this wavelength. However, if a high peak-power, pulsed laser is used (so that the mean power levels are moderate and do not damage the specimen), two-photon events will occur at the point of focus. At this point the photon density is sufficiently high that two photons can be absorbed by the fluorophore essentially and simultaneously. This is equivalent to a single photon with energy equal to the sum of the two that are absorbed. In this way, fluorophore excitation will only occur at the point of focus (where it is needed) thereby eliminating excitation of out-of-focus fluorophore and achieving optical sectioning. [37]

Three-photon excitation can also be used in certain circumstances. In this case three photons are absorbed simultaneously, effectively tripling the excitation energy. By this technique, UV excited fluorophores may be imaged with IR excitation. Because excitation levels are dependent of the cube of the excitation power, resolution is improved (for the same excitation wavelength) compared to the two photon excitation where there is a quadratic power dependence. It is possible to select fluorophores such that multiple labeled samples can be imaged by combination of 2- and 3 photon excitation, using a single IR excitation source.

Optical sections may be obtained from deeper within a tissue that can be achieved by confocal or wide-field imaging. There are three main reasons for this point: (1) the excitation source is not attenuated by absorption of fluorophore above the plane of focus; (2) longer excitation wavelengths suffer less scattering; and (3) fluorescence signal is not degraded by scattering within the sample as it is not imaged. When images of the optical sections that are deep within a light-scattering sample are obtained using confocal microscopy, the fluorescence signal is attenuated by light scatter. Furthermore, some fluorescence originating from the regions away from the point being instantaneously illuminated will be scattered such that this fluorescence will pass through the confocal pinhole, thereby increasing background. Confocal imaging therefore suffers deterioration in signal-to-background when obtaining images from deep within a sample. Multiphoton imaging is largely immune from these effects as little fluorescence is generated away from the point of illumination and all detected fluorescence photons may be used for imaging regardless of whether they have been scattered or not.

Recently, in order to improve the biomedical utilities of high-speed multiphoton microscopy, parallelizing the multiphoton imaging process which has been called multifocal multiphoton microscopy (MMM) was put forth. [38] In MMM, a specimen is scanned with multiple excitation foci instead of a single excitation focus. The emission light generated from the foci is collected simultaneously by a spatially-resolved detector such as a charge coupled device (CCD) camera. The detector integrates the emission photons from the specimen during 2D scanning of multiple

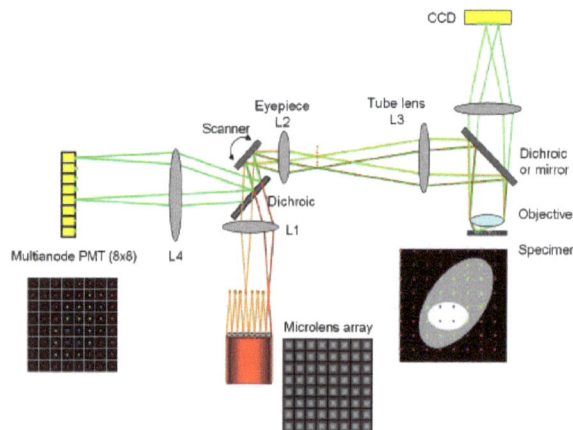

Fig. 8-18 A schematic of the multifocal multiphoton microscope based on a MAPMT. Excitation beams are depicted in red/orange colors and emission beams are in green color. In this figure, only two beam-lets are ray-traced. The excitation beam is splitted into 8 × 8 beam-lets via a microlens array. Multiple excitation foci (8 × 8) scan the specimen. The emission beam-lets are collected either by a CCD camera or a MAPMT which has 8 × 8 pixels. L1, L2, and L3 are lenses. The transmission of light through the microlens array and high NA objective causes appreciable pulse dispersion which can be corrected by pre-chirping using a pair of prisms. Reprinted with permission from Opt. Express 15, 11659-11678 (2007) with copyright © 2007 of Optical Society of America.

excitation foci in the sample plane. The imaging speed of enhancement of MMM is proportional to the number of excitation foci. However, for thick tissue applications, MMM has also an important shortage. Through many private communications with researchers, it has been noted that the imaging depth of MMM in tissues is significantly less than that of conventional single-focus multiphoton microscopy (SMM). In order to improve the imaging depth of MMM in turbid tissues, a multi-anode photomultiplier tubes (MAPMT) was put forth. [39] The MAPMT is composed of a cathode, a dynode chain and a segmented anode. Its structure is optimized to ensure the spatial distribution of photons incident upon the cathode to be reproduced faithfully as the signal at the anode. Similar to the conventional PMT, the MAPMT has a good quantum efficiency (over 20% in the blue/green spectral range), negligible read-out noise, and minimal dark noise with cooling. The anode of the MAPMT used in this instrument is a segmented, 8×8 pixel array (H7546, Hamamatsu, Bridgewater, NJ). A schematic of the new MMM design is depicted in Fig. 8-18. The light source used is a Ti-Sapphire (Ti-Sa) laser (Tsunami, Spectra-Physics, Mountain View, CA) pumped by a continuous wave Nd:YVO4 laser (Millenia, Spectra-Physics, Mountain View, CA). It outputs approximately 2 W of mode locked pulsed light at 800 nm wavelength. Excitation beam from the laser is expanded and illuminates a square micro-lens array with 12×12 micro-lenses. The micro-lenses are of 1 mm × 1 mm in size with 17 mm focal length (1000-17-S-A, Adaptive Optics, Cambridge, MA). The degree of beam expansion is set such that 8 ×8 beam-lets are produced after the micro-lens array. The beam-lets are collimated after a lens (L1) and reflected onto x-y galvanometric mirror scanners (6220, Cambridge Technology, Cambridge MA) which is positioned in the focal plane of L1. In this configuration, the beam-lets overlap each other on the scanner mirror. The beam-lets, reflected by the scanner, incident upon the epi-illumination light path of a microscope at different entrance angles (BX51, Olympus, Melville, NY). A lens combination of L2 and L3 expands the beam-lets to slightly under-fill the back aperture of an objective (approximately 0.75 of the aperture diameter) balancing the need to maximize excitation power at each foci and to utilize the high numerical aperture of the objective lens. The expanded beam-lets are reflected by either a dichroic mirror (for CCD setup) or a mirror (for MAPMT setup) toward the objective lens. The objective lens used here is a 20×water immersion lens with 0.95 NA (XLUMPLFL20XW, Olympus, Melville, NY). The effective *NA* due to under-filling is approximately 0.71. The objective lens generates an array of 8 × 8 excitation foci in the sample plane of the specimen. The image is formed by raster scanning the excitation foci across the specimen using the scanner mirrors. The excitation foci are separated each other in distance of 45 μm and each focus scans the area of 45 μm×45 μm in the sample plane. The frame size covered by the 8×8 array of foci is 360 μm×360 μm. Emission light is collected by the same objective lens forming an array of emission beam-lets. For signal collection using the CCD camera, the emission beam-lets are transmitted through a short-pass dichroic mirror (650dcxxr, Chroma Technology, Brattleboro, VT) and focused on the CCD camera via a lens. The CCD camera integrates the emission signal during 2D scanning of the foci in the sample

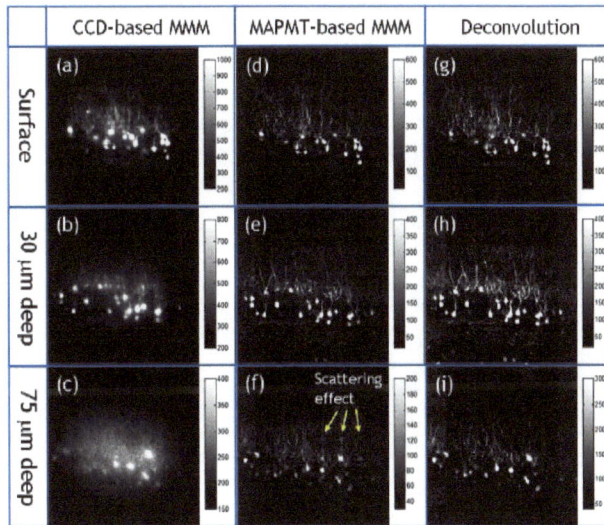

Fig. 8-19 Images of GFP expressing neurons in the *ex vivo* mouse brain acquired with CCD-based MMM (a-c) and MAPMT-based MMM (d-f) at different depth locations (surface, 30 um, and 75 um deep). For images with MAPMT-based MMM, a deconvolution algorithm was applied to remove the effect of emission photon scattering (g-i). The objectiveused is a 20× water immersion with NA 0.95. The input laser power is 300 mW at 890 nm wavelength. The frame rate is 0.3 frames per second with a frame size of 320 × 320 pixels. Reprinted with permission from "Ki Hean Kim, Christof Buehler, Karsten Bahlmann, Timothy Ragan, Wei-Chung, A. Lee, Elly Nedivi, Erica L. Heffer, Sergio Fantini, and Peter T. C. So, Opt. Express 15, 11659-11678 (2007)" with copyright © 2007 of Optical Society of America.

plane. For the MAPMT, the emission beam-lets are reflected on the mirror toward the scanner mirror retracing the excitation paths and are de-scanned via the scanner mirrors. The de-scanned emission beam-lets become stationary irrespective of the scanner motion. The emission beam-lets are reflected by a long-pass dichroic mirror (Chroma Technology, Brattleboro, VT) toward the MAPMT and are focused onto the MAPMT by a lens (L4). A short-pass filter (E700SP, Chroma Technology, Brattleboro, VT) blocks any strayed excitation light. The 8×8 emission beam-lets are focused at the centers of the corresponding MAPMT pixels. Since the emission beam-lets are detected after de-scanning, a loss of emission light (approximately 30 %) is expected compared to non-de-scanning systems. Both the MAPMT and the CCD detection light paths are implemented on the same system so that accurate comparison of these two detection methods is possible. The image in Fig. 8-19 shows that the MAPMT-based MMM can provide superior images as compared to the CCD-based MMM for tissue imaging.[38] However, additional improvements of this system are possible. First of all, since the MAPMT is positioned in the image plane, the location of each excitation focus corresponds to the center position of the matching pixel of the MAPMT. The effective detector area is scaled quadratically with the separation distance of the foci. Therefore, with wider foci separation, each pixel of the MAPMT will have better collection efficiency for scattered emission photons. Second, having lower signal collection efficiency is a disadvantage of a MAPMT-based MMM designs in comparison to a CCD-based MMM design. Third, the MAPMT used in this experiment is composed of bi-alkali cathodes with very low efficiency in the red. However, newer MAPMTs with multi-alkali cathodes have a more uniform spectral response having a quantum efficiency of 15-20% throughout the visible spectrum. Further, these MAPMTs have very low noise. 100 dark counts per second are typical and cooling can reduce this further by several orders of magnitude. Fourth, the photon sensitivity of each MAPMT pixel can vary up to 50%. Fifth, there is also electronic crosstalk among neighboring pixels of the MAPMT. The typical crosstalk is about 2 – 4% when the photons are collected at the center of each pixel. Finally, MMM tissue imaging requires high laser power.

Generally speaking, limitations of the multi-photon excitation includes: (1) slightly lower resolution with a given fluorophore when compares to confocal imaging. This loss in resolution can be eliminated by the use of a confocal aperture at the expense of a loss in signal; (2) thermal damage can occur in a specimen if it contains chromophores that absorb the excitation wavelengths, such as the pigment melanin; (3) only works with fluorescence imaging; and (4) currently rather expensive.

8.4 Summary

Two characterization approaches: geometrical and optical characterization, were described in this chapter. Currently, SPM technology is a crucial tool in the nanosciences. With the concept of SPM, three relevant techniques: AFM, STM, and NSOM were rapidly developed subsequently. For geometrical characterization in nanoscale, an AFM technique is introduced. For optical characterization, several methods were illustrated: NSOM/SNOM, SERS, confocal microscope, and multiphoton microscope. Amongst, NSOM and SERS are most important methods for characterization of the plasmonic structures and nanophotonic devices. For current commercial relevant equipments, some functions are integrated together to be two-in-one or three-in-one system in respect to different applications.

Reference

[1]   M. R. Jarvis,1 Rubén Pérez,2 and M. C. Payne, "Can Atomic Force Microscopy Achieve Atomic Resolution in Contact Mode?", Phys. Rev. Lett. 86, 1287-1290 (2001).

[2]   G Binnig, C Gerber, E Stoll, TR Albrecht, CF Quate, "Atomic Resolution with Atomic Force Microscope". Euro. Phys. Lett. 3, 1281 (1987).

[3]   G. Binnig, C. F. Quate, and C. Gerber, "Atomic Force Microscope", Phys. Rev. Lett. 56, 930 (1986).

[4]   K. Kimberlin, Edt.: Basic SPM training course. Digital Instruments Veeco, Inc., 2000.

[5]   Yongqi Fu, Wei Zhou, Lennie E.N. Lim, Chunlei Du, Haofei Shi, Changtao Wang, "Geometrical characterization issues of plasmonic nanostructures with depth-tuned grooves for beam shaping", Opt. Eng. 45 (10), 108001(2006).

[6]   William L. Barnes, Alain Dereux, and Thomas W. Ebbesen, "Surface plasmon subwavelength optics." Nature 424, 824-830 (2003).

[7]   H.J. Lezec, A. Degiron, E. Devaux, R.A. Linke, L. Martin-Moreno, F.J. Garcia-Vidal, T.W. Ebbesen, "Beaming light from a subwavelength aperture", Science 297, 820-822 (2002).

[8]   Tineke Thio, H.J. Lezec, T.W. Ebbesen, "Strongly enhanced transmission through subwavelength holes in metal films", Physica *B* 279, 90-93 (2000).

[9]   L. Martín-Moreno, F.J. García-Vidal, H.J. Lezec, K.M. Pellerin, T. Thio, J.B. Pendry, and T.W. Ebbesen, "Theory of extraordinary optical transmission through subwavelength hole arrays", Phys. Rev. Lett. 86, 1114-1117 (2001).

[10] L. Martin-Moreno, F.J. García-Vidal, H.J. Lezec, A. Degiron and T.W. Ebbesen, "Multiple path to enhance optical transmission through a single subwavelength slit", Phys. Rev. Lett. 90, 167401 (2003).

[11] K.J. Klein Koerkamp, S. Enoch, F.B. Segerink, N.F. van Hulst, and L. Kuipers, "Strong influence of hole shape on extraordinary transmission through periodic arrays of subwavelength holes", Phys. Rev. Lett. 92, 183901 (2004).

[12] Zhijun Sun and Hong Koo Kim, "Refractive transmission of light and beam shaping with metallic nano-optic lenses", Appl. Phys. Lett. 85, 642-644 (2004).

[13] F.J. García-Vidal, L. Martin-Moreno, H.J. Lezec and T.W. Ebbesen, "Focusing light with a single subwavelength aperture flanked by surface corrugations", Appl. Phys. Lett. 83, 4500-4502 (2003).

[14] Haofei Shi, Changtao Wang, Chunlei Du, Xiangang Luo, Xiaochun Dong, Hongtao Gao, "Beam manipulating by metallic nano-slits with variant widths", Optics Express 13, 6815-6820 (2005).

[15] H. Reather, Surface *Plasmons on Smooth and Rough surfaces and on gratings* (Springer, Heidelberg, 1988), Chapter 2.

[16] Lim Boon Hong, thesis of Master Degree, "Study of modification of atomic force microscope probe via focused ion beam milling", in Nanyang Technological University, Singapore (2001).

[17] Yongqi Fu, Ngoi Kok Ann Bryan, "Investigation of physical properties of quartz via focused ion beam Bombardment", Appl. Phys. B 80, 581–585 (2005).

[18] Ji-Young Kim, Vladimir P. Drachev, Hsiao-Kuan Yuan, Reuben M. Bakker, Vladimir M. Shalaev, "Imaging contrast under aperture tip–nanoantenna array Interaction", Appl. Phys. B 93 (1), 189-198 (2008).

[19] J. Prikulis,a) H. Xu, L. Gunnarsson, H. Olin, and M. Ka¨ ll, "Phase-sensitive near-field imaging of metal nanoparticles", J. Appl. Phys. 92, 6211-6213 (2002).

[20] R. Hillenbrand and F. Keilmann, Appl. Phys. B: Lasers Opt. 73, 239 (2001).

[21] Ellis, D.I. and Goodacre, R., "Metabolic fingerprinting in disease diagnosis: biomedical applications of infrared and Raman spectroscopy", Analyst, 131, 875-885 (2006).

[22] Khanna, R.K. *Evidence of ion-pairing in the polarized Raman spectra of a Ba_2+CrO doped KI single crystal.* John Wiley & Sons, Ltd, 1957.

[23] Jeanmaire, David L.; Richard P. van Duyne,, "Surface Raman Electrochemistry Part I. Heterocyclic, Aromatic and Aliphatic Amines Adsorbed on the Anodized Silver Electrode". Journal of Electroanalytical Chemistry 84: 1–20, (1977).

[24] K. Kneipp, H., Itzkan, I., Dasari, R. R., Feld, M. S., "Ultrasensitive chemical analysis by Raman spectroscopy". Chem. Phys. 247: 155–162 (1999)..

[25] Khanna, R.K.. *Theoretical and experimental resonance Raman intensities for the manganate ion.* John Wiley & Sons, Ltd. 1974.

[26] P. Matousek, I.P. Clark, E.R.C. Draper, M.D. Morris, A.E. Goodship, N. Everall, M. Towrie, W.F. Finney, A.W. Parker, "Subsurface Probing in Diffusely Scattering Media using Spatially Offset Raman Spectroscopy". Applied Spectroscopy 59: 393 (2005).

[27] Kneipp, K., Kneipp, H., Itzkan, I., Dasari, R. R., Feld, M. S. "Surface-enhanced Raman scattering: A new tool for biomedical spectroscopy". *Current Science* 77, 915-924 (1999).

[28] Kneipp, K.; Kneipp, H.; Itzkan, I.; Dasar, R.R.; and Feld, M.S., "Ultrasensitive chemical analysis by Raman spectroscopy". Chem. Rev. 99, 2957-2975 (1999).

[29] Moskovits M., "Surface-enhanced spectroscopy", Rev. Mod. Phys. 57, 783 (1985).

[30] Kambhampati, P.; Child, C.M.; Foster, M.C.; and Campion A., "On the chemical mechanism of surface enhanced Raman scattering: Experiment and theory," J. Chem. Phys. 108, 5013-5026 (1998).

[31] Michael J. Weaver, Shouzhong Zou and Ho Yeung H. Chan, "Peer Reviewed: The New Interfacial Ubiquity of Surface-Enhanced Raman Spectroscopy", Anal. Chem. 72, 38A-47A (2000).

[32] Garrell, R.L. "Surface-enhanced Raman spectroscopy", Anal. Chem. 61, 401A- 411A (1989).

[33] Zou, S.; Williams, C.T.; Chen, E. K.-Y.; and Weaver, M.J. J., "Surface-Enhanced Raman Scattering as a Ubiquitous Vibrational Probe of Transition-Metal Interfaces: Benzene and Related Chemisorbates on Palladium and Rhodium in Aqueous Solution", J. Phys. Chem. B 102, 9039-9049 (1998).

[34] M. V. Can˜amares, J. V. Garcia-Ramos, C. Domingo and S. Sanchez-Cortes, "Surface-enhanced Raman scattering study of the adsorption of the anthraquinone pigment alizarin on Ag nanoparticles", J. Raman Spectrosc. 35, 921–927 (2004).

[35] Denis Semwogerere, Eric R. Weeks, *Encyclopedia of Biomaterials and Biomedical Engineering*, 2005 by Taylor & Francis.

[36] Minsky, M., "Memoir on inventing the confocal microscope". Scanning 10, 128–138 (1988).

[37] Ammasi Periasamy and Alberto Diaspro., "Multiphoton Microscopy", J. Biomed. Opt., Vol. 8, 327 (2003).

[38] Ki Hean Kim, Christof Buehler, Karsten Bahlmann, Timothy Ragan, Wei-Chung, A. Lee, Elly Nedivi, Erica L. Heffer, Sergio Fantini, and Peter T. C. So, "Multifocal multiphoton microscopy based on multianode photomultiplier tubes", Opt. Express 15, 11659-11678 (2007).

[39] J. Bewersdorf, R. Pick, and S. W. Hell, "Multifocal multiphoton microscopy", Opt. Lett. 23, 655-657 (1998).

9 NANOHOLES ARRAY AND APPLICATIONS

Abstract: One of important types of plasmonic structures, nanoholes array was highlighted and presented in this chapter. Firstly, optical transmission properties of the nanoholes were introduced. Then polarization effect of elliptical nanoholes array was highlighted. Finally, one typical application of the nanohole array embedded on biochips for cancer cell detection was described briefly.

9.1 Brief introduction

Integrated photonic structures were found in a wide variety of flora and fauna [1]. The most basic type of such structures is a hole, and light transmission through the holes has been an object of scientific inquiry for centuries [2]. Bethe [3] studied theoretically the light transmission through a subwavelength hole in an infinitely thin, perfectly conducting metal screen. However, in terms of subwavelength optics, the results were not encouraging; Bethe found the transmission of light, $T(k)$, to scale with the ratio of hole radius to wavelength to the fourth power, $T(k) \propto (r/k)^4$. This dependence is not necessarily followed for holes in real metal films with a finite thickness. As a matter of fact, later experimental studies showed extraordinary transmission through arrays of subwavelength holes in silver and gold films [4]. The transmission was extraordinary in the absolute transmission efficiencies at peak wavelengths which were significantly higher than the light that impinged on the holes, and orders of magnitude higher than the predicted value calculated by the earlier theory. Subsequent works showed that the extraordinary transmission results from a combination of waveguiding effects within the holes and electromagnetic waves at the surface of the metal, such as surface plasmons (SP). This initial discovery has motivated many studies on the physics of enhanced transmission through nanohole arrays [5, 2], and toward their application in sensing [6], optical filters [7], imaging [8], nano-lithography[9], and photonic circuits [10]).

9.2 Transmission properties of nanoholes array [11]

9.2.1 Single holes

Bethe's predicted cutoff in the transmission of a single hole in a perfect-conductor thin screen as $(b/\lambda)^4$ is the leading-order term in the expansion of the transmission cross section in powers of b/λ [3]. Bethe's theory is too idealized to treat situations where surface modes are involved and where propagating or evanescent modes can additionally be excited inside the hollow aperture, thereby significantly under estimating the transmission efficiency. Subsequent higher-order analytical corrections, and eventually rigorous numerical calculations, demonstrated that the cross section lies below the holes area up to a radius of $b \approx 0.2\lambda$. These results found experimental corroboration down to the NIR regime, with new localized plasmon resonances showing up at shorter wavelengths. Two different mechanisms were, however, suggested to achieve enhanced transmission in a single hole: filling it with a material of high permittivity, thus creating a partially bound cavity mode that couples resonantly to incident light (see Sec. III. with paragraph E); and decorating the aperture with periodic corrugations in much the same way as highly directional antennas which are capable of focusing electromagnetic radiation on a central dipole element by means of concentric, periodically spaced metallic rings.

9.2.2 Holes array

The intensity of light passing through holes is boosted at certain wavelengths when we arrange them periodically. Pioneering calculations and microwave experiments showed zero reflection in thin films perforated by periodic arrays of small apertures with radius of $b \approx 0.36\lambda$. Further seminal experiments focused on the relation between hole arrays in thin metal screens and their complementary screens, putting Babinet's principle to a test in the far-infrared region. This was followed by numerous applied studies of hole arrays (regarded as frequency-selective surfaces) in the engineering community, including filters for solar energy collection and elements to enhance antennae performance. Ebbesen *et al.* [4] demonstrated in the optical domain an extraordinary light transmission for openings of the radius below cutoff wavelength of the first propagating mode in a circular waveguide, $b<0.29\lambda$. Since then, this phenomenon has been consistently observed for a varied list of metallic materials, over a wide range of wavelengths (*e.g.*, for microwaves, to which metals respond as nearly perfect conductors, in the infrared, and in the vacuum ultraviolet (VUV), using a good conductor in this regime like Al), and with different types of array symmetries, including 2D quasicrystal arrangements.

The influence of various geometrical and environmental factors was studied. In particular, the role of holes shape has been shown to yield nontrivial effects, such as larger enhancement and redshift of the transmission peaks with respect to the Rayleigh condition for the light polarized along the short axis of elongated apertures. Finite arrays exhibit interesting shifts in the transmission maxima as well, depending on the number of apertures. More exotic shapes like annular holes had also been simulated and measured, with the additional appeal that annular waveguides support always one guided mode at least. The transmission is exponentially attenuated with holes depth because it is mediated by evanescent modes of the apertures regarded as narrow subwavelength waveguides. However, strong signatures of interaction between both two metal interfaces at double sides have been reported, as well as high sensitivity to dielectric environment, so that maximum transmission is achieved when the permittivity is the same on the double sides of the metal film. Extraordinary optical transmission has expanded to a wide range of phenomena, like the interaction of holes arrays with molecules for potential applications in biosensing and all-optical switching, and the demonstration of the quantum nature of plasmons through photon entanglement preservation after traversing a holes array. More detailed theoretical analysis can be seen in Ref. [11].

9.3 Polarization effect of elliptical nanoholes array [12]

This topic was put forth in focused ion beam (FIB) fabrication point of view. FIB fabrication technique was described in Chapter 6 Sec.2 of this book. As well known, like conventional optical system having imaging and focusing aberration, the focused ion beam has also aberration in the ion column such as chromatic, spherical aberration, coma, and stigmation. Stigmation always exists and accompanies the entire milling and deposition process. It will be changed due to many unforeseen factors such as power fluctuation, vacuum, and environmental vibration even if it is well corrected by operators before starting milling. As a consequence, shape of the focused ion beam will be changed to be ellipse instead of the theoretically designed circular shape. The elliptical beam spot causes deformation of the hole during drilling. It is especially apparent for nanoholes drilling, as shown in Fig. 9-1. It can be seen that for the central micron holes, the holes deformation is slight in comparison to that of the nano-elliptical holes distributed at outer rings due to

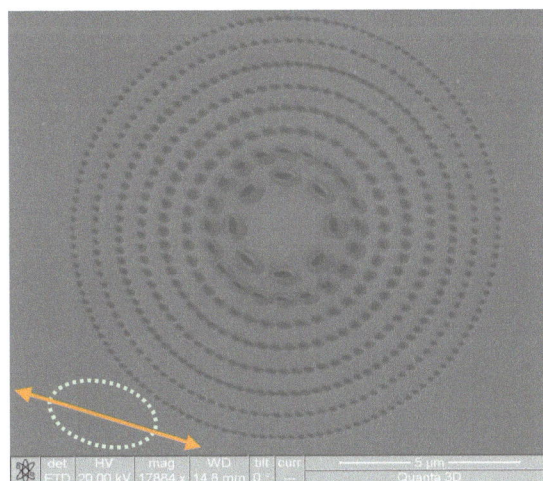

Fig. 9-1 Holes ranging from micron to nanoscale were fabricated using FIB direct fine milling. The arrow indicates orientation of the focused ion beam spot shown as the elliptical dot line. Stigmation exists and causes the beam spot shape to be ellipse instead of the normal circular.

the size matter. However, for the nanoholes drilling, the spot shape of ion beam directly influences the nanoholes shape. Orientation of the deformed nanoholes strongly depends on that of the ion beam spot. For current commercial FIB machines, the focused beam quality can be judged by operator's nuke eyes only. Therefore, the stigmation exists to a certain extent no matter how fine adjustment and correction of the stigmation is done. It means that the stigmation cannot be completely eliminated by the operators in practical use.

Polarization effect of the elliptical nanoholes array was firstly reported by R. Gordon in 2004. Accidentally, elliptical nanoholes instead of circular holes were obtained when the FIB fabrication was completed due to focusing aberration existing in the ion column. Significant polarization effect was surprisingly found while tested transmission properties after the fabrication. The following analyses were cited from the author's report regarding the polarization effect.

The nanoholes were created using focused Ga-ion beam milling (30 keV) on a 100-nm thick gold film coated on a

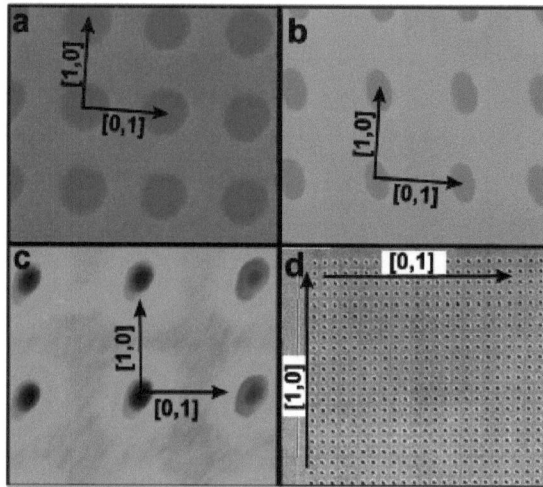

Fig. 9-2 Scanning electron micrographs of square nanohole arrays in gold. (a) Nearly circular holes. (b) Elliptical holes, 0.6 aspect ratio and major axis at -12° to the [1; 0]. (c) Elliptical holes, 0.6 aspect ratio and major axis at 33° to the [1; 0] axis. (d) An expanded view of (c) showing the full 16.1 μm wide array of 529 holes (holes spaced by 704 nm).

Reprinted with permission from "R. Gordon, A.G. Brolo, A. McKinnon, A. Rajora, B. Leathem, and K. L. Kavanagh, Phys. Rev. Lett. 92, 037401(2004)" with copyright © 2004

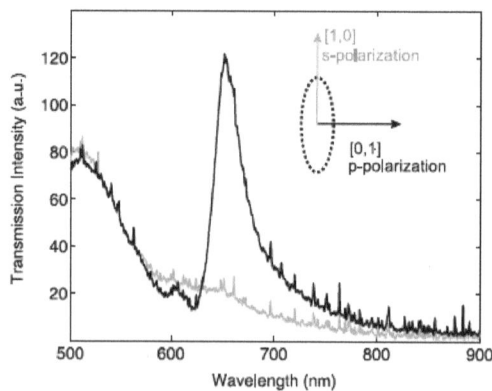

Fig. 9-3 Transmission spectra through elliptical nanohole array for two orthogonal linear polarizations, with a 0.3 aspect ratio between the minor and major axes of the ellipse. The *p* polarization is parallel to the [0; 1] direction. Reprinted with permission from "R. Gordon, A.G. Brolo, A. McKinnon, A. Rajora, B. Leathem, and K. L. Kavanagh, Phys. Rev. Lett. 92, 037401(2004)" with copyright © 2004 of The American Physical Society.

glass substrate. Several square arrays of 23 × 23 holes with different periodicities were fabricated. The ellipticity of the holes was controlled by means of an astigmatic ion beam. The orientation of the major axis of the ellipse was also varied. Figure 9-2 shows SEM micrographs of three samples, demonstrating the ability to create circular and elliptical holes, as well as the ability to control the orientation of the ellipse's major axis relative to the array lattice. The optical transmission spectra of normally incident unpolarized light through all the arrays utilized in this work showed the previously reported enhanced resonances, and is in agreement with the Bragg condition for the SP modes [13]. Figure 9-3 shows the transmission spectra for two orthogonal linear polarizations of the incident white light. The major axis of the elliptical holes is parallel to the orientation of [1; 0] axis of the lattice, the aspect ratio of the ellipse is 0.3, and the spacing between the holes is 500 nm. The peak at 655 nm agreed well with the SP resonance at the gold-glass interface. There is a dramatic reduction in this peak as the polarization is rotated from the *p*-polarization to the *s*-polarization (with the *p*-polarization defined along the [0; 1] direction as shown in Fig. 9-2). It is noted that the [0; 1] direction was selected to be most perpendicular to the major axis of the ellipse. The resonant transmission is enhanced when the direction of electric field polarization is perpendicular to the major axis of the elliptical holes.

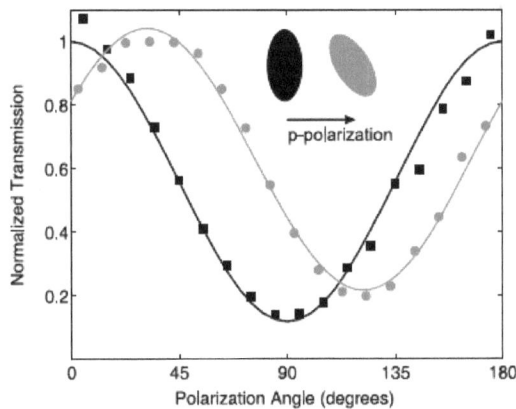

Fig. 9-4 Polarization dependence of transmission at the [0;1] resonance peak, normalized to the maximum. Transmission shows cosine dependence when the major axis of the ellipse is oriented both at 0° and 33° to the [1; 0] axis of the array. The *p* polarization, along the [0; 1] direction of the lattice, corresponds to a polarization angle of zero degrees. The maximum transmission occurs for polarization perpendicular to the broad side of the ellipse. Reprinted with permission from "R. Gordon, A.G. Brolo, A. McKinnon, A. Rajora, B. Leathem, and K. L. Kavanagh, Phys. Rev. Lett. 92, 037401 (2004)" with copyright © 2004 of The American Physical Society.

Figure 9-4 shows the polarization dependence of the transmitted intensity at the resonant wavelength, which follows a cosine function, with the maximum for *p*-polarized light (along the [0; 1] lattice direction). The same cosine dependence of the polarization was observed for ellipses oriented parallel and tilted 33° to the [1; 0] axis, respectively. When the ellipse is tilted, the orientation of the electric field polarization corresponding to the maximum transmission follows the orientation of the ellipse. This result was verified on five different ellipse orientations. The maximum transmission is for the *p*-polarization, along the direction that is perpendicular to the major axis of the ellipse. Ellipses with different aspect ratios were tested, each one was aligned with the major axis along the [1; 0] direction of the lattice, to within 20°. The transmission ratio had a squared dependence of the aspect ratio—a log-log fitting produced a slope of 2.07. The squared plot is shown with a solid line.

The strong polarization dependence of the transmission as a function of the ellipse orientation resembles previous observations of transmission through a subwavelength slit surrounded by surface corrugations [12] and also the transmission through metallic nanowire gratings [14]. In those experiments, it was shown that the enhanced transmission occurred only for the *p*-polarization, where the electric field is perpendicular to the long axis of the slits. The enhanced transmission for *p*-polarization was also found in simulations for a 1D array of grooves in a metal film

[15]. If we consider an array of elliptical holes with their length of major axis equal to the lattice spacing itself, and aligned along the [1; 0] direction, then the holes become a series of parallel grooves, as shown in Fig. 9-5 (a). Therefore, in the limiting case we recover a geometry that resembles the 1D case, and the enhanced polarization is expected to be

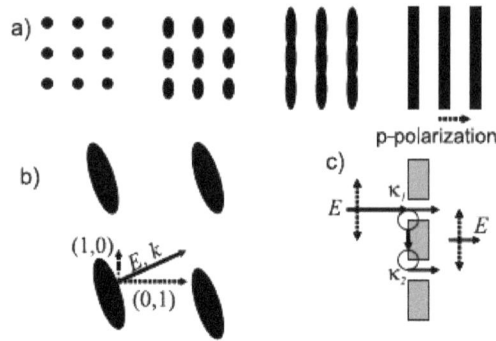

Fig. 9-5 (a) Elliptical holes approaching limiting case of slits, which only how enhanced transmission for *p* polarized light. (b) Enhanced SP excitation perpendicular to major axis of ellipse leads to preferential excitation of the [0; 1] resonance with respect to the [1; 0] resonance. (c) Coupling both into and out of the SP mode is required to obtain enhanced transmission from periodic array of holes, which results in a squared dependence of the enhanced transmission on the coupling strength. Reprinted with permission from "R. Gordon, A.G. Brolo, A. McKinnon, A. Rajora, B. Leathem, and K. L. Kavanagh, Phys. Rev. Lett. 92, 037401(2004)" with copyright © 2004 of The American Physical Society.

perpendicular to the major axis of the ellipse, as was observed in Fig. 9-2. Both the influence of the orientation and the aspect ratio dependence can be understood by noting that the direction of the SP mode propagation is parallel to the electric field [16], and by considering that the coupling into and out of the SP modes occurs at the edges of the holes. Since the SP modes propagate parallel to the electric field polarization, the Bragg resonance from the periodic array is aligned with the optical polarization. For example, the [0; 1] Bragg resonance will be excited by a polarization along the [0; 1] direction. Therefore, if the ellipse, by virtue of its orientation, enhances the coupling to the grating parallel to the [0; 1] direction, as shown in Fig. 9-5 (b), then the transmission will be enhanced for the polarization along the [0; 1] direction as well, as was observed in Fig. 9-4.

9.4 Applications for cancer cell detection
9.4.1 Nanohole Arrays as SPR based sensors [17]

Incident light on nanometer scale metallic structures such as nanoholes can also generate surface plasmon resonance (SPR) [2, 5]. In transmission mode, incident light on a metallic film with nanoholes is initially scattered into SPs that penetrate the holes and are again scattered on the other side of the film. The extent of SP generation and the degree of transmission, depend on the incident light wavelength, holes geometry/periodicity and material/medium dielectric constants. The approximate SPR wavelength is given by:

$$\ddot{e}_{SP}(i,j) = p(i^2 + j^2)^{-1/2}(\frac{\varepsilon_m \varepsilon_d}{\varepsilon_m + \varepsilon_d}) \tag{9-1}$$

where the wavelength for coupling (λ_{SP}) is governed by the periodicity p of the array, ε_d is the relative permittivity of the dielectric, ε_m is the dielectric constant of the metal, and i and j are integers related to the scattering order of the grating. The central role of SPs in the enhanced transmission motivated the application of nanohole arrays as surface based chemical and biological optofluidic sensors. In the context of on-chip analysis, nanohole arrays have several unique advantages [18]: (1) in contrast to reflective mode SPR, transmission mode operation at normal incidence simplifies alignment, facilitates the use of high numerical aperture optics [19], and permits eventual device level miniaturization and integration of supporting optics; (2) the footprint of a nanohole array is small relative to that typically required in reflective mode SPR [20], enabling miniaturization and integration into microfluidic architectures, multiplexed analysis, higher spatial resolution; (3) in contrast to other local surface plasmon strategies based on colloidal nanoparticles, or roughened surfaces, nanohole arrays can be fabricated with high reproducibility; (4) the sensitivity of the optical response to hole shape, periodicity, and lattice versus basis orientation provides a large variety of handles with which to control array/sensor response; and (5) the geometry differentiating each array element is fixed within the structure, and is generally more robust than an adsorbed probe.

The transmission spectrum of asymmetric structures contains two sets of peaks, and each set belongs to one of the surfaces. In many applications, hole arrays of a finite size were used for practical reasons. If the arrays contain small numbers of holes then the periodicity is not well defined and the contribution from the edges becomes significant, changing the spectrum and leading to unusual reemission patterns [21].

9.4.2 Field and spectroscopic enhancements with nanohole arrays

The confinement offered by a simple microfluidic channel, as shown in Fig. 9-6 (a), was proved useful in the delivery of solutions to nanohole arrays [19, 22-24]. The incorporation of arrays of nanohole arrays within a microfluidic framework for spatial and temporal detection on-chip is a natural progression of this technology.

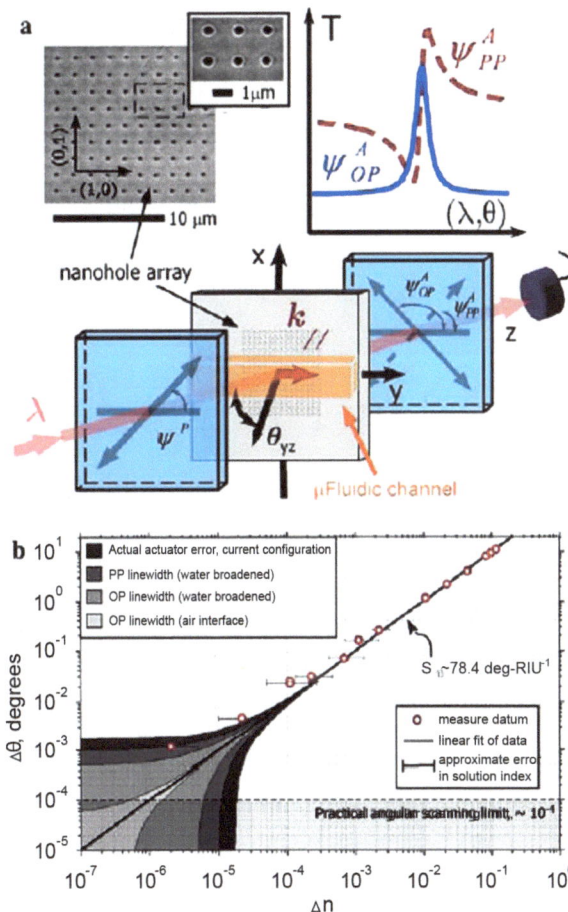

Fig. 9-6 The high-resolution SPR sensor: a conceptual diagram of the nanohole-array-based sensor with optical and fluidic interactions indicated. The input and output polarization states of a tunable laser provided variable spectral or angular Fano type profiles. A microfluidic channel transported analyte to the sensing area; b a plot of resonance peak position shift determined through angular interrogation as a function of solution refractive index change. Shaded regions indicate error bounds for the conditions indicated inset (PP indicates parallel polarizer-analyzer and OP indicates orthogonal polarizer-analyzer pair). Reprinted with permission from "Kevin A Tetz, Pang L, Fainman Y, Opt. Lett. 31, 1528–1530 (2006)" with Copyright © 2006 of Optical Society of America.

$$t(\zeta) = t_b(\zeta) + \sum_n \frac{b_n \Gamma_n e^{i\phi_n}}{\zeta - \zeta_n + i(\gamma_n + \Gamma_n)}$$

(9-2)

where t_b is the slowly varying background amplitude, ζ_n is the real wave vector/frequency of SPP on either interface, ϕ_n is the phase of SPP wave, γ_n and Γ_n is the non-radiative and radiative damping, respectively, $\zeta \equiv \{k_\parallel, \omega\}$. The item $\exp(i\phi_n)$ is SPP resonance.

A high spectral resolution, 2D nanohole-array-based SPR sensor that operates at normal or near normal incidence and facilitates high spatial resolution imaging was presented (see Fig. 9-6 (a)). The angular and spectral transmittance of the structure was modified from a Fano type to a pure Lorentzian line shape with a parallel and orthogonal polarizer–analyzer pair. This change leads to a linewidth narrowing that maximizes the sensor resolution. Measurements were carried out using a simple setup, as shown in Fig. 9-6 (a), where a collimated, tunable laser source (1520–1570 nm, 6 dBm) of ~1 cm in diameter was used to excite a SPP field in the 2D nanohole array. We referred to two polarization states in our measurements. (i) Parallel polarizer–analyzer (PP): polarizer and analyzer axes were parallel and oriented at $+\pi/4$ with respect to the [1, 0] direction of the nanohole array yielding equal electric field amplitudes in the x- and y-directions. (ii) Orthogonal polarizer–analyzer (OP): the polarizer (analyzer) axis was oriented at $+\pi/4$ $(-\pi/4)$ with respect to the [1, 0] direction. Resonant transmittance through the 2D nanohole array depends on the interrogation angle and wavelength of radiation and has a Fano-type line shape for PP and a Lorentzian shape for OP. There have been a number of studies that have investigated and explained the effects of the various geometric parameters on the shape of the resonant transmission (*e.g.*, hole size, metal film thickness, and optical properties of the metal), and we noted that the critical feature (assume a relatively "thick" film) is the hole diameter, which increases the scattering rate and hence broadens the resonance linewidth.[14] This resonant transmission mechanism, involves coupling to an SPP mode, evanescent transmission through the below-cutoff waveguide hole, and scattering of radiation again from the hole array to produce propagating modes in free-space. The surface wave is excited by a projection of the incident electric field polarization in the propagation direction, and the reradiated field is again projected onto the analyzer. This effect was explored previously with 1D gratings and utilized in imaging SPPs excited on 2D grating couplers.[19]

For the SPP wave propagation on nanoholes as shown in Fig. 9-7, it can be regarded as on 2D cubic hole gratings with SPP Bloch mode. Corresponding phase matching condition can be written as

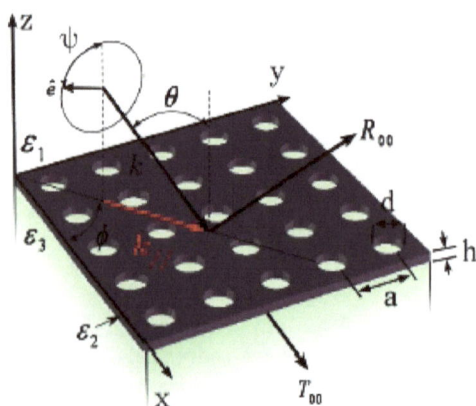

Fig.9-7 Coordinate system of the nanoholes with geometrical parameters and dielectric constants.

$$\vec{k}_{SP} = \vec{k}_\parallel \pm i \vec{K}_G^x \pm j \vec{K}_G^y$$

(9-3)

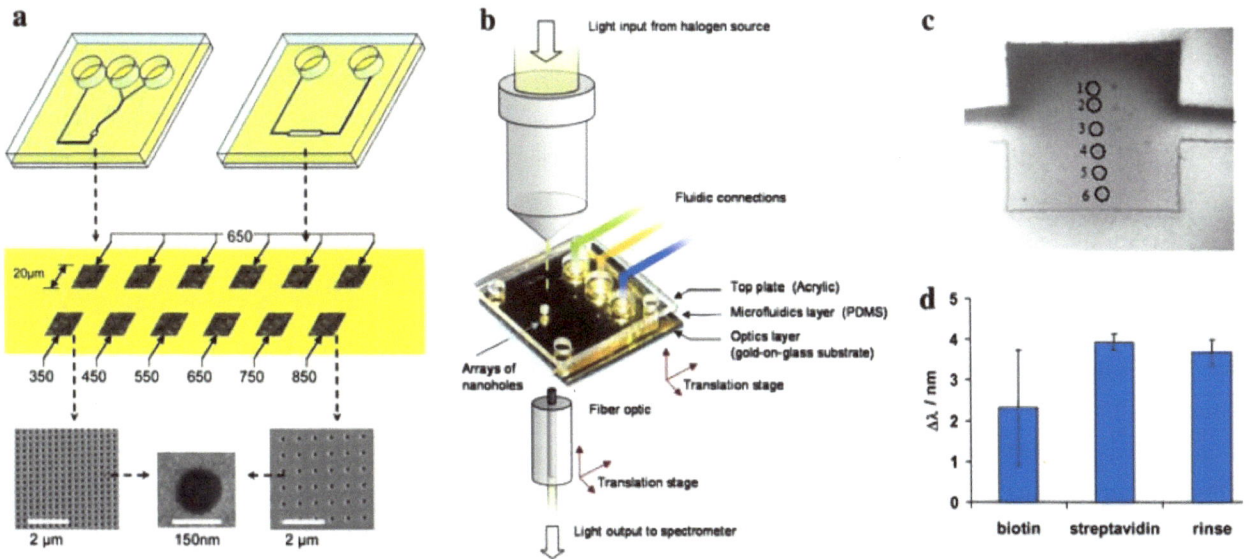

Fig. 9-8 On-chip integration and application of nanohole array sensing elements: a schematic of device at range of relevant lengthscales; b schematic of optical and fluidic setup with an image of the device; c image showing six cross-stream nanohole sensors sampling a crossstream microfluidic concentration gradient; and d results of on-chip biosensing tests showing relative peak red-shifts accompanying biotin and stretavidin absorption to the cysteamine monolayer Reprinted with permission from "De Leebeeck et al Anal Chem 79: 4094-4100 (2007)" with Copyright © 2007 of American Chemical Society.

$$\vec{k}_{\parallel} = \vec{k}_x + \vec{k}_y = k_0[x\sin\theta\cos\phi + y\sin\theta\sin\phi] \tag{9-4}$$

where K_G^x, K_G^y is the grating vector in x- and y-directions. Assuming small modulation (d<<a), and no coupling between adjacent sides (planar case), then

$$|\vec{k}_{1-2,2-3}^{SP}| = k_0\sqrt{\frac{\varepsilon_{1,3}\varepsilon_2}{\varepsilon_{1,3} + \varepsilon_2}} \tag{9-5}$$

We demonstrated nanohole arrays as discrete, 20 μm × 20 μm, optofluidic elements which are capable of both chemical sensing and bio-molecular adsorption monitoring for spatial/temporal measurements in a microfluidic chip platform. Figures 9-8 (a) and (b) show schematic diagram of the chip and the optical/fluidic setup employed, respectively. A set of six arrays was employed to spatially resolve a cross-stream concentration gradient, as shown in Fig. 9-8 (c). A light absorbing dye was used to illustrate the cross-stream gradient relative to the arrays and in comparison to the nanohole array-based measurements. The sensors were applied for the purpose of detecting surface-binding in the assembly process of a cysteamine monolayer-biotin linker-streptavidin protein system with SPR peak shift results, as shown in Fig. 9-8 (d). Specific adsorption was indicated at the set of arrays with average shifts in resonance wavelength of 3.5–4.0 nm which was calculated relative to the SPR peak of the cysteamine layer. The sensitivity of the device in resonance units (RU), whereas one RU refers to a picogram of biomaterial per square millimeter of the sensor surface, was estimated to be *6,500 RU [18] as compared to the commercial SPR devices at *70,000 RU. Opportunities to increase the sensitivity of that nanohole array device include using a laser source, increasing the sensitivity of the detection system, optimizing hole geometry, and tuning the array periodicity to match the SPR resonance corresponding to the laser source.

9.4.3 Optical trapping with nanoholes

Optical trapping with nanohole arrays is an optofluidic application more in the "optical devices that employ fluids" category. Optical trapping has been used to manipulate a wide range of biological (including cells, viruses, and DNA) and non-biological (including colloidal spheres) particles in fluids. It is natural to consider the optical force for manipulation of particles within the microfluidic context and nanohole arrays, and present several possibilities for optical manipulation. For example, fluorescence-based sorting was demonstrated from laser bars using polydimethylsiloxane (PDMS) flow channels. In addition, fiber optic traps were used to isolate particles into linear arrays in a microfluidic channel. Optical trapping uses the force transmitted to the particle from the refractive or scattering momentum change of photons. It is convenient to separate the optical force into intensity gradient and scattering components. The gradient force is particularly interesting for optical trapping using SPs at metal surfaces because of the exponential decay in the intensity of light away from the surface. Recently, the optical forces from SPs were measured using coupling through a metal film. Surface plasmon forces also allow for the unprecedented possibility of nanometric control of the position of particles in the microfluidic environment. It is possible to envision optical trapping using nanohole arrays. Two orientations are possible: the light transmits through the holes into the particle/fluidic medium; or the light transmits firstly through the particle/fluidic medium and then through the nanoholes. Regarding the first orientation, the optical gradient force arising from the SP generation at the surface should overcome the scattering force of light transmitted through the arrays. Thereby, an optical trap that is sensitive to particle size may be formed, but only in the proximity of the metal surface. The critical trapping distance may be derived by equating the repulsive scattering force of the transmitted light to the attractive gradient force of the SP, and using the well known field distribution for the SP. The critical trapping distance from the surface, d, may be expressed as [25]:

$$d = \frac{\lambda}{4\pi} \sqrt{\frac{n_m^2 + \varepsilon_m}{-n_m}} \ln\left(\frac{3\lambda^3 n_m (m^2 + 2I_x(0)}{16\pi^3 a^3 I_z (m^2 - 1)\sqrt{-(n_m^2 + \varepsilon_m)}}\right) \tag{9-6}$$

where k is the wavelength of the light source, n_m is the refractive index of the medium in which the particles are suspended, ε_m is the real part of the dielectric constant of the metal, a is the particle radius, $m = n_p/n_m$ is the refractive index contrast, with n_p being the refractive index of the particle, $I_x(0)$ is the surface plasmon intensity at the surface of the metal, and I_z is the intensity of the transmitted light through the nanoholes. If the surface plasmon intensity is slightly higher than the transmitted light intensity, a trapping distance of 0.5 μm may be obtained for micron sized polystyrene spheres in water and excitation at λ=632 nm. With such an arrangement, micron-sized spheres will be selectively trapped, whereas larger particles will be washed away. In the alternative orientation, light transmits firstly through the particle/fluidic medium and then through the nanoholes. In that case, the forward-scattering force and the SP-generated surface normal gradient force act together. While selectively trapping based on particle size would not be possible, this configuration can be used to increase the effective sampling volume of the array in a microfluidic environment. Specifically, it would be possible to concurrently generate SP and direct particles toward surface of the nanohole array.

9.5 Summary

This chapter introduced basic transmission property of nanoholes arrays firstly. Then polarization effect and influence on transmission passing through the elliptical nanoholes were illustrated. After that, some typical applications of the metallic nanoholes arrays such as cancer cell detection and optofluidic biochips were described. Resonant SPP excitation in nanoholes arrays were demonstrated and used for sensing applications. It is a promising approach combining with optofluidic biochips for biosensing such as cancer cell detection and imaging.

Detection of cancer cells in blood is another approach that would provide information about the metastatic potential of the cancer, and therefore clinical outcomes. The treatment for metastasis is much more severe to the patient than the treatment for those of no metastasis, and avoiding the former treatment is a significant benefit to the patient. The subwavelength optics-based nanotechnology approaches that are being taken toward beating cancer that were described above were considered to have a likelihood of success for many applications. On the other hand, some cancers are drug resistant or mutate after initial therapy and treatment is mostly ineffective. This is where drastic approaches and new nanomedicines are needed to combat cancer in the future.

Reference

[1] Vukusic P, Sambles JR, "Photonic structures in biology". Nature 424, 852–855 (2003).

[2] Genet C, Ebbesen TW, "Light in tiny holes". Nature 445, 39–46 (2007).

[3] Bethe H. A, "Theory of diffraction by small holes". Phys. Rev. 66, 163–182(1944).

[4] Ebbesen, T. W., H. J. Lezec, H. F. Ghaemi, T. Thio, and P. A. Wolff, "Extraordinary optical transmission through subwavelength hole arrays," Nature 391, 667–669 (1998).

[5] Barnes WL, Dereux A, Ebbesen TW, "Surface plasmon subwavelength optics". Nature 424, 824–830 (2003).

[6] Brolo AG, Gordon R, Leathem B, Kavanagh KL, "Surface plasmon sensor based on the enhanced light transmission through arrays of nanoholes in gold films". Langmuir 20, 4813–4815 (2004).

[7] DiMaio JR, Ballato J, "Polarization-dependent transmission through subwavelength anisotropic aperture arrays". Opt Express 14, 2380–2384 (2006).

[8] Smolyaninov II, Elliott J, Zayats AV, Davis CC, "Far-field optical microscopy with a nanometer-scale resolution based on the in-plane image magnification by surface plasmon polaritons". Phys Rev Lett 94, 057401 (2005).

[9] Srituravanich W, Fang N, Sun C, Luo Q, Zhang X, "Plasmonic nanolithography". Nano Lett 4, 1085–1088 (2004).

[10] Yin LL, Vlasko-Vlasov VK, Pearson J, Hiller JM, Hua J, Welp U, Brown DE, Kimball CW, "Subwavelength focusing and guiding of surface plasmons". Nano Lett 5, 1399–1402 (2005).

[11] F. J. García de Abajo, "Colloquium: Light scattering by particle and hole arrays". Rev. Mod. Phys. 79, 1267-1286 (2007).

[12] R. Gordon, A.G. Brolo, A. McKinnon, A. Rajora, B. Leathem, and K. L. Kavanagh, "Strong Polarization in the Optical Transmission through Elliptical Nanohole Arrays", Phys. Rev. Lett. 92, 037401 (2004).

[13] T.W. Ebbesen, H. J. Lezec, H. F. Ghaemi, T. Thio, and P. A.Wolff, "Extraordinary optical transmission through sub-wavelength hole arrays". Nature 391, 667 (1998).

[14] G. Schnider, J. R. Krenn, W. Gotschy, B. Lamprecht, H. Ditlbacher, A. Leitner, and F. R. Aussenegg, "Optical properties of Ag and Au nanowire gratings". J. Appl. Phys. 90, 3825 (2001).

[15] F. J. Garcı`a-Vidal and L. Martı`n-Moreno, "Transmission and focusing of light in one-dimensional periodically nanostructured metals", Phys. Rev. B 66, 155412 (2002).

[16] H. Raether, *Surface Plasmons,* Springer-Verlag, Berlin, 1988.

[17] David Sinton, Reuven Gordon, Alexandre G. Brolo, "Nanohole arrays in metal films as optofluidic elements: progress and potential", Microfluid Nanofluid 4:107–116 (2008).

[18] De Leebeeck A, Kumar LKS, de Lange V, Sinton D, Gordon R, Brolo AG, "On-chip surface-based detection with nanohole arrays". Anal Chem 79, 4094–4100 (2007).

[19] Kevin A Tetz, Pang L, Fainman Y, "High-resolution surface plasmon resonance sensor based on linewidth-optimized nanohole array transmittance". Opt. Lett. 31, 1528–1530 (2006).

[20] Stark PRH, Halleck AE, Larson DN, "Short order nanohole arrays in metals for highly sensitive probing of local indices of refraction as the basis for a highly multiplexed biosensor technology". Methods 37:37–47 (2005).

[21] Degiron,A.,Lezec,H.J.,Barnes,W.L.&Ebbesen,T.W. Effects of hole depth on enhanced light transmission through subwavelength holearrays. Appl. Phys. Lett. 81, 4327-4329(2002).

[22] R. Rokitski, KA. Tetz, Y. Fainman, "Propagation of Femtosecond Surface Plasmon Polariton Pulses on the Surface of a Nanostructured Metallic Film: Space-Time Complex Amplitude Characterization". Phys. Rev. Lett. 95, 177401 (2005).

[23] Liu Y, Bishop J, Williams L, Blair S, Herron J, "Biosensing based upon molecular confinement in metallic nanocavity arrays". Nanotechnology 15, 1368–1374 (2004).

[24] Rindzevicius T, Alaverdyan Y, Dahlin A, Hook F, Sutherland DS, Kall M, "Plasmonic sensing characteristics of single nanometric holes". Nano Lett 5, 2335–2339 (2005).

[25] David Sinton, Reuven Gordon, Alexandre G.Brolo, Nanohole arrays in metal films as optofluidic elements: progress and potential. Microfluid Nanofluid, 4, 107-116 (2008).

10 PLASMONIC STRUCTURES FOR IMAGING AND SUPERFOCUSING

Abstract: Superlens for imaging was introduced firstly. Two types of the superlenses were presented: negative refraction-based perfect lens, and metamaterials-based silver superlens. After that, several plasomonic structures such as the subwavelength metallic structures with depth tuning and width tuning, variant periods metallic structures, and funnel-shaped array for beaming and superfocusing were illustrated. A recently developed light source, radial polarization illumination was cited as a new concept of polarization for illumination. Then the radial polarization illumination-based plasmonic structures for superlensing were described. Finally, a nanoplasmonic waveguide for nanofousing was described.

10.1 Introduction

The electromagnetic diffraction limit predicts that one can not get an image better resolved than the wavelength of light. The initial idea of beating the diffraction limit can be dated back to the era of Synge in 1928, and it was proposed that an effective way to realize the subwavelength imaging was to extract the near field evanescent information. Such an idea is nowadays widely used in the commercially available scanning near-field optical microscopy (SNOM) for subwavelength imaging. However, the SNOM requires a point-by-point scanning and thus the process is extremely slow. Most recently, the extensive research interests in exploring the real time subwavelength imaging method led to several promising technologies such as the perfect lens and time reversal mirror, which have the potential to overcome the drawbacks of the SNOM.

The use of a perfect lens was firstly proposed by J. B. Pendry in 2000. [1] It is an attractive way to obtain subwavelength images at once without point-by-point scanning. The idea was put forth on the basis of negative refraction and evanescent wave amplification, one of special assets possessed by left-handed metamaterials, and, therefore, the quality of the restored subwavelength imaging is largely determined by the properties of physically available artificial materials. Unfortunately, in practice, constructing lossless and broadband metamaterials is challenging at both microwave and optical frequencies, and hence the perfect lens degenerates to the super lens that operates at limited frequency range with limited resolution.

A so-called "time reversal technology" was initially demonstrated at underwater acoustic communication and the broadband super resolution was readily obtained due to the expansion of the effective aperture. Recently, it was experimentally validated for the electromagnetic wave that the time reversal operation can also refocus the electromagnetic energy at its source and has the potential resolution of one thirtieth of the wavelength. The interesting refocusing effect at its initial source is critical for the applications in complex environment. However, various practical applications such as imaging require the far field image, which is difficult to fulfill by the current time reversal configuration.

The recent breakthrough by J. B. Pendry [2] linked the concept of negative refraction and time reversal, which provides us a new avenue to realize negative refraction without typical metamaterials and thus has the potential to build more refined super lenses. Pendry emphasized that a time reversal interface is subject to the wavelength limitation on resolution, and this limitation can be overcome by the addition of further time reversal interface to the system. Recently, M. Fink group's reported their work on the time reversal super focusing experiment.[3] They distributed random scatterers in the near field of the focusing point, and used the far-field time-reversal mirror to build the time-reversed wave field, which interacts with the random scatterers to regenerate not only the propagating waves but also the evanescent waves required to refocus them below the diffraction limit. This indicates that only one time reversal surface can restore the far field subwavelength information with the aid of micro-structured scatterers. Further experimental results for practical far-field imaging still need in-depth study in the future.

In this chapter, metamaterials-based structures for imaging were described firstly, including silver film-based superlens and hyperlens. After that, plasmonic structures-based superfocusing at near-field were studied recently. It may be useful in the areas of data storage, nanolithography, and nanometrology. This chapter highlights the two functions of plasmonic structures: imaging beyond diffraction limit, and near-field superfocusing.

Yongqi Fu (Ed.)

10.2 Superlens

10.2.1 Negative refraction-based perfect lenses

J. B. Pendry put forth a perfect lens theoretically in 2000,[1] as shown in Fig. 10-1. The optical material has a negative refractive index with dielectric constant of $\varepsilon=-1$ and $\mu=-1$. It can realize imaging with spatial resolution beyond diffraction limit due to diffraction free for the negative refraction. Figure 10-2 is his designed perfect lens with calculated two-dimensional (2D) electrostatic profile in comparison to that of the imaging system with conventional diffraction limit. This is achieved by recognizing that the recently discovered negative refractive index material restores not only the phase of propagating waves but also the amplitude of evanescent states. For very short working distances

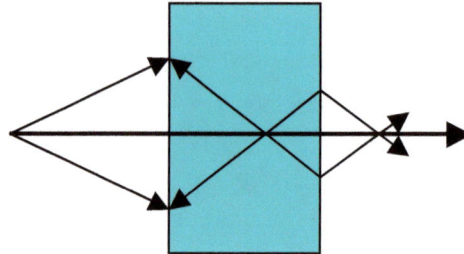

Fig. 10-1 A negative refractive index medium bends light to a negative angle with the surface normal. Light formerly diverging from a point source is set in reverse and converges back to a point. Released from the medium the light reaches a focus for a second time.

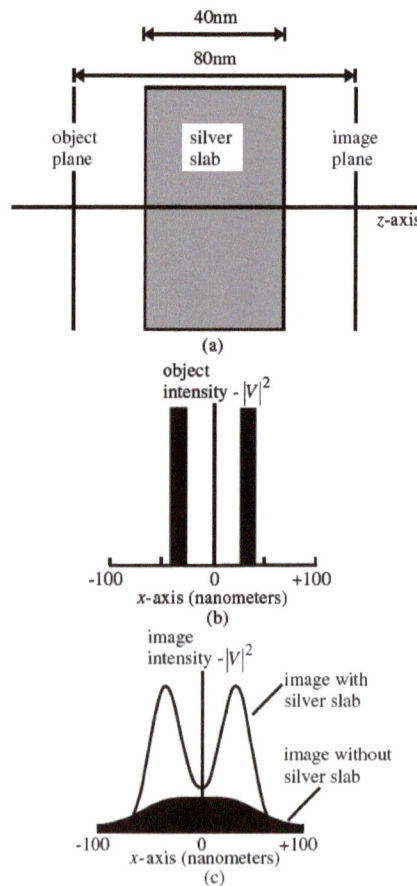

Fig. 10-2 (a) Plan view of the new lens in operation. A quasi-electrostatic potential in the object plane is imaged by the action of a silver lens. (b) The electrostatic field in the object plane. (c) The electrostatic field in the image plane with and without the silver slab in place. The reconstruction would be perfect were it not for finite absorption in the silver. Reprinted with permission from J. B. Pendry, Phys. Rev. Lett. 85, 3966-3969 (2000) with copyright © 2000 of The American Physical Society.

the electrostatic or magnetostatic limits apply, enabling a practical implementation to be simulated in the form of a slab of silver. This device focuses light tuned to the surface plasma frequency of silver and is limited only by the resistive losses in the metal. Here we discussed limitations on subdiffraction imaging by use of a negative refractive index slab.

The values of the electromagnetic parameters and the spatial periodicity render the experimental sample distinct from the idealized perfect lenses. We explored the inherent limitations associated with realizable materials, and the expected impact on the focusing properties of a slab. When μ is positive, the first term in brackets will dominate the behaviour of the transmitted wave for sufficiently large d and $|k_x|>\omega/c$, with the evanescent components decaying exponentially through the slab. When both ε and μ are equal to -1, however, the normally dominant solution vanishes, and $\tau_s=1$ for every component from the source field: homogeneous or inhomogeneous, thus exactly reproducing the source field in the image plane. But the balance is delicate; any deviation from the perfect lenses condition, even if very small, will result in an imperfect image that degrades exponentially with slab thickness d, until the usual diffraction limit is reached. This sensitivity was noted by other researchers.[4] While losses severely limit the obtainable resolution of the focus for a negative index slab, periodicity, as exists for the examples in structured metamaterials or in the sampling associated with finite-difference simulations, also imposes a significant resolution limitation. The effect of periodicity is minimal for propagating plane waves which wavelength is much larger than the repeated unit-cell size. However, periodicity has a significant effect on the recovery of the nonpropagating components having large transverse wave number. We assumed that the periodic variation in the index is only in the transverse (x) direction. Under these assumptions, $n^2(x)=n^2(x+a)$, where a is the repeat distance. In the limit case $k_0\Delta/g\ll1$, where reciprocal lattice vectors $g=2\pi/a$, the limitation that periodicity imposes on the resolution enhancement can be expressed as [4]

$$R \equiv \frac{\lambda}{\lambda_{\text{min.}}} = \frac{1}{2\pi} \ln(\frac{\lambda^2}{a^2\Delta^4})\frac{\lambda}{d} \tag{10.1}$$

where d is the thickness of slab, a is the repeat distance, and $\lambda_{\text{min.}}$ is the minimum resolvable feature which will be on the scale of $\lambda_{\text{min.}}=2\pi/k_x$.

A similar resolution enhancement that is limited to that obtained by a variation in material parameter is thus imposed by the introduction of periodic modulation. As an example, for $R=10$ and slab with $\lambda/d=10$ and $\Delta\sim1$ requires a periodicity of $\sim\lambda/20$. While admittedly of limited quantitative use, Eq. (10.1) shows that the inherent periodicity in metamaterials will impose a limitation on the resolution of the lens. This same limitation will result also from numerical methods that model a continuous material by evaluating the fields at a finite number of sampling points periodically spaced. This limitation has undoubtedly complicated numerical attempts to observe the superfocusing effect.

10.2.2 Metamaterials-based Silver superlens

Superlens as a type of nanolens with imaging resolution beyond diffraction limit was firstly reported by X. Zhang's group.[5-8] It was deigned on the basis of the negative refractive index (NFI) materials (also named "metamaterials"). Considering fabrication limitation of the metamaterials for the optical systems working at visible wavelength, an approximate metamaterial: Ag thin film was adopted to replace the ideal NFI materials. A sandwiched PMMA-Ag-photoresist structure was designed to realize superfocusing. The words "NANO" were written on Cr film firstly using focused ion beam milling, and clearly imaged and recorded into the backside photoresist layer, as shown in Fig- 10.3. In this structure, the Ag film was approximately regarded as a negative refractive index material assuming that imaginary part of the Ag dielectric constant ε''_{Ag} can be ignored because $|\varepsilon'_{Ag}|>>|\varepsilon''_{Ag}|$, and thus $\varepsilon\approx\varepsilon'_{Ag}$. Lateral imaging resolution as small as 89 nm were obtained by this structure, as shown in Fig. 10-4. They claimed that the silver superlens can also image arbitrary nanostructures with sub–diffraction-limited resolution. Only the scattered TM evanescent waves from the object are coupled into the surface plasmon resonance (SPR) of the silver film, and they become a primary component for restoring a sub–diffraction-limited image. The photoresist layer for record herein was designed tightly attached on topside of the superlens. Thus the imaging was still carried out at near-field region. Their further presentations [9] reported that the superlens imaging can be extended into far-field. The higher resolution imaging was realized by surface excitation at the negative index medium. A so-called "far-field optical superlens" (FSL) that is capable of imaging beyond the diffraction limit was put forth. Conventional optical systems working at far field is fundamentally limited by diffraction, which typically is about half of the wavelength because evanescent waves carrying small scale information from an object that fades away in the far field. The FSL significantly enhances the

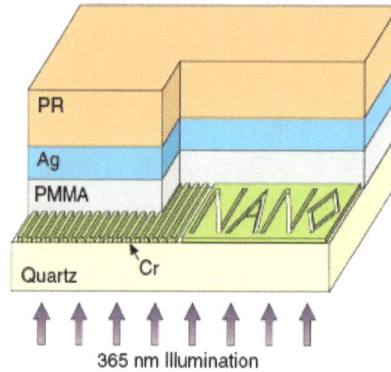

Fig. 10-3 Optical superlensing experiment. The embedded objects are inscribed onto the 50 nm-thick Cr; at left is an array of 60 nm-wide slots of 120 nm pitch, separated from the 35-nm-thick silver film by a 40-nm PMMA spacer layer. The image of the object is recorded by the photoresist on the other side of the silver superlens. Reprinted with permission from "N. Fang, H. Lee, C. Sun, X. Zhang, Science 308, 534 (2005)" with copyright ©2005 of American Association for the Advancement of Science.

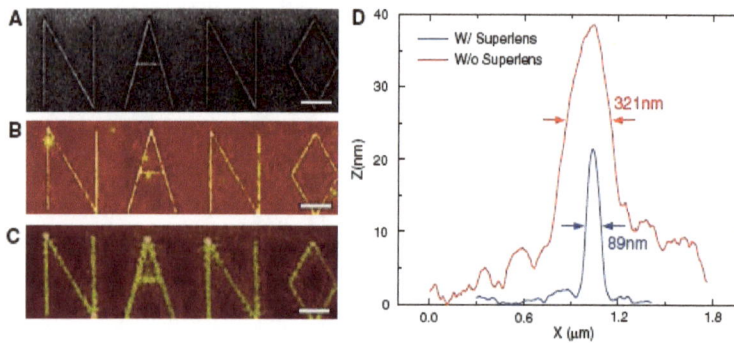

Fig. 10-4 An arbitrary object "NANO" was imaged by silver superlens. (A) FIB image of the object. The linewidth of the "NANO" object was 40 nm. Scale bar in (A) to (C), 2 mm. (B) AFM of the developed image on photoresist with a silver superlens. (C) AFM of the developed image on photoresist when the 35-nm-thick layer of silver was replaced by PMMA spacer as a control experiment. (D) The averaged cross section of letter "A" shows an exposed line width of 89 nm (blue line), whereas in the control experiment, we measured a diffraction-limited full width at half maximum. Reprinted with permission from "N. Fang, H. Lee, C. Sun, X. Zhang, Science 308, 534 (2005)" with copyright ©2005 of American Association for the Advancement of Science.

evanescent waves of an object and converts them into propagating waves that are measured in the far field. A subwavelength object consisting of two 50 nm wide lines separated by 70 nm working at 377 nm wavelength can be imaged, as shown in Fig. 10-5. Working wavelength of a FSL can be tuned by changing either the metal or the dielectric. For example, the working wavelength of a silver-structured FSL can move to visible range if a very high refractive index dielectric is used. Essentially, function of imaging devices lies in its ability to convert the larger wave vector information to a smaller one that can be detected either optically or electronically. The superlens takes advantage of the short wavelength of surface plasmons and effectively enables one to image at subdiffraction limit.

However, there is no working distance between the superlens and object plane. Moreover, the planar superlenses have no amplifying function. An ideal imaging device will avoid this problem: it would not only capture evanescent fields to retrieve subwavelength information, but will also allow for their processing with standard optical components. According to this point of view, an ideal device will convert evanescent waves to propagating waves for ease of

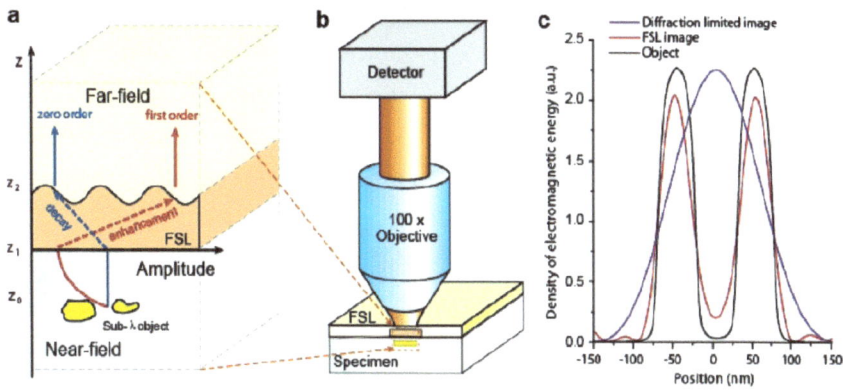

Fig. 10-5 Far-field superlens (FSL) for subwavelength imaging. (a) A FSL is constructed by adding a subwavelength grating onto a thin silver slab. It has two major functions: first, it selectively enhances the evanescent waves from the object; second, it converts evanescent wave into propagating waves. (b) Far-field superlens optical microscope can be realized by insertion of a FSL between the specimen and objective of a regular optical microscope. (c) Subwavelength object with two line sources of 50 nm width separated by a 50 nm gap and its far-field image by FSL calculated for p-polarized normally incident laser light at a wavelength of 377 nm. Calculation shows a unique subdiffraction-limited image can be obtained by FSL. A diffraction limited image from a conventional optical microscope (NA) 1.5) is also shown as comparison. Reprinted with permission from "Zhaowei Liu, Ste´phane Durant, Hyesog Lee, Yuri Pikus, Nicolas Fang, Yi Xiong, Cheng Sun, and Xiang Zhang, Far-field optical superlens. Nano Lett. 7, 403-408 (2007)" with copyright ©2007 of American Chemistry Society.

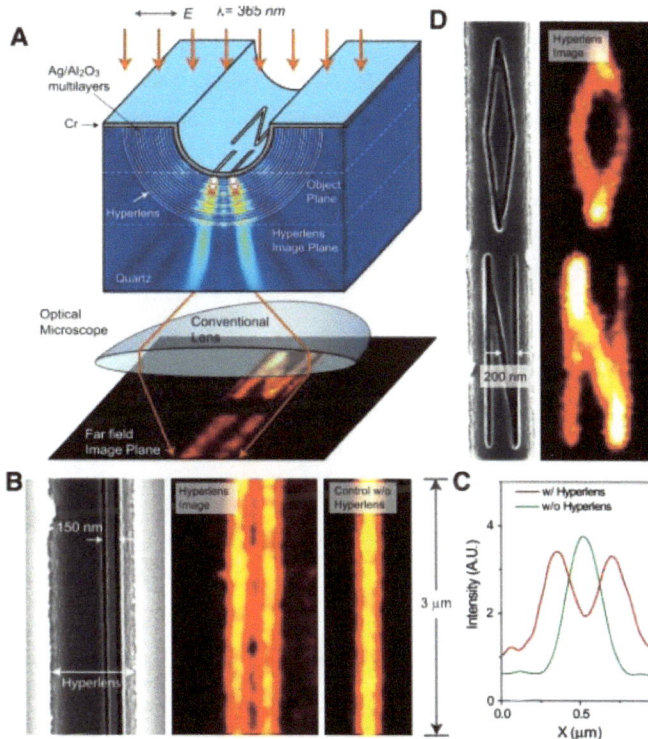

Fig. 10-6 Magnifying optical hyperlens. (A) Schematic of hyperlens and numerical simulation of imaging of sub-diffraction-limited objects. (B) Hyperlens imaging of line pair object with line width of 35 nm and spacing of 150 nm. From left to right, scanning electron microscope image of the line pair object fabricated near the inner side of the hyperlens, magnified hyperlens image showing that the 150-nm-spaced line pair object can be clearly resolved, and the resulting diffraction-limited image froma control experiment without the hyperlens. (C) The averaged cross section of hyperlens image of the line pair object with 150-nm spacing (red), whereas a diffraction-limited image obtained in the control experiment (green). A.U., arbitrary units. (D) An arbitrary object "ON" imaged with subdiffraction resolution. Line width of the object is about 40 nm. The hyperlens is made of 16 layers of Ag/Al_2O_3. Reprinted with permission from "Zhaowei Liu, Hyesog Lee, Yi Xiong, Cheng Sun, Xiang Zhang, Science 315, 1686 (2007)" with copyright ©2007 of American Association for the Advancement of Science.

detection and processing; these waves do not mix with the propagating waves emanating from the object. This can be accomplished by transferring the information carried by evanescent fields into a portion of the propagating spectrum. Following the conversion, these propagating waves can be detected and processed in the far field by methods similar to those of conventional imaging. Considering this, another superlens called "hyperlens" was presented, as shown in Fig. 10-6.[10] This device relies on recently proposed strongly anisotropic metamaterials that featured opposite signs of the two permittivity tensor components, $\varepsilon_\parallel$ and $\varepsilon_\perp$. Such metamaterials were theoretically shown to support propagating waves with very large wave numbers [11, 12] (in ordinary dielectrics, such high-k modes undergo evanescent decay). The hyperlens utilizes cylindrical geometry to magnify the subwavelength features of imaged objects so that these features are above the diffraction limit at the hyperlens output. Thus, the output of the hyperlens consists entirely of propagating waves, which can be processed by conventional optics. Moreover, the material losses do not appreciably degrade the performances of the proposed devices due to its non-resonant nature. In the case of uniaxial anisotropy, dielectric permittivity can be characterized by two values: $\varepsilon_\square$ along the optical axis of the crystal, and $\varepsilon_\perp$ transverse to the optical axis. Propagating modes, in turn, can be decomposed into two polarization states: the ordinary (TE) and extraordinary (TM) waves. For ordinary (TE) waves, the electric field vector is transverse to the optical axis and produces the same dielectric response (given by $\varepsilon_\perp$) independent of wave propagation direction. However, for the extraordinary (TM) waves, the electric field vector has components both along and transverse to the optical axis (given by $\varepsilon_\square$). Accordingly, both $\varepsilon_\square$ and $\varepsilon_\square$ play a role in the dielectric response and in the dispersion relation, given by [13]

$$\frac{k_\perp}{\varepsilon_\parallel} + \frac{k_\parallel}{\varepsilon_\perp} = \frac{\omega^2}{c^2} \tag{10.2}$$

where $k_\parallel$ and $k_\perp$ denotes the wave numbers along optical axis of the crystal and transverse to the optical axis, respectively, ω and c is angular frequency and light speed in vacuum, respectively. In the case of strong anisotropy where $k_\parallel$ and $k_\perp$ are of opposite signs, the dispersion relation (10.2) becomes hyperbolic. For extraordinary waves (TM modes) in a bulk medium with strong cylindrical anisotropy where dielectric permittivities have different signs in the tangential and radial directions ($\varepsilon_\theta > 0$, $\varepsilon_r < 0$), Eq. 10-2 changes to be

$$\frac{k_r}{\varepsilon_\theta} - \frac{k_\theta}{\varepsilon_r} = \frac{\omega^2}{c^2} \tag{10.3}$$

As the tangential component of the wave vector increases towards the center, the radial component increases also. There is no caustic issue, and thus the field of high angular momentum states has appreciable magnitude close to the center as long as the effective medium description is valid. The cylindrical anisotropy causes a high angular momentum state to penetrate toward the center. Such a high angular momentum state can act as a subwavelength probe for an object placed inside the core. In the medium under consideration, these states are propagating waves. They can carry information with respect to the detailed structure of the object to the far field region. The proposed hyperlens enables extra information channels for retrieving the object's subwavelength structures. In the absence of the anisotropic structures, the high angular momentum modes representing these channels do not reach the core and no information about the object as such carry. Because outer radius R_{outer} is always larger than the hollow core radius R_{inner} due to the geometrical relationship, the image along tangential direction is constant (normal dielectric material $\varepsilon_\theta > 0$) and amplified along radial direction ($\varepsilon_r < 0$). Therefore, it can produce amplified image by the hyperlens composed of alternative dielectric-Ag multilayers structured in parabolic shape. The resolution of the hyperlens is determined by the effective wavelength at the core and is given by $\Delta \propto R_{\text{inner}} \lambda / R_{\text{outer}}$. But there is still no working distance for the hyperlens. The object with etched letters of "NANO" is tightly attached to surface of the hollow core sidewall of the hyperlens.

Apart from the Ag film-based planar superlens, Zhang *et. al.* reported their Al_2O_3-Ag configured nanowires array-based artificial negative refractive index material recently, as shown in Fig. 10-7. [14] By their superlens, the incidence at visible wavelength of 660 nm and 780 nm can be negatively refracted experimentally. A porous alumina template was prepared by electrochemical anodization, through which silver nanowires were electrochemically deposited. A 1-µm-wide slit, etched through a 250 nm thick silver film coated on the metamaterials, was illuminated by a collimated diode laser beam at different incident angles. The transmitted light was mapped by scanning a tapered optical fiber at the bottom surface of the metamaterial.

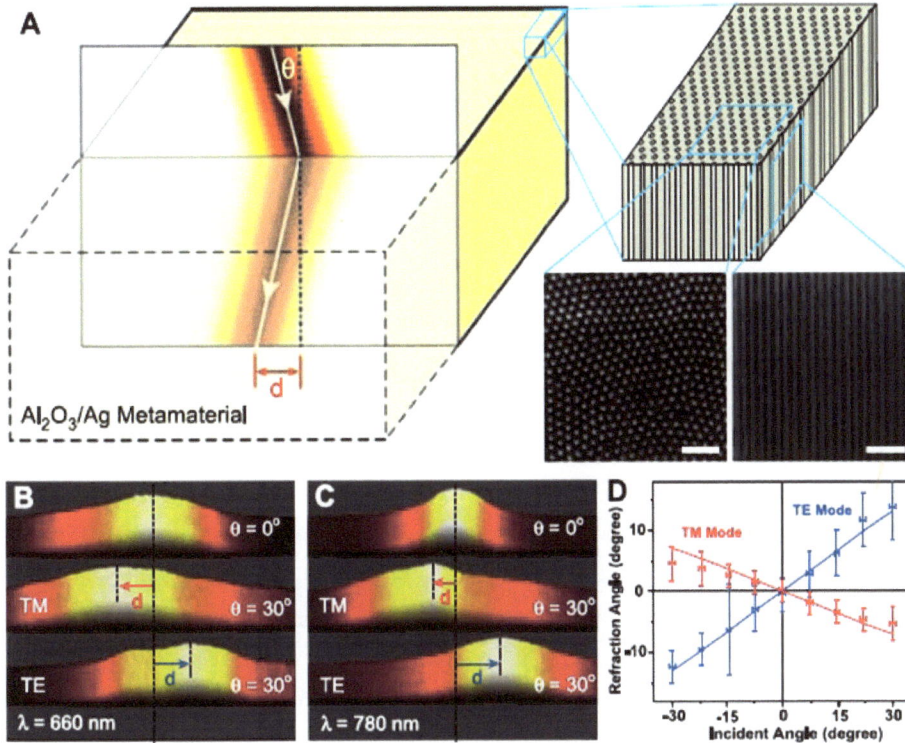

Fig. 10-7 Negative refraction in bulkmetamaterial at visible frequencies. (A) (Left) Schematic of negative refraction from air into the silver nanowire metamaterials. (Right) Nanowires embedded in an alumina matrix, as well as scanning electron microscopy images showing the top and side view of the nanowires (60-nm wire diameter and 110-nmcenter-to-center distance). The scale bars indicate 500 nm. Measured beam intensity at the existing surface of the metamaterial slab at the wavelength of 660 nm (B) and 780 nm (C). The lateral displacement (d) of TM polarized light shows the negative refraction in the metamaterial at both wavelengths, whereas TE light undergoes positive refraction. The horizontal sizes of (B) and (C) are 5 mm and 12 mm, respectively. (D) The dependence of refraction angles on incident angles and polarizations at 780-nm wavelength. The negative refraction occurs for broad incident angles. The experiment data agree well with calculations (solid curves) using the effective medium theory. The sample thicknesses in (B) and (C) are 4.5 mm and 11 mm, respectively. Reprinted with permission from "Jie Yao, Zhaowei Liu, Yongmin Liu, Yuan Wang, Cheng Sun, Guy Bartal, Angelica M. Stacy, Xiang Zhang, Science 321, 930 (2008)" with copyright ©2008 of American Association for the Advancement of Science.

In addition, a nano-optical negative refraction lens with reconstructed image of the transmitted field in the air cavity at 1530 nm and focal length of ~12 µm away from the concave face was realized. The FWHM of the beam spot size is in the order of ~0.68λ. It is an ultrashort focal length plano-concave microlens in an InP/InGaAsP semiconductor two-dimensional (2D) photonic crystal with negative index of refraction (−0.7) at λ=1.5 µm. For more information, please see Fig. 4-11 in Chapter 4.

It is reasonable to believe that the next generation optical system will be the hybrid negative-positive refraction systems. The aberrations of a spherical lens composed of left-handed materials were studied in Ref.15. Five Seidel aberrations (spherical, coma, astigmatism, field curvature, and distortion) as a function of the refractive index n and shape factor q of the lens were considered. The numerical calculations showed that the negative refractive index gives much larger windows of small values of aberrations than that of the positive index, which will significantly enhance the flexibility for the design of an optical lens. Two possible regions with optimized aberrations were proposed: $n=-1$, $q=-2.2$ and $n=-0.81$ and $q=0.83$. Furthermore, D. R. Smith *et. al.* also examined the Seidel aberrations of the thin spherical lenses composed of media with refractive index not restricted to be positive.[16] They found that consideration of this expanded parameter space allows for the reduction or elimination of more aberrations than that of the lens with positive index media only. In particular, they found that spherical lenses possessing real aplanatic focal points are possible only with a negative index. They performed ray tracing using a custom code which relies only on Maxwell's equations and conservation of energy, and confirmed the results of the aberration calculations.

10.3 Plasmonic structures for imaging and superfocusing

Essentially, both the superlens and hyperlens can function as one-dimensional imaging only because the object of "NANO" consists of single lines instead of two-dimensional patterns. Thus it is still not a real engineered optical lens. Considering this, plasmonic metallic structures flanked with subwavelength corrugations/grooves/nanoslits were explored. They can realize beaming, imaging and superfocusing by means of phase modulation and near-field interference. Here two methods of phase modulation were introduced: (1) depth tuning; and (2) width tuning. After that, microstructures with variant periods and a funnel-shaped array for superfocusing were presented.

10.3.1 Depth tuned method

Figure 10-8 (a)-(c) illustrates a typical slit-groove-based focusing structure, which was formed by a single subwavelength metallic slit surrounded by a finite array of grooves at the output surface. When the TM polarized light is launched in the slit from the left side, it couples to SPPs in the metallic slit and then is diffracted to the output surface as well as the forward region in the air/free space. The diffracted light wave on the output surface will propagate along the silver-air interface and scatter into radiation light at the groove region with specific phase and amplitude. The interference of the light emerging from the slit and grooves will create a focus at the point where the values of phase difference between the emissions are the integer multiples of 2π. Thus, the relative phase profile distribution at the output surface basically determines the focus position. [17]

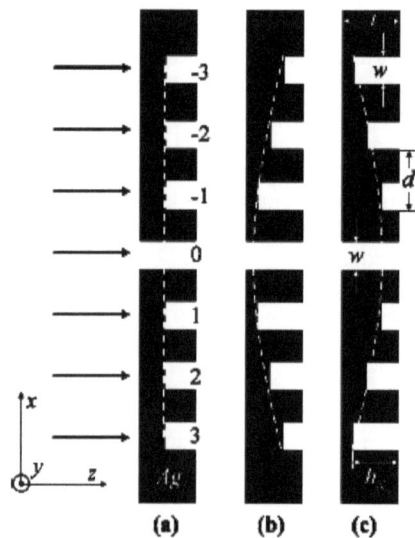

Fig. 10-8 Schematic view of the structures formed by a single subwavelength slit surrounded by grooves with traced depth profile. *t* denotes the thickness of silver film, *w* denotes the width of the slit and all grooves, *d* denotes the groove period, and h_N denotes the depth of grooves with the serial number of *N*. kd denotes the depth difference between the adjacent grooves, and kd=0, kd<0, kd>0 represent the cases shown in (a), (b), and (c), respectively. Reprinted with permission from "Haofei Shi, Chunlei Du, and Xiangang Luo, Focal length modulation based on a metallic slit surrounded with grooves in curved depths. Appl. Phys. Lett. 91, 093111 (2007)" with copyright © 2007 of American Institute of Physics.

The SPPs wave is a special type of electromagnetic field, which can propagate along the metallic surfaces while keep bounded near the surface without radiating away. Considering two closely placed parallel metallic plates, the SPPs of each surface will be coupled and propagate in the form of a waveguide mode, mainly for TM polarized set (E_x, H_y, and E_z). The dispersion relation between the effective refractive index and slit depth implies a potential way of phase modulation by simply tuning the slit depth. When TM polarized incident plane wave impinges the slit entrance, it will excite SPPs wave. Then the SPPs wave couples with the incident wave and propagates along the slits region until it reaches the exit where the coupled wave radiates into the beam in free space. The phase of light ϕ transmitted through the nanoslits can be expressed as

$$\phi = \phi_{01} + \phi_{12} + n_1 kh(x) - \theta \tag{10.4}$$

where ϕ_{01} and ϕ_{12} are the phase changes at the entrance and exit interface, respectively. The last term θ, originating from the multiple reflections between the entrance and exit interfaces, can be calculated using the following equation

$$\theta = \arg(1 + \rho_{01}\rho_{12}e^{i2kh(x)}) \tag{10.5}$$

where k is the wave vector of the light at dielectric/Ag interface, and $h(x)$ is the slit depth. Subscripts 0, 1, 2 denote the media before, inside, and after the nanoslit array, respectively. The dispersion relation between the effective refractive index and slit width implies a potential way of phase modulation by simply tuning the slit depth. The wave vector k for the coupled surface plasmons in the interface can be expressed as

$$\hat{e} = k_0[1 + \tfrac{1}{2}\eta^2(1 + \sqrt{1 + \tfrac{4}{\eta^2}(1 + |\varepsilon|)})] \tag{10.6}$$

where $\eta = 2/(k_0 w|\varepsilon|)$, w is the width of slit defined in the structure, and $\varepsilon = \varepsilon_m\varepsilon_d/\varepsilon_m + \varepsilon_d$.

The target is that design a nanostructure which can act as a "nano-lens" to realize beam shaping such as collimation or focusing for the future possible usage of detection and inspection. A higher spatial resolution can be expected through this type of lens. Figure 10-8 shows the designed enhanced SPPs-based nanostructure, which consists of 7 slits. The central slit is thoroughly penetrated through the Ag thin film. Depth distribution of the other 6 slits is symmetrical to the central slit, and their depths gradually decrease from center to edge. Outline of the grooves bottom is designed forming a parabolic curve. Beam shaping can be realized theoretically by a way of controlling phase distribution along the direction perpendicular to the grooves. Required phase distribution of the emitted light along x-axis can be derived readily according to the equal optical path length principle

$$\phi(x) = \frac{2m\pi}{\lambda}h(x) = 2m\pi + \frac{2\pi f}{\lambda} - \frac{2\pi\sqrt{f^2 + x^2}}{\lambda} \tag{10.7}$$

where m is the arbitrary integer number, and f is the focal length of the probe which is a function of slit width, period, metal thickness, and depth distribution curve as $f = f(a, \Lambda, h, d(x))$. The corresponding phase delay due to the modulated depth defined using the standard parabolic function $d(x) = ax^2 + bx + c$ causes redistribution of the peak transmission and convergence of the coupled SPPs wave in the grooves by which the beam is shaped. Considering fabrication possibility, we discretized the continuous phase distribution into 7 interval steps and transferred them into 7 corresponding slits with different depths. Period and width of the slits is 500 nm and 200 nm, respectively. Incident wavelength and excited SPPs wavelength is $\lambda_{in} = 527$ nm (for white light source), and $\lambda_{SP} = 525$ nm, respectively. The nanostructure will be fabricated on the 200 nm Ag thin film coated on the quartz substrate.

Calculated E_x relative to the phase distribution profiles at groove regions along the x-axis at the exit surface is shown in Fig. 10-9 (a). The circle marks denote the data of the control structure for the case of $kd = 0$, and the square and triangle marks denote the data of the structure for the cases of $kd = -50$ nm and $kd = 50$ nm, respectively. Obviously, the curvature of the relative phase profile at the exit surface varies with the distribution profile of the groove depth as expected; the deeper of the groove depth, the larger the relative phase value will be. Figures 10-9 (b)–(d) illustrate the corresponding $|H_y|^2$ intensity distributions for the cases shown in Figs. 10-9 (a)–(c). The $|H_y|^2$ intensity distribution results show that the energy emerging from the structure overlaps the x-axis within several microns, concentrating most of the energy in an extremely small region. For example, the trace case of flat groove depth shown in Fig. 10-9 (b) reveals the focal length of 1.49 μm and the beam spot size at site of full width at half maximum (FWHM) of 0.62 μm. If the trace of groove depth is designed to spatially decrease with their distance from the central slit for the case of $kd =$ −50 nm, the focal length will be reduced to be 1.22 μm with the decreased focal spot of 0.55 μm. On the contrary, the trace of groove depth for the case of $kd = 50$ nm shows an opposite phenomenon which the focal length and spot size increases to be 1.99 μm and 0.67 μm, respectively. The focal length of the slit-groove-based focusing structures can be tailored in certain value if the groove depths are arranged in a traced profile. With the regulation of the groove depth profile, it is possible to modify the focus position in the precision of nanoscale level without increasing size of the nanodevices.

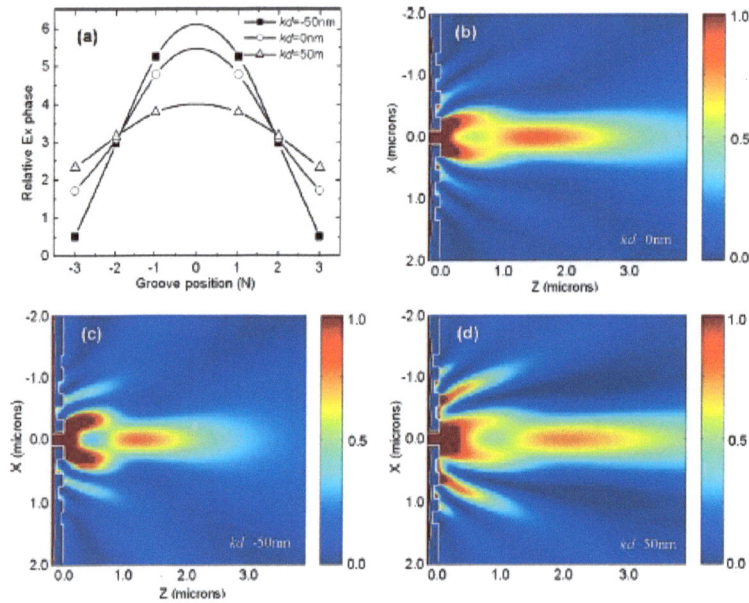

Fig. 10-9 FDTD simulation results for modulating the focal length with trace profiles. (a) Relative phase Ex distributions for the three depth trace profiles shown in Fig. 1 (b) $|H_y|^2$ distributions for the cases shown in Fig. 1 (a) with $kd=0$, (c) $|H_y|^2$ distributions for the cases shown in Fig. 1 (b) with $kd=-50$ nm, and (d) $|H_y|^2$ distributions for the cases shown in Fig. 1 (a) with $kd=50$ nm. The other parameters are set as $t=200$ nm, $w=200$ nm, and $d=420$ nm. Reprinted with permission from "Haofei Shi, Chunlei Du, and Xiangang Luo, Focal length modulation based on a metallic slit surrounded with grooves in curved depths. Appl. Phys. Lett. 91, 093111 (2007)" with copyright © 2007 of American Institute of Physics.

Figures 10-10 (a)~(d) are optical characterization results using a near-field scanning optical microscope system.[18] To characterize the functionalities of the slits, a commercial system of s-SNOM (Russian, NT-MDT) was employed. At a plane approximately 10 nm away from the metal surface, a strong focusing region with a width of roughly 200 nm and a length of about 10 μm produced by the morphological profile of the slit and grooves were observed. Since the film thickness (200 nm) is several times larger than the skin depth of gold (~20 nm at 632.8 nm incident wavelength), the observed high contrast optical signal is solely generated from the excitation of SPPs rather than the pure interference with directly transmitted light beam. Apart from this strong signal at the central area, a low intensity region over an area of ~1.7×10 μm² was also be observed on both sides of the slit, which clearly demonstrates the propagation of the excited SPPs along the incident polarization direction. The Near-field measurement reveals unambiguously the light interaction with the slits and confirms the functionalities of the plasmonic lens. The simple plasmonic lens demonstrated in this paper can find broad applications in ultra-compact photonic chips particularly for biosensing and high-resolution imaging.

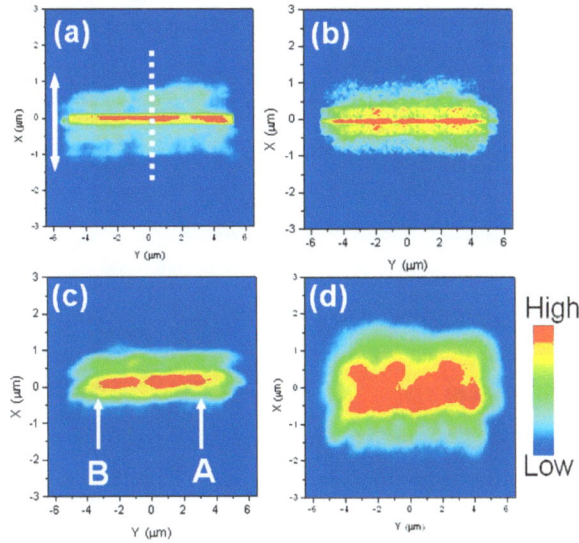

Fig.10-10 Intensity distributions of transmitted light through the flat slit at planes with distance at (a) z=10 nm, (b) z=50 nm (c) z=1600 nm and (d) z=4000 nm to the slit surface. The arrow in (a) indicates the incident polarization direction.

10.3.2 Width tuned method

A novel method was proposed to manipulate beam by modulating light phase through a metallic film with arrayed nano-slits, which have constant depth but variant widths.[19-21] The slits transport electro-magnetic energy in the form of SPPs wave propagating in nanometric waveguides and provide desired phase retardations of the beam manipulating with variant phase and propagation constant.

To illustrate the above idea of modulating phase, a metallic nano-slits lens was designed.[19] The parameters of the lens were defined as follows: $D = 4$ μm, $f = 0.6$ μm, $\lambda = 0.65$ μm, $d=0.5$ μm, where D is the diameter of the lens aperture, f the focus length, λ the wavelength, and d the thickness of the film. Media of both sides of the lens are air. The schematic diagram of lens is given in Fig. 10-11, where a metallic film is perforated with a great number of nano-slits

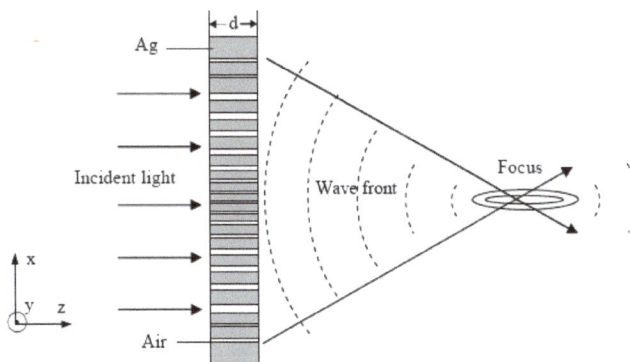

Fig. 10-11 A schematic of a nano-slit array with different width formed on thin metallic film. Metal thickness in this configuration is d, and each slit width is determined for required phase distribution on the exit side, respectively. A TM-polarized plane wave (consists of E_x, H_y and E_z field component, and H_y component parallel to the y-axis) is incident to the slit array from the left side. Reprinted with permission from "Haofei Shi, Changtao Wang, Chunlei Du, Xiangang Luo, Xiaochun Dong, Hongtao Gao, Opt. Express 13, 6815-6820 (2005)" with copyright © 2005 of Optical Society of America.

with specifically designed widths. Transmitted light from the slits is modulated and converged in free space. The required phase distribution of emitted light at x-axis can be obtained readily according to the equal optical length principle, where n is an arbitrary integer number.

$$\phi(x) = 2n\pi + \frac{2\pi f}{\lambda} - \frac{2\pi\sqrt{f^2 + x^2}}{\lambda} \qquad\qquad (10.8)$$

After FDTD calculation, the resulting Poyinting vector is obtained and shown in Fig. 10-12 (a). A clear-cut focus appears about 0.6 micron away from the exit surface, which agrees well with that of the designed. The cross section of focus spot in *x*-direction is given in Fig. 10-12 (b), indicating a full-width at half-maximum (FWHM) of 270 nm. The extraordinary light transmission effect of SPPs through sub-wavelength slits was also observed in the simulation with a transmission enhancement factor of about 1.8 fold.

Another application of the width tuning is beam deflector designed on the basis of SPPs wave effect. The nanoslits with different width and depth can lead to different phase retardations while the SPPs wave passing through the structure. A beam splitting effect with the splitting angle of near 90° can be achieved by adjusting the width and depth of the slits accordingly. The splitting angle can be modulated precisely from 0° to ~90° by changing the position and width of each nanoslit. The structure, with its miniaturized size, can be applied in the fields like optical beam control, optical switch, imaging, and micro-mechanical-electronic-systems (MEMS) etc.

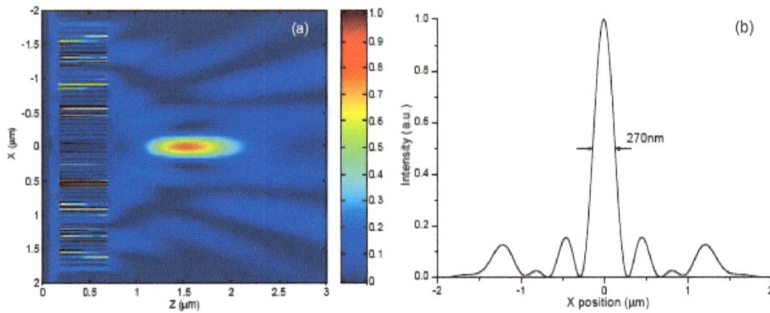

Fig. 10-12 (a) FDTD calculated result of normalized Poynting Vector Sz for designed metallic nano-slits lens. Film thickness is 500nm, and the total slits number is 65. The structure's exit side is posited at z=0.7 μm. (b) Cross section of the focus at z=1.5 μm. Reprinted with permission from "Haofei Shi, Changtao Wang, Chunlei Du, Xiangang Luo, Xiaochun Dong, Hongtao Gao, Opt. Express 13, 6815-6820 (2005)" with copyright © 2005 of Optical Society of America.

Considering the surface plasmon mode in metallic grooves for true metals, H. Shi *et. al.* proposed a revised version of the analytical and numerical model on the light diffraction from a subwavelength slit surrounded by finite number grooves on a metallic film.[19] The revised model indicates that the optical property of true metals, even for good metals such as Ag and Au, may possess great influence on the light diffraction as the size of subwavelength structure is comparable to the depth of light penetration into the surface of metal films.

In summary, the modulation ability is limited to a certain extent by means of either pure depth tuning or width tuning only. Combination of both tuning method together may be a more effective option for the purpose of further improvement of superfocusing and imaging. Further study is necessary to verify the combination approach.

10.3.3 Plasmonic structures with variant periods

10.3.3.1 Metallic subwavelength slits

A new superlens, a plasmonic structure with variant periods was put forth.[22, 23] It is different from the concept of metameterials-based superlens reported by X. Zhang *et. al.*.[9] It can realize converting the enhanced SPP wave to propagating waves so as to increase the effective propagation distance. Liu *et. al.* theoretically studied that diffractive elements may work at near-field in the scanning near-field optical microscope (SNOM) systems to replace the

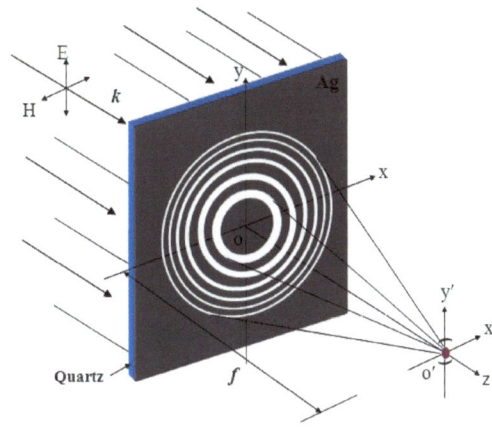

Fig. 10-13 Schematization of the plasmonic microzone plate superfocusing with focal length f. It is illuminated by a plane wave with 633 nm incident wavelength. The Ag film has permittivity $\varepsilon = -17.6235 + 0.4204i$ at $\lambda_0 = 633$ nm. In our FDTD simulations the perfectly matched layer (PML) boundary condition was applied at the grid boundaries.

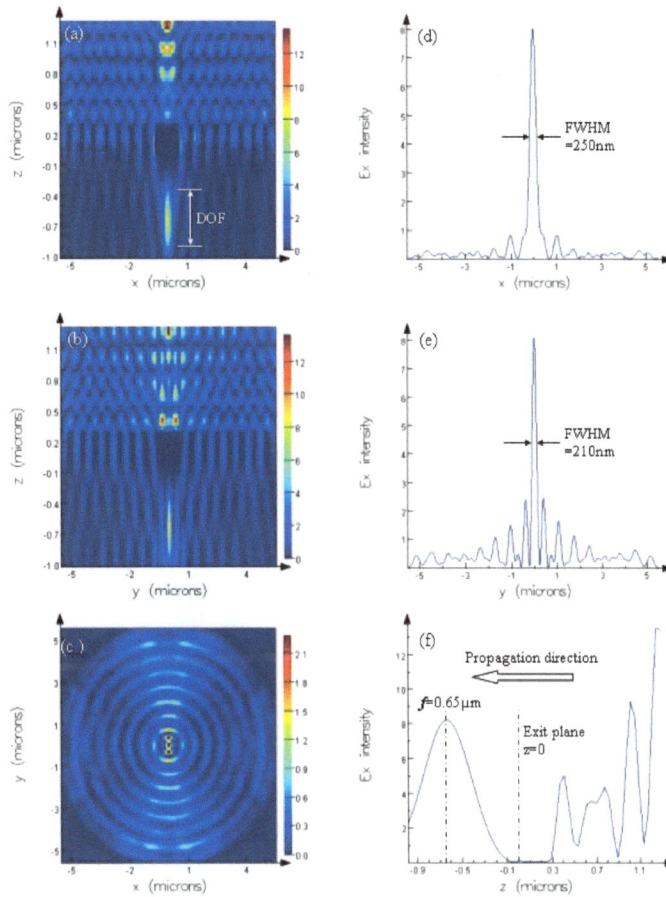

Fig. 10-14 Example of a plasmonic microzone plate (negative) with outer diameter, Ag film thickness, and working wavelength of 10.34 μm, 300 nm, and 633 nm, respectively. Electric field distribution results were calculated using FDTD algorithm. Propagation direction is z. Electric field E_x intensity in (a) x-z plane, (b) y-z plane, and (c) x-y plane. Electric field transmission in line $z = -0.65$ μm (calculated focal plane) in (d) x-z plane, $y=0$; (e) y-z plane, $x=0$; and (f) x-y plane, $z=0$. Designed focal length, outer diameter, and outmost zone width using scalar theory are $f = 0.5$ μm, 10.53 μm, and 28 nm, respectively. Calculated depth of focus (*DOF*) is ~500 nm (scalar theory designed value is 2.58 nm). The site of $z=0$ is the exit plane of the Ag film.

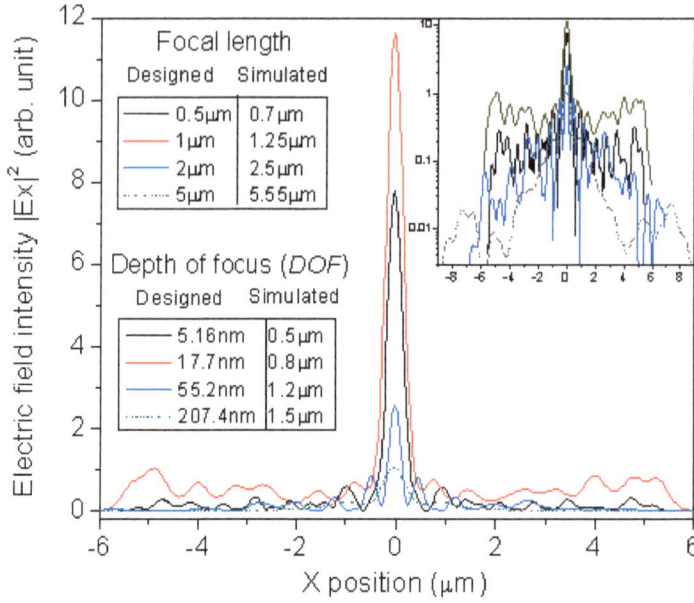

Fig. 10-15 Electric field intensity distribution $|E_x|^2$ vs x in focal planes for the lens with designed four different focal lengths. Simulated focal length and *DOF* are greatly larger than those of the designed values. The inset is the corresponding plot with logarithmic scale in longitudinal axis. Corresponding beam spot sizes (FWHM) in the x direction for the four designed focal lengths of 0.5, 1, 2, and 5 μm are 250, 370, 280, and 550 nm, respectively.

conventional optical fiber probes. [24] But controlling constant working distance l ($l<\lambda/10$) between the planar diffractive element and sample surface is difficult in practice. Configuration of our proposed lens is constructed with chirped circular nanoslits which were corrugated through an Ag thin film supported on quartz substrate. The lens is an asymmetric structure with variant periods in which a thin film of Ag is sandwiched between air and glass, as shown in Fig. 10-13. The chirped circular nanoslits, herein means that slit widths and periods of the rings are changed. Unlike the conventional zone plates with metal film of Cr, Al, or Ni, our structure is a device that a quartz substrate coated with Ag thin film which is embedded by a zone plate-like structure with the zone number $N<10$. Optical length difference for the adjacent zones is λ instead of $\lambda/2$ of the conventional zone plate. For an evanescent wave with a given wave vector k_x, we have $k_{zj}=+[\varepsilon_j(\omega/c)^2-k_x^2]^{1/2}$ for j=1 (air) and j=3 (glass); and $k_{zj}=+i[k_x^2-\varepsilon_j(\omega/c)^2]^{1/2}$ for j=2 (Ag film). Superfocusing requires regenerating the evanescent waves through the lens. Thus the lens needs to be operated in the condition of $|k_{z1}/\varepsilon_1+k_{z2}/\varepsilon_2||k_{z2}/\varepsilon_2+k_{z3}/\varepsilon_3|\rightarrow0$. Physically, this will require exciting a surface plasmon at either the air or the glass side. For p-polarized light (SPPs wave is excited for TM mode only), a negative permittivity is sufficient for focusing evanescent waves if thickness of the metal film and object are much smaller than the incident wavelength. Because electric permittivity $\varepsilon<0$ occurs naturally in silver and other noble metals at visible wavelengths, a thin metallic film can act as an optical superlens. In the electrostatic limit, the p-polarized light, in dependence of permeability μ is eliminated and only permittivity ε is relevant.

The super lens can image at propagation distance in micron scale (around $\lambda\sim5\lambda$ and even longer, we call this range a "quasi-far-field region". Both scalar theory and vector theory are applicable in this region.) along the propagation direction. The conversion from SPP wave to propagation waves in the quasi-far-field region takes place by diffraction from the subwavelength zones of the lens. Focusing characteristics of the lens is quite different from the conventional Fresnel zone plates. The simulated focal length of the lens f and depth of focus (*DOF*) are larger than that of the designed value using the classical equations: $DOF=\pm2\Delta r^2/\lambda$, where n=1,2,3,…, f_{FZP} is the designed principal focal length of Fresnel zone plates and given in terms of radius R of the inner ring and incident wavelength by $f_{FZP}=R^2/\lambda$, whereas Δr is the outmost zone width, and λ is the incident wavelength. This phenomenon is apparent especially for the *DOF*, as shown in Fig. 10-14 (a) ~ (f), and Fig. 10-15. It may attribute to the SPP wave coupling through the cavity mode and is involved for contribution of the beam focusing. The focusing is formed by interference between the SPP wave and the diffraction wavelets originating from the slits. The interference can exist within a coherence length Lc of the "source", but not beyond it. The source is equivalent to a source with central wavelength of λ_{in}=633 nm, and bandwidth $\Delta\lambda=\lambda_{in}-\lambda_{SP}$, and thus $DOF \leq Lc$, where $Lc \approx \lambda^2/(\lambda_{in}-\lambda_{SP})$. For the SPP interference, the wave vector k_x of the incident optical wave projected on the plane parallel to the surface of the metal film must equal to k_{SP}.

Corresponding FDTD analysis results reveal that it has unique focusing performance of the elongated focal length and *DOF* with a focused spot size beyond diffraction limit in comparison to the conventional zone plates. In addition, it can work at longer working distance ranging from $\sim\lambda$ to 5λ or more that gives more flexibility for the relevant application systems working in this quasi-far-field region. This performance will be helpful for the lens being used as optical probes for high resolution imaging and detection.

A plasmonic lens with metallic chirped circular nanoslits corrugated on Au film supported on quartz substrate for the purpose of superfocusing was put forth and fabricated by means of focused ion beam direct milling technique. [25] Topography of the lens was imaged using an AFM. After that a NSOM was employed for optical characterization of focusing performance of the lens. Our experimental results verified the focusing performance and further demonstrated that they are in agreement with the theoretical calculation results. Focusing performance is significantly improved in comparison to that of the non-chirped lens. The lenses are possible to be used for the applications of bioimaging, detection, and inspection in submicron scale resolution. It is the interference of the SPPs wave from the multiple metallic slits that produce the intensity enhancement. Focusing performance of the proposed plasmonic lens was demonstrated in detail on the basis of our experimental study presented below.

In order to further demonstrate and verify the focusing performance of the plasmonic lens with the chirped circular slits experimentally, optical characterization by means of a NSOM (MultiView 2000[TS] from Nanonics Inc.) was carried out recently in our laboratory for the newly designed plasmonic lens with total 8 chirped circular slits which were etched through an Au film supported with glass. Width of the outmost circular slit (8th) is 95 nm. The wave field at the focal point was mapped using the NSOM. For the NSOM measurement, near-field intensity distributions at different

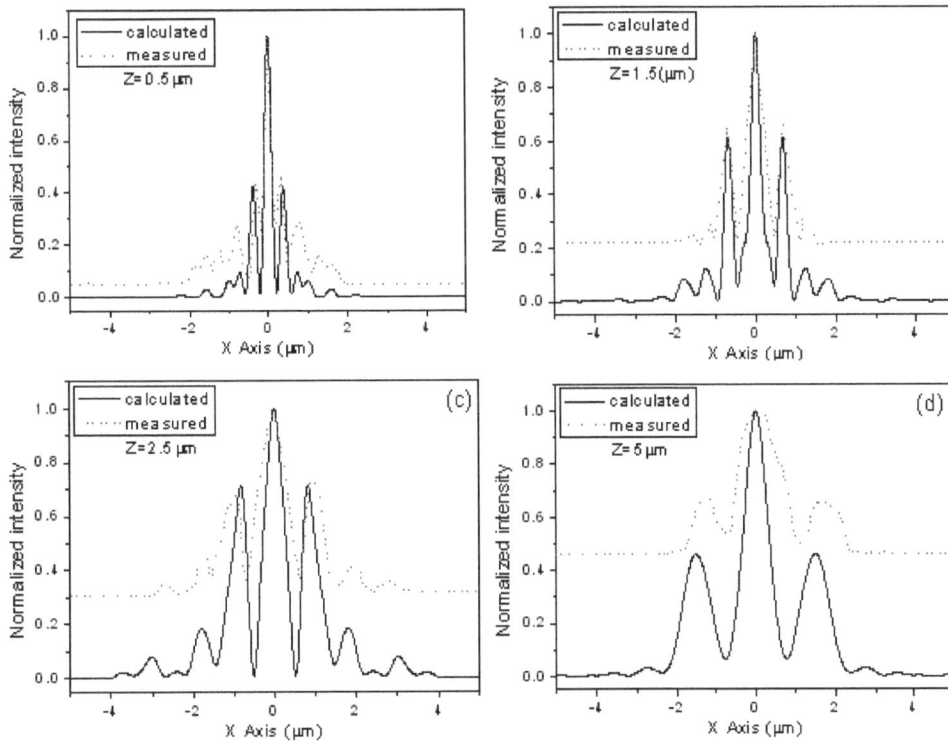

Fig. 10-16 Comparison between the calculated and measured E-field intensity profiles probed at x-z plane at propagation distance of (a) 0.5 μm; (b) 1 μm; (c) 2.5 μm, and (d) 5μm.

horizontal planes vertical to the optical axis have been obtained and compared with simulation results calculated using the finite difference time domain (FDTD) algorithm. The NSOM probing results were found to be in agreement to the theoretical calculation results. To further compare the measured results with the theoretically calculated results, we plotted E-field intensity profiles at x-axis together with that of the numerical calculation, as shown in Fig. 10-16 (a)-(d). In the 3D FDTD calculations, simulation time and mesh size were defined as 150 fs, and $\Delta x = \Delta y = \Delta z = 5$ nm, respectively. The optical field is p-polarized monochromatic wave with 532 nm wavelength in the air. At this wavelength, the Au layer has the refractive index of 0.452+i2.451. It can be seen that variation tendency of the E-field intensity of the measured is in agreement to the theoretically calculated results. As can be seen from Fig. 3, the measured FWHM of the central lobe is slightly larger than the calculated value for cases (a)-(c), and the difference is large (5 μm) for the case (d) in near-field region. It can be attributed to the background noise signal which is stronger gradually with increasing the probing distance from 0.5 μm to 5 μm. It directly leads to the increase of base intensity and causes degradation of the signal-to-noise ratio of the detected optical signal. Thus the difference of FWHM between the measured and calculated data is larger as the probing distance is getting far away from the exit plane.

Figure 10-17 is a re-plotted 3D image of the NSOM measured intensity profiles along x-axis which was probed at the different propagation distance z ranging from 5 nm to 5 μm. It intuitively shows the intensity distribution along propagation distance z. It can be seen that the peak intensity is significantly enhanced from 0.01 μm to 1 μm, and then degraded gradually in near-field region because of SPP-enhanced wave propagation on Au surface vanished in free space when z >1 μm. The interference-formed beam focusing region exists in near-field region only. In comparison to the theoretical calculated results, our experimental results demonstrated that the measured results are in agreement to the calculated results.

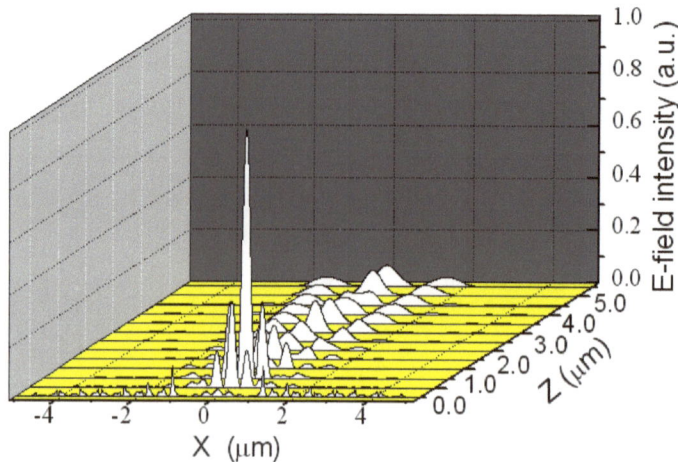

Fig. 10-17 Measured 3D E-field intensity distribution of the plasmonic lens vs. lateral x and propagation distance z using NSOM. The figure was replotted using the NSOM probed data.

10.3.3.2 Radically distributed nanopinholes

The structures with metallic subwavelength slits were described in above section. However, sidelobes for the previous lens structures are high. It occupies a lot of energy and degrades intensity of main lobe. To solve this problem, a novel nanostructure was put forth that is composed of pinholes with micron scale dimension and subwavelength diameter, as shown in Fig. 10-18 [26]. We referred to this structure as nanopinhole-based plasmonic structure (NPPS) and explored its application as a plasmonic lens. Influence of the cut-off wavelength effect on propagation and transmission properties were analyzed for the purpose of revealing optical performance and physical picture of the structure in near-field region.

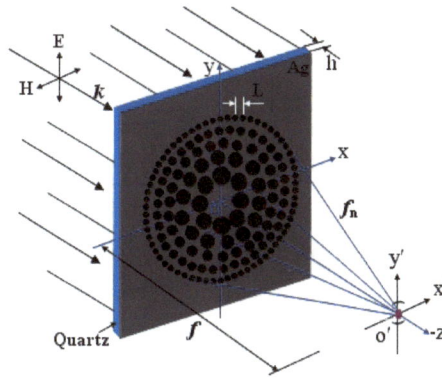

Fig. 10-18 Schematic of the pinhole array with focal length *f*. Lateral central distance *L* determines wave coupling between the neighboring holes. The pinholes are uniformly distributed in the zones. It is illuminated by a plane wave with 633nm incident wavelength and p-polarization (transverse-magnetic field with components of E_x, H_y, and E_z). The Ag film has permittivity $\varepsilon = -17.6235 + 0.4204i$ at $\lambda_0 = 633$ nm. The perfectly matched layer (PML) boundary condition was applied at the grid boundaries in the 3D FDTD simulation.

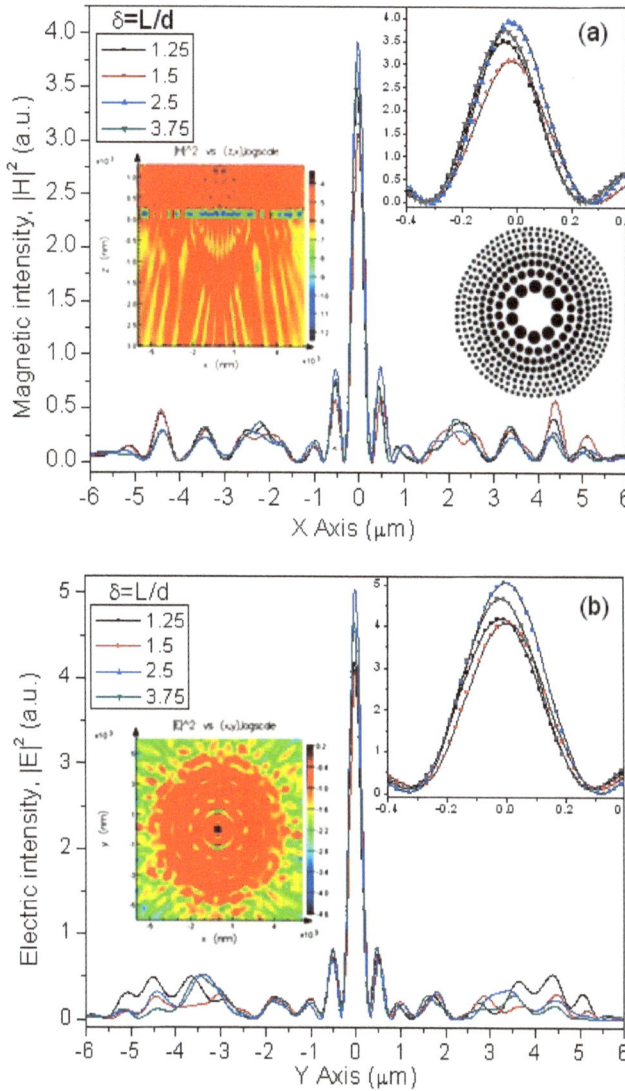

Fig. 10-19 Results of electromagnetic field analysis for (a) *H*-field intensity $|H|^2$ distribution at *x-z* plane. Inset is 2D image of *H*-field intensity $|H|^2$ distribution at *x-z* plane for incident wavelength of 633nm and layout of the pinhole structure. (b) *E*-field intensity $|E|^2$ distribution at *y-z* plane. Inset is 2D image of *E*-field intensity $|E|^2$ distribution at *x-y* plane.

The proposed plasmonic lens here is an asymmetric structure in which a 200 nm Ag thin film is sandwiched between air and quartz substrate. The pinholes with different diameters are uniformly and symmetrically distributed along the open subwavelength zones with variant periods. It works at visible wavelength regime and propagates at near-field with the outer diameter $D<13$ μm and zone number $N<10$. The structure consists of 8 rings of pinholes with the following diameters from outer to inner (see inset pattern in Fig. 10-18 (a)): 158 nm (8th ring), 177 nm (7th ring), 202 nm (6th ring), 234 nm (5th ring), 280 nm (4th ring), 349 nm (3rd ring), 467 nm (2nd ring), and 735 nm (1st ring), respectively. For the wavelength $\lambda_{in.}=633$ nm at normal incidence, the pinhole diameters at the outer four rings are less than $\lambda/2$. The total dimension of the structure is 12.07 μm. All the pinholes have a fixed ratio $K=d/w=3.0$, where d is the diameter of pinholes and w is the width of corresponding open zones in a zone plate. For the transmission with the SPPs excitation ($E_\perp$ field), the electric intensity $|E|^2$ distribution at x-z plane and magnetic intensity $|H|^2$ distribution at y-z plane were calculated using the FDTD algorithm respectively, as shown in Fig. 10-19 (a) and (b). It can be seen that an apparent focal region was observed for both components $|H|^2$ and $|E|^2$. Spot size at site of full-width and half-maximum (FWHM) increases slightly with increasing the pinholes periodicity L. As can be seen, the sidelobes (±1 orders) are significantly suppressed.

If we replace the circular pinholes to be elliptical pinholes, as shown in Fig. 10-20 (a), the propagation intensity distribution in free space will be significantly tailored. The focusing region is ultra-elongated with increasing the ratio δ of long-axis to short axis gradually from 1 to 10.[27] Figure 10-20 (b) shows the E-field intensity distribution at y-z plane in the case of $\delta=5$ (length of long-axis is 3 times of the corresponding annular width) under the beaming produced from SPP wave excitation and coupling as well as interference. It can be seen that an ultra-long extended focusing region appears. Further study is necessary to reveal its physical mechanism.

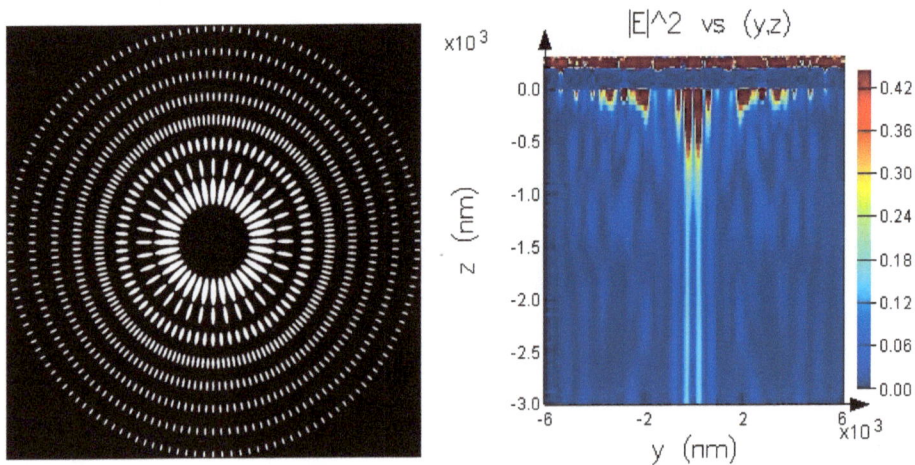

Fig. 10-20 (a) Designed pattern with elliptical pinholes with orientation along radial direction for $\delta=5$, length of long-axis is 3 times of the annular width. (b) image of E-field intensity $|E|^2$ distribution at y-z plane for linear polarized incident beam with working wavelength of 633 nm.

10.3.3.3 Hybrid structures with metallic subwavelength slits

The plasmonic structures with pure Ag for superfocusing were introduced in above sections. However, most of them were designed on the basis of Ag thin film metallic nanostructures. Corrosion-induced electrochemical damage on surface of the Ag film exists at ambient atmosphere, especially in the period of time after micro/nanofabrication. The corrosion originates from oxidation and sulfuration which is well known for bulk Ag. But for the Ag thin film, dielectric constant of the Ag thin film will be definitely changed due to the oxidation and sulfuration. Optical performance of the nanophotonic devices varies accordingly. To overcome this problem, we further put forth a hybrid Au-Ag subwavelength structures with Au thin film covered on the Ag film surface, as shown in Fig. 10-21.[28,29] Therefore, the Au film acts as both a protector and modulator here. On the one hand, it can protect Ag film surface from oxidation. On the other hand, it can modulate beaming and propagation properties of the devices. But it is a tedious work to study the Au film modulation here by experiments only due to complex thin film coating process required. Considering this, a theoretical study was performed firstly by means of computational numerical calculations for the purpose of revealing physical picture of the hybrid Au-Ag film modulation in the subwavelength structures. A FDTD algorithm was adopted

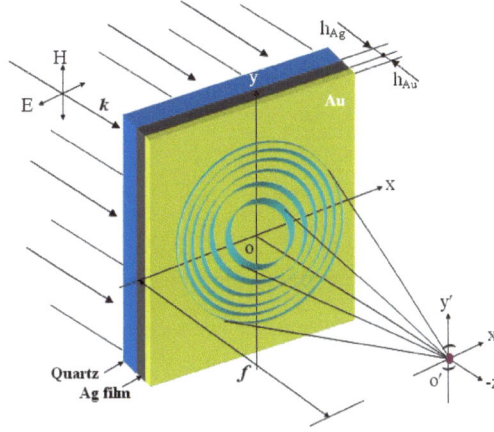

Fig.10-21 Schematic of the plasmonic zone plate-like structure for super-focusing with focal length *f*. The rings penetrate both Ag and Au thin films. It is illuminated by a plane wave with 633 nm incident wavelength. In our FDTD simulations the perfectly matched layer (PML) boundary condition was applied at the grid boundaries.

here for the computational calculation and numerical simulation. Corresponding calculation results further show that thicknesses of both the Au and Ag thin films have significant tailoring function due to their great contribution on superfocusing and transmission. Improved focusing performance and enhanced transmission can be obtained if h_{Au} and h_{Ag} match each other.

The hybrid structure-based device was designed as a zone plate-like structure with total 8 rings penetrating through both the Au and Ag films and reaching the supported glass substrate, as shown in Fig. 10-21. The rings act as cavity mode while the incident light beam propagates through them. Fabry-Pérot-like phenomenon as the SPPs wave coupling and passing through the slits occurs during the propagation [9]. The superfocusing is formed on the basis of beam interference between the coupled SPP wave and diffraction wavelets generated from each ring. Resonance capacity of the surface plasmon can be tailored by the corrugated subwavelength metallic structures [10]. For our variant period-based subwavelength structures, the SPP wavelength can be expressed as

$$\lambda_{SP} = \frac{\Lambda_{eff}}{m} \mathrm{Re}[(\frac{\varepsilon_m}{1+\varepsilon_m})^{1/2}] \ , \tag{10.9}$$

For normal incident where *m* is the nonzero relative integer, which is deduced from the SP momentum matching condition, Λ_{eff} is the effective period of the structures, and ε_m is the dielectric constant of hybrid metal films Au-Ag, $\varepsilon_m = f(n_{eff})$, and $n_{eff} = \beta/k$, whereas β is the propagation constant, and *k* is the wave number. The variant rings period Λ_i can be expressed as

$$\Lambda_i = \sqrt{2 f \lambda_{SP} (i+1) + \lambda_{SP}^2 (i+1)^2} - \sqrt{2 i f \lambda_{SP} + i^2 \lambda_{SP}^2} \ , \tag{10.10}$$

where *i* is the integer number of variant periods, *f* is the focal length of the structure. The effective period can be written as $\Lambda_{eff} = f(\Lambda_i)$.

We assume that the device is illuminated by a plane wave in *p*-polarization. The dispersion relationship of *p*-polarization in the metallic-dielectric-air guide can be written as [2]

$$k_d h_d = a \tan(\frac{\varepsilon_d k_a}{\varepsilon_0 k_d}) + a \tan(\frac{\varepsilon_d k_m}{\varepsilon_m k_d}) \ , \tag{10.11}$$

where k_d and k_a is the wave vector of quartz substrate and air, respectively, ε_d and ε_a is the dielectric constant of quartz and air, respectively, h_d is the thickness of quartz. The final intensity at the focal point is synthesized by iteration of each slit focusing and interference each other, and can be expressed as

$$I = \alpha \sum_{i=1}^{N} C \, I_{i0} I_{SP} \frac{4r_i}{\lambda_{SP}} e^{-(r_i/l_{SP})} , \qquad (10.12)$$

where I_{i0} is the intensity of diffractive wavelet at i[th] zone, I_{SP} is the intensity of the SPP wave passing through the ith slit, r_i is the radius of each zone, i is the number of the zones, l_{SP} is the propagation length for the SPP wave, α is the interference factor, and C is the coupling efficiency of the slits. C is a complicated function of the slit geometry and will likely have a different functional form when the slit width is much larger or much smaller than the incident wavelength. The interference phenomena is theoretically described in Eq. (10.12) as the term of "$I_{i0}I_{SP}$" which originates from both the subwavelet I_{i0} and surface plasmonic wave I_{SP}.

Superfocusing performance is improved accordingly by this approach. Sidelobes play a negative role on the focusing and can be suppressed greatly in the case of h_{Au}=50 nm and h_{Ag}=10 nm. Our target of suppression of sidelobes and enhancement of the main lobe can be realized by optimization of the heterostructure by means of tuning the bilayer thicknesses. Generally speaking, thinner film results in a better focusing performance. Considering our computational results and metal film coating techniques, the estimate values of the thicknesses for Au and Ag are ~50 nm and ~10 nm, respectively. Figures 10-22 (a) and (b) show E-filed intensity distribution at x-z plane for the components of E_x, E_y, and E_z, respectively, for the case of $h_{Au} = h_{Ag} = 50$ nm and $\alpha = 1$. It can be seen that the calculated focal length is longer than the designed value of 1 µm. Figure 10-22 (b) shows the component intensity $|E_y|^2 = 0$. It attributes to the p-polarization effect. Both E_x and E_z have significant focusing performance.

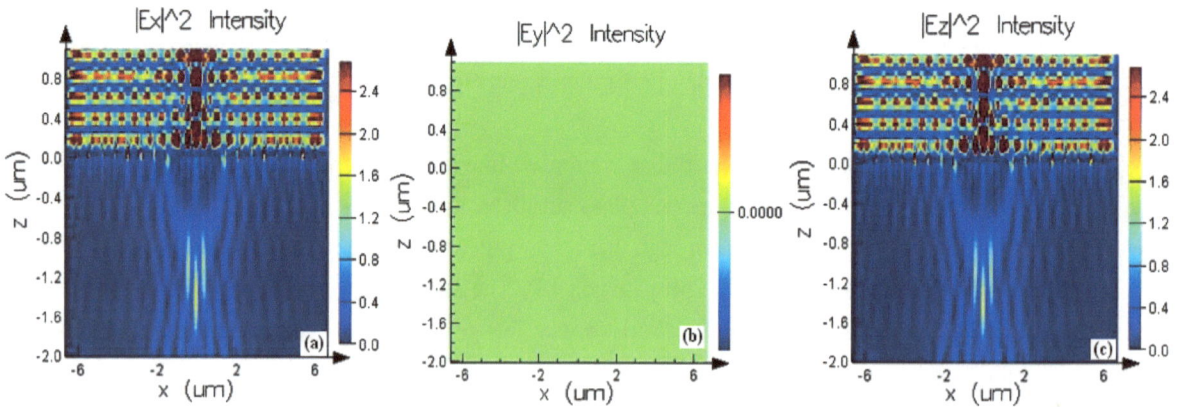

Fig. 10-22 2D images of E-field intensity distribution at x-z plane for components of (a) E_x (b) E_y, and (c) E_z for the case of $h_{Au} = h_{Ag} = 50$ nm and $\alpha = 1$.

The superfocusing performance strongly depends on thicknesses matching between the Au and Ag thin films. Optimization is necessary for designing the hybrid structures. Compared to the reported structure of pure Ag or Au metal film, this hybrid structure can function as the performance of corrosion free and tuning focusing. However, nanofabrication of the elliptical pinholes structures is quite challenging for current nanotechnologies. Currently, focused ion beam directly milling is most suitable to fabricate the structures. But orientation of the nanopinholes especially for the pinhole-based structures with pinholes distributed at outer rings is too difficult to ensure their fabrication quality due to inherent stigmation during the process. Stigmatism is hardly completely eliminated because of lack of quantitatively controlling approach for the current commercial focused ion beam machines.

10.3.3.4 Influence on V-shaped metallic slits

Currently, the FIBM technique is a commonly used approach to fabricate such nanostructures due to its unique advantages of one-step fabrication, nanoscale resolution, and no materials selectivity etc. However, the FIBM process has its own problems also. Regarding the FIBM fabricated corrugations and subwavelength aperture, there are two issues that have not been addressed at present: (1) V-shaped structuring effect; (2) variation of optical property of the metal film (optical property will be changed due to ion implantation of the Ga^+ after the FIBM). These two issues have not been mentioned in the existing theoretical literature. The second issue will be further investigated via both theoretical and experimental study and reported in an individual section later. This section extends the prior works from theoretical study of the ideal rectangular-based corrugations to the practical nanofabrication-related issue: the V-shaped structuring effect during formation of the corrugations. In cross-section view of the FIBM fabricated structures (see Fig. 10-23), a V-shaped grooves instead of the grooves ideal rectangular shape (rectangular grooves were assumed in all the previous reports which were cited in the Reference.) appear while the FIB mills the structures designed with high aspect ratio (depth to width), as shown in Fig. 10-23.[30] For the designed nanostructures with rectangular shape in the cross-section view, a shape evolution from the designed rectangular to the "V" shape is unavoidable after the FIBM. Normally, geometrical characterization of the V-shaped nanostructures is carried out using an AFM working in tipping mode. However, it is too difficult to accurately measure the dimension of the nanostructures due to inherent shortages of the AFM. Therefore, it is quite challenge to experimentally study and evaluate their influence on optical performance which is an important issue for designing the metallic nanostructures. Considering this, we addressed this problem, and analyzed the influence on the transmission theoretically by means of computational calculation method: numerical simulation. Hopefully, exploring the influence on the light behavior in the nanostructures by the theoretical simulation can provide a better understanding of the beam propagation for the relevant researchers. As an example, we studied the case of 200 nm Ag film coated on quartz substrate and designed a structure with a thoroughly-penetrated central slit and 16 surrounded grooves at the exit plane. The 16 grooves having dimensions of 80 nm in depth and 200 nm in width are symmetrically surrounded the central slit, as shown in Fig. 10-23. For the grooves with the aspect ratio of 1.5, the final average width of the grooves after the FIBM is empirically estimated to be ~150 nm.

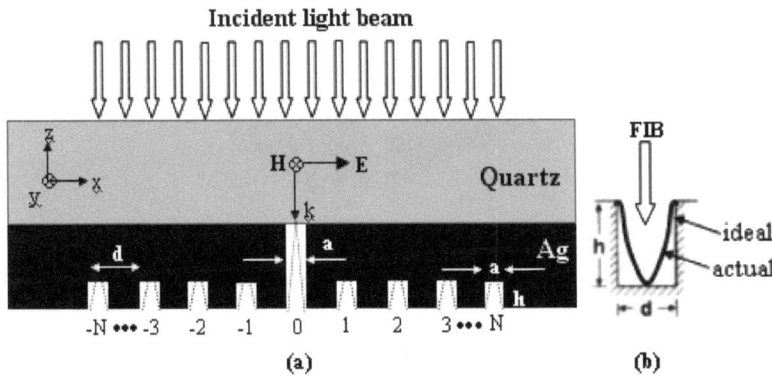

Fig.10-23 Design configuration of the enhanced plasmonic nanostructure. Side view of the designed structure. The grooves in the corrugation have the same depth and width. The dotted line in the grooves means the actual shape after the FIBM. Period of the corrugation is *d*=500 nm.

It was demonstrated by our simulation results that the sharp V-shaped central slit plays a positive dominant role in transmission to a large extent. It helps improving beam shaping significantly. The V-grooves cause a red-shift of the peak wavelength and broadening of the cut-off wavelength in transmission along the propagation direction due to the shape resonance.[31, 32] Moreover, it causes beam diverging in the far-field region. Thus it plays a negative role in the propagation process. However, the influence from the V-grooves on the transmission spectrum can be ignored in the mixture case of both the V-shaped central slit and grooves existing. As a matter of fact, the sharp edge of the slit / grooves with small size, *e.g.*, d_2=25 nm, is difficult to be intentionally and directly milled using the FIBM. However, it is possible to be formed naturally due to the inherited characteristics of the FIBM process mentioned above.

The V-shape effect is not a specific issue for the FIBM technique only. Some other techniques such as e-beam lithography, also has the V-shaped effect while the exposed pattern is transferred from the PMMA resist into substrate. Therefore, the simulation results are also applicable for other nanofabrication techniques.

In addition, for nanofabrication using focused ion beam (FIB) technique, Ga^+ implantation will be a negative factor that will change dielectric constant of the scanned region of the substrate or thin films.[33] Normally, for FIB milling under 30 keV ion energy, induced ion implantation depth is ~30 nm. The implanted region distributes at subsurface of the substrate. Of course, optical performance will be varied accordingly in this case. For more information, please read the relevant papers.[34] Figure 10-24 is SEM micrograph of the plasmonic structure which has variant periods and total

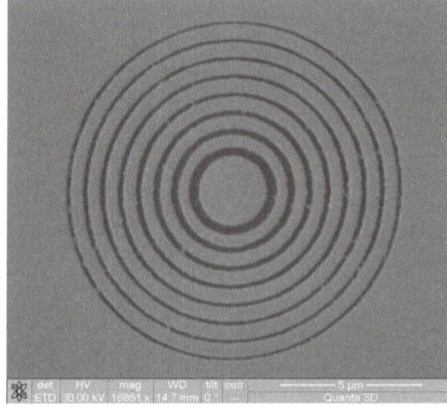

Fig.10-24 Micrograph of the FIB fabricated plasmonic structure.

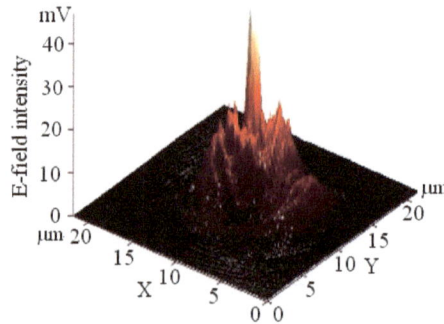

Fig.10-25 Three-dimensional intensity distribution of the focused beam
in x-y plane measured at propagation distance of 2 μm using NSOM.

8 rings which was fabricated on 200 nm Ag film using FIB direct milling. The plasmonic structure functions as superfocusing. Figure 10-25 is an optical characterization result measured using NSOM for the Ag film-based plasmonic lens. It can be seen that a sharp central peak transmission surely exists at the propagation distance of 2 μm away from the exit plane. It is in agreement to the theoretical calculation results. But FWHM is slightly larger than the designed value due to fabrication error and the Ga^+ implantation.

10.3.4 Superfocusing with radial polarization [35]

Recently radial polarization was also discovered to be the ideal light source for SPPs excitation with axially symmetric metal/dielectric structures. Most past research dealt with spatially homogeneous states of polarization (SOPs), such as linear, elliptical, and circular polarizations. For these cases, the SOP does not depend on the spatial location in the beam cross section. The radial polarization herein is the light beam with spatially variant states of polarization (SOPs). Spatially arranging the SOP of a light beam, purposefully and carefully, is expected to lead to new effects and phenomena that can expand the functionality and enhance the capability of optical systems. One particular example is

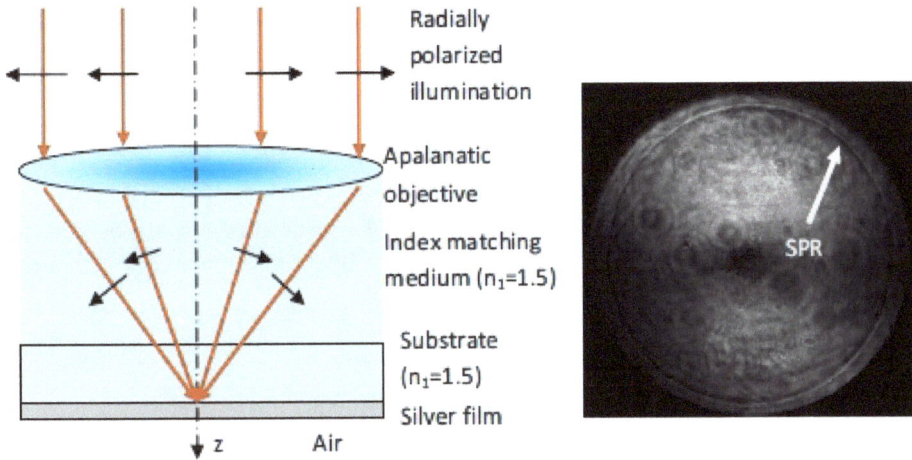

Fig. 10-26 Diagram of SPR excitation with highly focused radial polarization. The entire beam is p polarized with respect to the silver/dielectric interface. A pupil plane image of the reflected beam is shown to the right. A full dark ring is observed that is due to the SPR excitation by radial polarization. The nonuniformity in the beam profile is caused by the difference in reflection coefficients for the s- and p-polarization components of a beam splitter that steers the reflected beam to the CCD camera. Reprinted with permission from Qiwen Zhan, Adv. Opt. Photon. 1, 1-57 (2009) with copyright © 2009 of Optical Society of America.

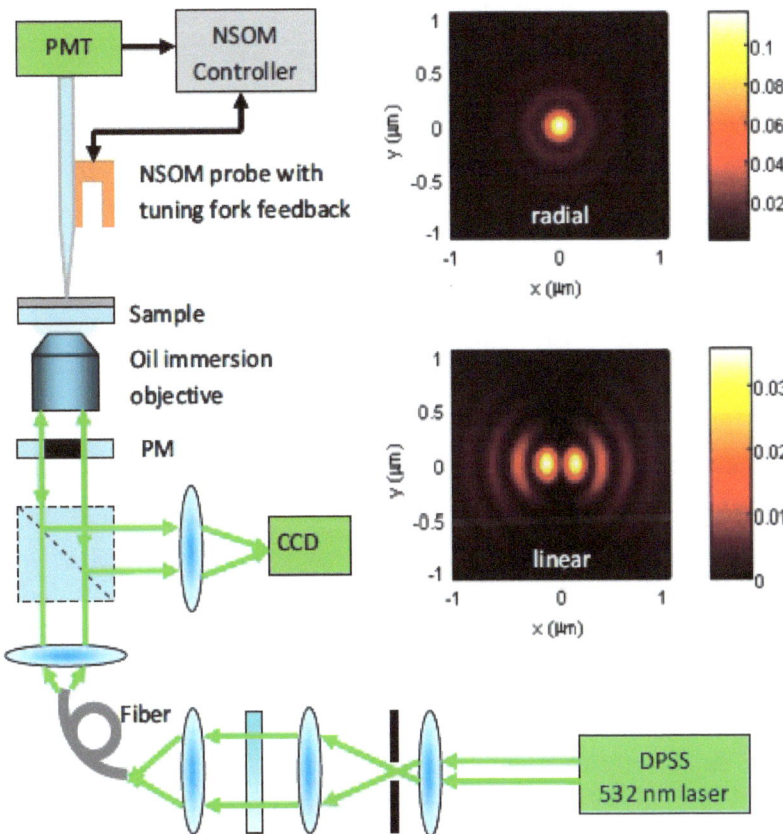

Fig. 10-27 Experimental setup for the generation and detection of plasmonic focusing with highly focused radial polarization. PMT, photomultiplier tube; PM, photomask. A Veeco Aurora 3 NSOM with a nanoaperture fiber probe is used to detect the plasmonic field. The insets show the calculated plasmon focal fields at the metal surface for radial and linear polarization incidence, respectively. For linear polarization, because of the mismatch of polarization and the destructive interference of counter-propagating plasmon waves, the field at the focal plane splits into two lobes with much lower peak intensity. Reprinted with permission from Qiwen Zhan, Adv. Opt. Photon. 1, 1-57 (2009) with copyright © 2009 of Optical Society of America.

laser beams with cylindrical symmetry in polarization, the so-called cylindrical vector (CV) beams. Because of their interesting properties and potential applications, there has been a rapid increase of the number of publications on CV beams [36-38].

Plasmon excitation has a strong dependence of excitation polarization. For example, in an attenuated total reflection configuration, only *p*-polarization can excite SPR. This leads to an interesting application of radial polarization in plasmonic wave-based focusing. It was pointed out that optimal plasmonic focusing can be obtained using a radial polarization for a rotationally symmetric illumination setup [39–41]. When a radially polarized beam is launched into these dielectric/metal plasmonic lens structures, the entire beam is *p*-polarized with respect to the dielectric/metal interface (see Fig. 10-26), providing an efficient way to generate a highly focused surface plasmon through constructive interference and create an enhanced local field. In contrast, if highly focused linear polarization is used to excite the surface plasmon, because of the destructive interference between counter-propagating surface plasmon waves, the focal field has a minimum value at the geometric focus. Figure 10-25 illustrates an experimental setup for the generation and detection of the plasmon field near the silver interface. The illumination was created by highly focused radial polarization [42]. Examples of a plasmonic focal field due to radial and linear polarizations were shown in the insets of Fig. 10-25 for comparison. An aplanatic oil immersion lens (*NA*=1.25) focuses the radially polarized beam onto a glass/silver interface. Immersion oil with a refractive index matched to the glass substrate (n_s=1.516) is filled between the lens and the substrate. A 50 nm silver film (ε=−0.18−0.824i) is deposited onto the glass substrate. The optical excitation wavelength was chosen to be 532 nm. The center part of the illumination corresponding to the incident angle below the critical angle θ_c=sin^{-1}(n_m/n_s)=41.27° is blocked by an annular photo-mask. The field right after the silver film can be calculated with the following integrals:

$$E_r(r,\phi,z) = 2A \int_{\theta_{min}}^{\theta_{max}} P(\theta)t_p(\theta)\sin\theta\cos\theta J_1(k_1 r \sin\theta)\exp[iz(k_2^2$$
$$- k_1^2 \sin^2\theta)^{1/2}]d\theta, \tag{10.13}$$

$$E_z(r,\phi,z) = i2A \int_{\theta_{min}}^{\theta_{max}} P(\theta)t_p(\theta)\sin^2\theta J_0(k_1 r \sin\theta)\exp[iz(k_2^2 \tag{10.14}$$
$$- k_1^2 \sin^2\theta)^{1/2}]d\theta,$$

where θ_{min} and θ_{max} is the minimum and maximum incident angles respectively, on the glass/silver interface corresponding to the annular illumination, $P(\theta)$ is the pupil apodization function, $t_p(\theta)$ is the transmission coefficient of *p*-polarization at the incident angle of θ, $J_{m(x)}$ is the *m*th-order Bessel function of the first kind, and k_1 and k_2 is the wave-number in the glass and air, respectively. Numerical results of the plasmonic field originating from highly focused radial polarization excitation were shown in Fig. 10-28 (a)–(c). From these plots, we can see that the plasmonic focal field generated by this way is dominated by the *z* component. A NSOM in collection-mode with a nanoaperture (aperture size is between 50 and 100 nm) fiber probe was used to map out the plasmonic field distribution near the silver/air interface. For the metal-coated fiber probe with a nanoaperture, the detected signal is proportional to $|\nabla_\perp E_z|^2$ owing to the symmetry of the HE$_{11}$ mode propagating in the fiber core of the probe [43]. The expected NSOM signal is calculated and shown in Fig. 10-28 (d), which actually shows a donut pattern. This was confirmed by the experimental measurements shown in Fig. 10-29 (a). Multiple concentric rings corresponding to SPPs wave propagation were observed. The plot was drawn in logarithmic scale for better visualization of those outer rings. A comparison of the measured and calculated transverse profiles of the NSOM signal was shown in Fig. 10-29 (b). Very good agreement between the experimental result and the simulated result was obtained. The locations of main lobe and sidelobe of the experimental and calculated results match each other quite well. Because of the finite aperture size of the probe, a certain amount of radial components will also be detected even if the probe is placed at the center of the focus, and leads to elevated NSOM signals at those minimum locations.

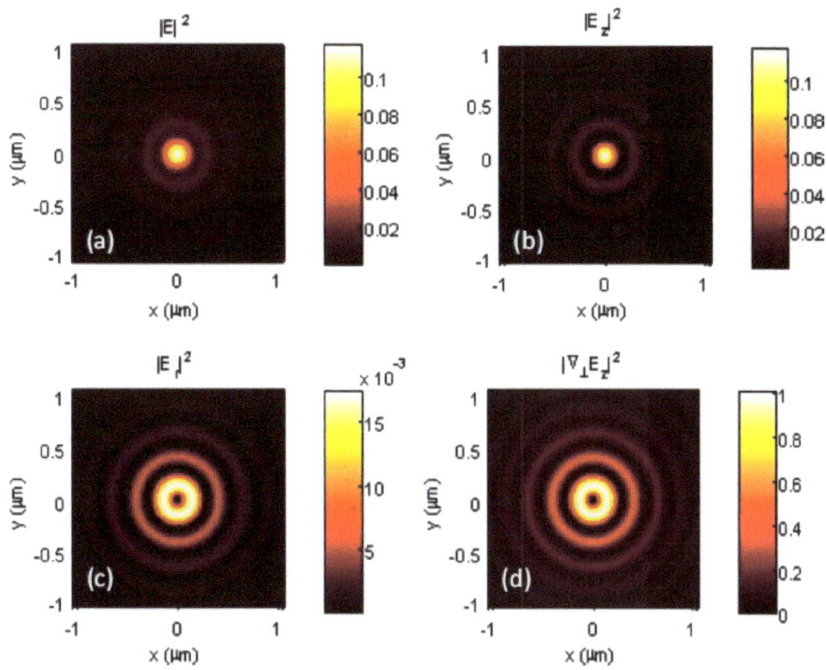

Fig. 10-28 Calculated surface plasmon intensity distribution at silver/air interface with radially polarized illumination. (a) Total intensity $|E|^2$; a homogeneous spot with a strongly enhanced field at the center is obtained. (b) Longitudinal component $|\nabla_\perp E_z|^2$, which is much stronger than $|E_r|^2$ and dominates the total field distribution. (c) Radial component $|E_r|^2$, which has a donut shape. (d) $|E_z|^2$ distribution, which is proportional to the NSOM signal detected by an apertured fiber probe. Reprinted with permission from Qiwen Zhan, Adv. Opt. Photon. 1, 1-57 (2009) with copyright © 2009 of Optical Society of America.

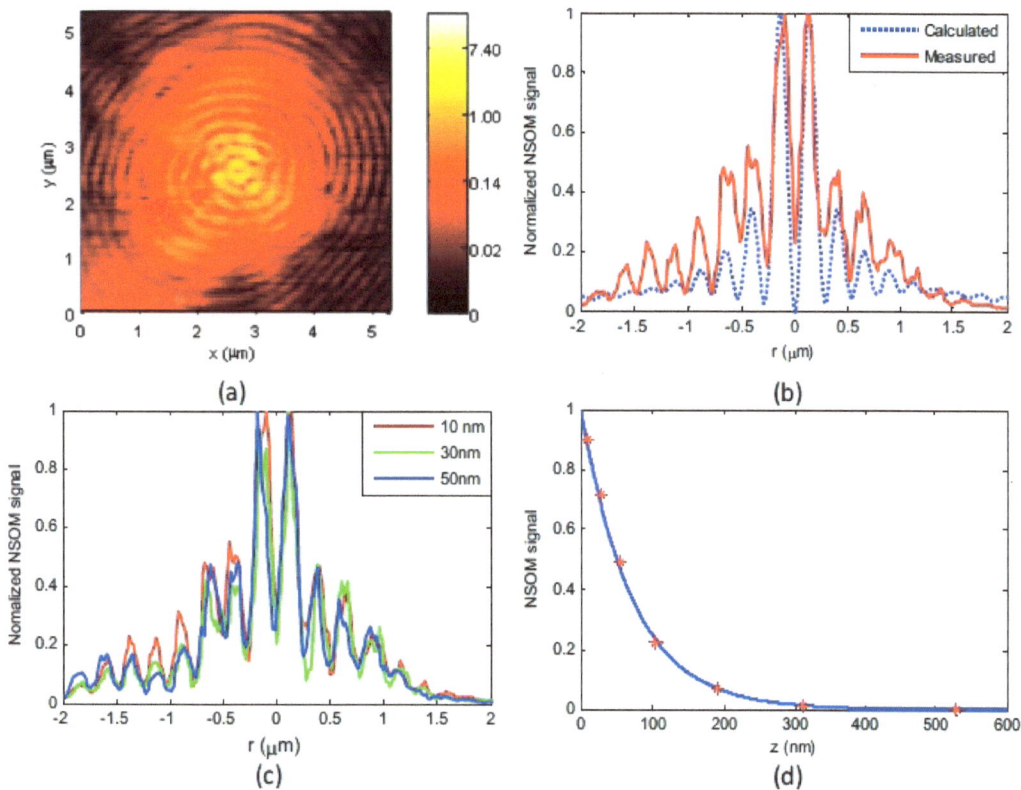

Fig. 10-29 Experimental confirmation of evanescent Bessel beam creation via surface plasmon excitation. (a) Logarithmic plot of measured NSOM signal in the near field. (b) Comparison of the measured and calculated transverse profiles of the NSOM signal $|\nabla_\perp E_z|^2$ in the near field. (c) Measurements at different planes to reveal the nondiffracting nature of the evanescent Bessel beam. (d) Exponential decay of the evanescent Bessel beam. Reprinted with permission from Qiwen Zhan, Adv. Opt. Photon. 1, 1-57 (2009) with copyright © 2009 of Optical Society of America.

More interestingly, the focal field created this way is an evanescent Bessel beam because the sharp SPR resonance effectively acts as an axicon [41]. This axicon-like function can be seen from the very narrow dark ring corresponding to the SPR, as shown in Fig. 10-26. The excitation angle is measured to be 45.51° with a FWHM angular width of 1.28°. Only the portion of the incident beam corresponding to the dark ring is coupled into surface plasmon modes. Thus the SPR excitation with highly focused radial polarization performs a rotationally symmetric angular filtering function for the transmitted field that mimics an axicon device, and allows the creation of an evanescent Bessel beam. This phenomenon was also confirmed by the experimental results. The normalized transverse profiles of intensity at different distances from the sample surface were plotted in Fig. 10-29 (c). It can be seen that all curves almost overlap with each other, indicating the nonspreading property of the beam. Then the probe was moved to the peak of the innermost ring, and the signal was measured at a series of distances away from the surface. The evanescently decaying nature of the Bessel beam is clearly shown in Fig. 10-29 (d), with a decay length of 143 nm obtained through the curve fitting.

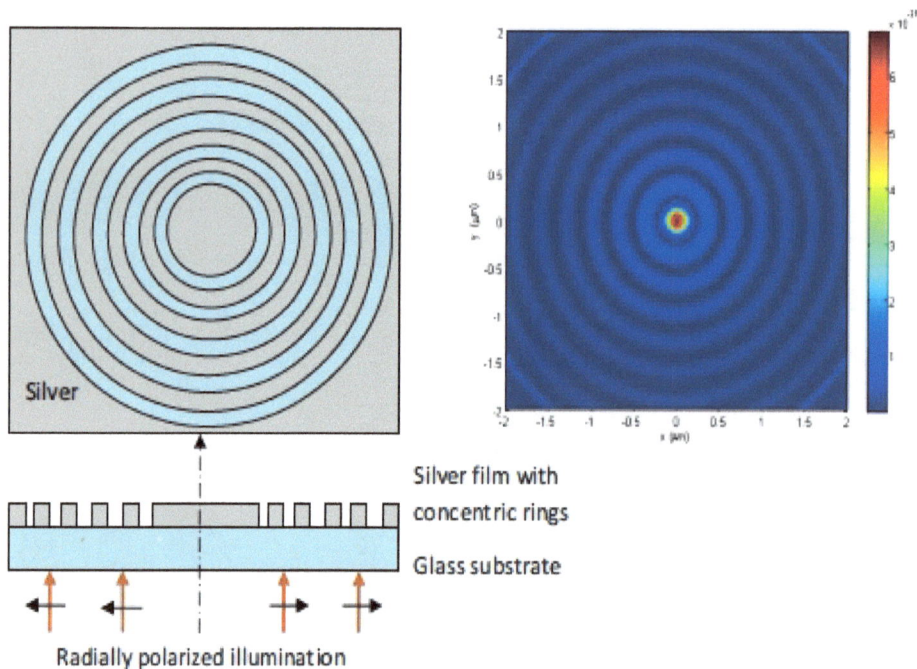

Fig. 10-30 Plasmonic lens structure with multiple concentric rings illuminated by radial polarization. Surface plasmons generated by each of the rings propagate to the center and constructively interfere to produce a tight plasmonic focus (shown to the right). Reprinted with permission from Qiwen Zhan, Adv. Opt. Photon. 1, 1-57 (2009) with copyright © 2009 of Optical Society of America.

These experimental results confirm that an evanescent Bessel beam is generated via surface plasmon excitation with a highly focused radially polarized beam. A spot size as small as 195 nm ($\lambda_0/2.728$) can be obtained by use of 532 nm optical excitation. The combination of symmetric radial excitation and the SPR angular selectivity eliminates the requirement for a conical device and significantly simplifies the alignment procedure simultaneously. The evanescent Bessel beam can be used as virtual probes for the near-field imaging and biosensing. In addition, the strong confinements along the longitudinal and transverse directions may be useful for application of optical tweezing.

The optimal plasmonic focusing generated by radial polarization illumination further applies to other structures such as multiple rings [38] and conical probes [44–46]. An exemplary multiple-ring structure is shown in Fig. 10-30. Rings with 250 nm width were etched into 200 nm silver film deposited on quartz substrate. For a 632.8 nm excitation wavelength, FWHM of about 222 nm ($\lambda/2.85$) can be obtained. The peak intensity can be increased by adding more rings to the structure, but the spot size remains almost unchanged with increasing the number of rings.

A conical tip structure illustrated in Fig. 10-28 has also been investigated numerically. The entire glass tip is coated with 50 nm silver film, and the tip radius is 5 nm. The numerical results showed that for a radial polarized illumination at 632.8 nm, the FWHM of less than 10 nm with an intensity enhancement higher than 105 can be obtained. This type of apertureless probe is very useful for near-field optical microscopy, near-field Raman mapping, and metrology.

10.3.5 Funnel-shaped array for superfocusing

A funnel-shaped array of single Ag nano-cylinders was proposed on the basis of transmission characteristics of single noble nano-cylinders.[47] Computational numerical study was performed by means of FDTD algorithm for the analysis of electromagnetic field propagation at the facet of the "funnel" as well as coupling effect between the localized SP waves and the metal nanocylinders. It demonstrated that super-focusing in near-field region can be realized through such a structure.

Figure 10-31 shows the top view of the proposed funnel-shaped waveguide. The structure consists of two-pathway Ag nanocylinders array with the gradually changed diameters, and discrete spacing forms the funneled-shaped waveguide with nano-scale dimension. The surrounding medium is simply set to be air. The geometrical parameters of the Ag nanocylinder array are denoted with the radius R and the transmission length D from the incident position to the funneled facet, the arm length L, the funnel obliquity θ, and the focus field width d, were denoted respectively in Fig.

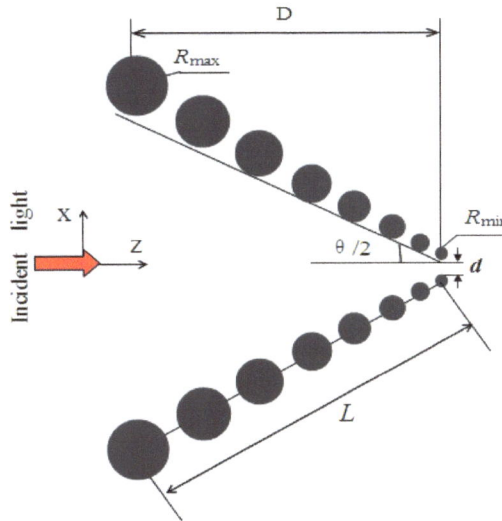

Fig. 10-31 The vertical view of the funnel-shaped waveguide with diameter gradually changed and discretized metal nanocylinders.

10-31. Corresponding FDTD simulations demonstrated that the proposed structures with different interval spaces and funnel angles have versatile characteristics of light propagation. The incident light exciting the SPP wave in this structure is a transmission mode (TM)-polarized (the magnetic field is parallel to y-axis) Gaussian beam with a spot size of 200 nm (FWHM) at λ_{SP}=437 nm, which is the resonance wavelength of the structure. Here only the TM mode consisted of E_x, E_z, and H_y components were discussed due to their obvious plasmon excitation on the metal surface. The dielectric constant of the Ag is -5.5773 + 0.2185i which was calculated using the Drude model

$$\varepsilon_m(\omega) = \varepsilon_\times - \frac{\omega_p^2}{\omega(\omega + i\gamma)} \tag{10.15}$$

where w_p is the plasma frequency, g is the absorption, ω is the frequency, and $e_\check{v}$ comes from the contribution of the bound electrons to the polarizability, for the Ag used here, w_p=1.346×10^{16}Hz, g=9.617×10^{13}Hz, and $e_\check{v}$=4.2.

FDTD-based numerical simulation of the funnel-shaped waveguide with the discrete Ag nanocylinders was performed for the purpose of further studying its optical property numerically.

Two-dimensional (2D) FDTD simulations were performed. The cell sizes were set to be $\delta=\Delta_x=\Delta_z=1$ nm，and the time step is $\Delta t=\delta/(2c)$. Figure 10-32 (a) and (b) present the typical $|E|^2$ and the time-averaged Poynting vector S_z distribution of the SPP passing through the proposed structure at the x-z plane with focal spot size of 10 nm at the funnel facet. It can be seen that the SPP energy is confined mainly in the inner of the funnel-shaped guide, especially confined along the inner edge of the discrete Ag cylinders. A strong focusing effect appears at the funnel facet of the funnel-shaped nanocylinder waveguide structures (FDNWS).

To fully understand the dependence of the focusing effect on geometry of the FDNWSs, the E-field intensity $|E|^2$ and Poynting vector S_z distribution for the structures with different geometrical parameters were calculated, whereas L denotes the gradient length from the centre of the largest Ag metal cylinder to the minimum one, and S denotes the length along the same line with the Ag occupied. As the structure is symmetrical, we defined the size of focusing spot width d as the central distance between the adjacent minimum nanocylinders at the funneled facet of the FDNWSs. The localization of focusing field distribution at the funneled facet and the localization of focusing field distribution of the FDNWS as function of the transmission distance were shown in Fig. 10-33 (a)-(d). It can be seen from Fig.10-33 (a)

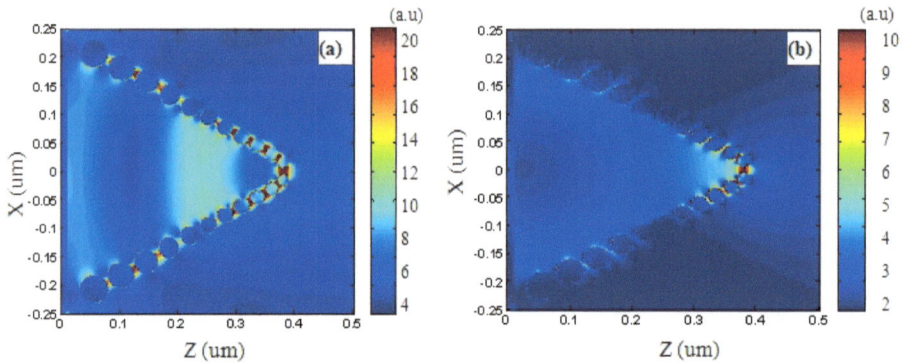

Fig. 10-32 Typical FDTD simulation distribution results of E-field intensity $|E|^2$ (a) and Poynting vector S_z (b) at the x-z plane of the structure shown in Fig.10-29 with the focusing width of 10 nm.

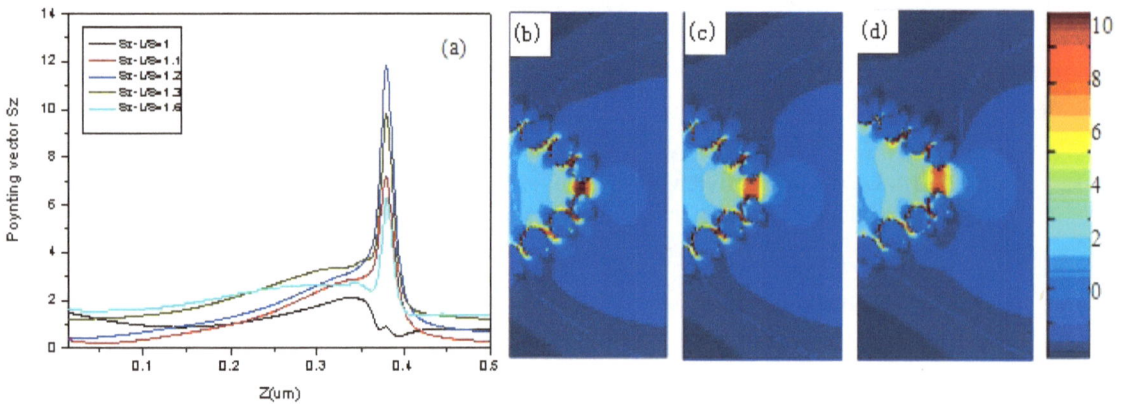

Fig. 10-33 The Localization Focus field distribution of the structure at the funneled month. (a) The Poynting vector Sz with different *L/S* which denote on the figure along the Z direction, at site X=0. (b) *L/S*=1.2 and focusing field width is 10 nm, (c) *L/S*=1.2 and focusing field width is 15 nm, (d) L/S=1.2 and focusing field width is 20 nm.

that for filling factor *L/S*=1.2, the localization of the focusing is maximized at the funneled facet. For the other ratio *L/S*, the focusing energy decreases, especially for the case of the Ag nanocylinders arranged without interval spacing along the *L* gradient array (a fixed ratio *L/S* corresponds to the case that the nanocylinders in both arms of the funnel contacts each other), almost no focusing appearing at the funneled facet but the intensity of the energy is relatively strong in front of the facet. These phenomena may be attributed to the optical energy coupling and scattering between the two arms which have not effectively transmitted through the funneled facet yet. Figure 10-31 (b), (c), and (d) show the focusing energy distribution at the funneled facet with L/S=1.2 and the size of the focusing spot size of 10 nm, 15 nm, and 20 nm, respectively. The observed focusing spots are clear and apparent. We defined the focusing efficiency S_z as the intensity of the light at the funneled facet plane divided by the one at the input plane. It can be seen that, with a fixed D=360 nm, and a fixed 10 nm focusing width, as filling factor L/S changes from 1 to 1.1, 1.2, 1.3, and 1.6, respectively. The corresponding focusing efficiency is 9.6%, 19.2%, 38.6%, 17% and 10%, respectively. However, for the structures with L/S=1.2, if we change the localization of the focusing width to be 15 nm, and 20 nm, the focusing efficiency will increase to be 41.5% and 80.4 %, respectively. Figure 10-33 shows the S_z distribution for the structures with L/S=1.2 at the funneled facet, and the focusing width varies from 10 nm to 15 nm and 20 nm. As a supplement to Fig. 10-34, it shows that the localization of the focusing is obvious at the funneled facet.

Changing the focusing efficiencies of the proposed structures for the funnel angles was studied also. For the filling factor L/S=1.1 and funnel angle varying from 50°, 60°, 70°, to 80°, and fixing the 10 nm focusing width, the corresponding focusing efficiency is 18.1%, 19.2%, 18.2%, and 17%, respectively. When the funnel angle for the ratio L/S=1.2 was set to be 50°, 60°, 70°, and 80°, while the focusing width is fixed to be 10 nm, and the corresponding focusing efficiency turns to be 35.3%, 41.5%, 35.3% and 32%, respectively. When the focusing width is fixed to be 20 nm, the corresponding efficiency will be 71.2%, 80.4%, 71.2%, and 65%, respectively. It shows that the best focusing

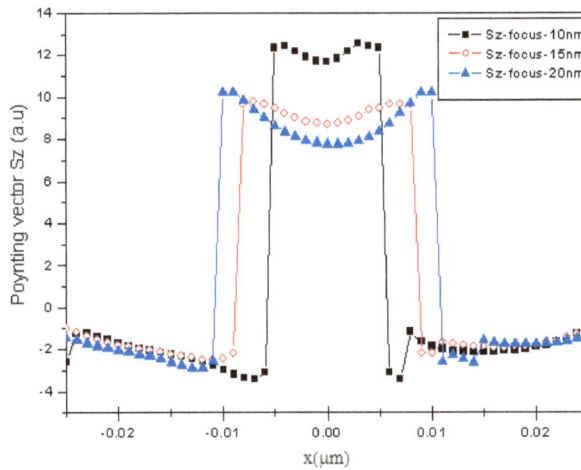

Fig. 10-34 Focusing intensity of S_z at the funneled facet with different focusing field widths.

is reached at around funnel angle of 60°. It is the reason that the changes to the smaller angles induce stronger coupling between the two arms in which the optical intensity is attenuated along the Z direction. The larger funnel angles result in higher E-field intensity distribution in the waveguide structure which decays the E-field intensity along the Z direction simultaneously.

Furthermore, we changed FWHM of the incident Gaussian beam from 50 nm to 100 nm, 150 nm, 200 nm, 250 nm, and 300 nm. The maximum focusing efficiency of our structures keeps around the value ranging from 75% to 82%. It shows that the diversifications of the transmissions are not apparent and significant. But it changes the absolute focusing intensity at the funneled facet.

These structures can focus the incident Gaussian beam (200 nm FWHM) to be the size of 10 nm, 15 nm, and 20 nm, which corresponds to the transmission efficiency of 38%, 40%, and 80%, respectively. These simulation results imply that the SP-based discrete array with the metal nanowire may make some nanophotonic devices such as high-intensity SP source and biophotonics into reality in the near future.

Another structure that can realize nanofocusing was theoretically reported. [48] SPPs propagating toward the tip of a tapered plasmonic waveguide are slowed down and asymptotically stopped when they tend to the tip apex, and never actually reach it (the travel time to the tip is logarithmically divergent). This phenomenon causes accumulation of energy and giant local fields at the tip. Focusing of fundamental cylindrical SPP wave is formed at apex of the taper tip, as shown in Fig. 10-35. Figure 10-36 displays the amplitudes of the local optical fields in the cross section of the system for the normal and longitudinal (with respect to the axis) components of the optical electric field. As SPP's move

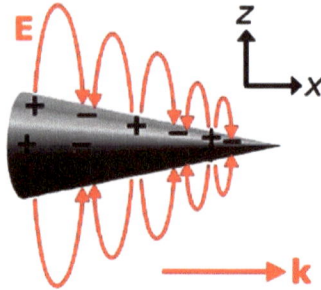

Fig. 10-35 Geometry of the nanoplasmonic waveguide. The radius of the waveguide gradually decreases from 50 to 2 nm.

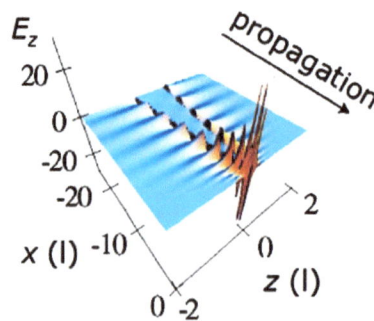

Fig. 10-36 Snapshot of instanteneous fields (at some arbitrary moment $t=0$): longitudinal component Ez of the local optical electric field are shown in the longitudinal cross section (xz) plane of the system. The fields are in the units of the far-zone (excitation) field. Reprinted with permission from "Mark I. Stockman, Phys. Rev. Lett. 93, 137404 (2004)" with copyright © 2004 of American Society of Physics.

toward the tip, the SPP wave starts to localize at the metal surface, and simultaneously, their wavelength is progressively reducing and amplitude growing. The field magnitudes grow significantly at small $|Z|$. The transverse x component grows by an order of magnitude as the SPP wave approaches the tip of the waveguide, while the longitudinal z component, which is very small far from the tip, grows relatively much stronger. The 3D energy concentration occurs at the tip of a smoothly tapered metal nanoplasmonic waveguide. This causes the local field increasing by 3 orders of magnitude in intensity and four orders in energy density.

10.4 Summary

A negative refractive index medium-based superlens and hyperlens for imaging at both near- and far-field regions was illustrated. They can realize imaging in resolution beyond the diffraction limit and even reach to sub-100 nm level (one sixth incident wavelength). Although a silver far-field superlens concept was demonstrated here, a more practical and better performing far field superlens microscope can be realized by further development of negative materials.

Producing flawless images has been a lens maker's aspiration for many decades. The far-field superlens optical imaging has great potential for many exciting applications in optical imaging, electronics manufacturing, and biomedical sensing.

The plasmonic structures for superfocusing were described in detail. They can be classified in terms of the tuning methods as: depth, width, period, and material (hybrid structure). Radial polarization illumination-based plasmonic structures for superfocusing were described also. Finally, the tapered plasmonic structure for nanofocusing was presented in this chapter. They provided some potential applications in data storage, photolithography, nanometrology, and communication etc.

It is reasonable to believe that the next generation optical system will be hybrid negative-positive refraction system. It will be extensively applied in industry, military and life science etc. With rapid development of metamaterials, the time of negative refractive index-dominated optics will come soon.

Reference

[1]　J. B. Pendry, "Negative refraction makes a perfect lens", Phys. Rev. Lett. 85, 3966-3969 (2000).

[2]　J.B. Pendry, "Time reversal and negative refraction", Science 322, 71-73 (2008).

[3]　Mathias Fink, "Time reversal of ultrasonic fields". II. Experimental results, Ultrasonics, Ferroelectrics and Frequency Control, IEEE Transactions on 39(5), 567 – 578 (1992).

[4]　David R. Smith, David Schurig, Marshall Rosenbluth, and Sheldon Schultz, S. Anantha Ramakrishna and John B. Pendry, "Limitations on subdiffraction imaging with a negative refractive index slab", Appl. Phys. Lett. 82, 1506-1508 (2003).

[5]　N. Fang, Z.W. Liu, T.J. Yen, X. Zhang, "Regenerating evanescent waves from a silver superlens", Opt. Express 11, 682-687 (2003).

[6]　N. Fang, X. Zhang, "Rapid growth of evanescent wave by a silver superlens", Appl. Phys. Lett. 82, 161-163 (2003).

[7]　N. Fang, H. Lee, C. Sun, X. Zhang, "Sub–Diffraction-Limited Optical Imaging with a Silver Superlens", Science 308, 534-537 (2005).

[8]　Zhaowei Liu, Nicholas Fang, Ta-Jen Yen, and Xiang Zhang, "Rapid growth of evanescent wave by a silver superlens," Appl. Phys. Lett. 83, 5184-5186 (2003).

[9]　Zhaowei Liu, Ste´phane Durant, Hyesog Lee, Yuri Pikus, Nicolas Fang, Yi Xiong, Cheng Sun, and Xiang Zhang, "Far-field optical superlens". Nano Lett. 7, 403-408 (2007).

[10]　Zhaowei Liu, Hyesog Lee, Yi Xiong, Cheng Sun, Xiang Zhang, "Optical hyperlens: far-field imaging beyond diffraction limit", Science 315, 1686 (2007).

[11]　V. A. Podolskiy, and E. E. Narimanov, "Strongly anisotropic waveguide as a nonmagnetic left-handed system," Phys. Rev. B 71, 201101 (2005).

[12]　A. A. Govyadinov and V. A. Podolskiy, "Meta-material photonic funnels for sub-diffraction light compression and propagation," Phys. Rev. B **73**(15), 155108 (2006).

[13]　Zubin Jacob, Leonid V. Alekseyev and Evgenii Narimanov, "*Optical Hyperlens*: Far-field imaging beyond the diffraction limit", Opt. Express 14, 8247-8256 (2006).

[14]　Jie Yao, Zhaowei Liu, Yongmin Liu, Yuan Wang, Cheng Sun, Guy Bartal, Angelica M. Stacy, Xiang Zhang, "Optical Negative Refraction in Bulk Metamaterials of Nanowires", Science 321, 930 (2008).

[15]　Jiajun Chen, Cosmin Radu, and Ashok Puri, "Aberration-free negative-refractive-index lens", Appl. Phys. Lett. 88, 071119 (2006).

[16]　D. Schurig and D. R. Smith, "Negative index lens aberrations", Phys. Rev. E 70, 065601(R) (2004).

[17]　Haofei Shi, Chunlei Du, and Xiangang Luo, "Focal length modulation based on a metallic slit surrounded with grooves in curved depths". Appl. Phys. Lett. 91, 093111 (2007).

[18]　Baohua Jia, Haofei Shi, Jiafang Li, Yongqi Fu, Chunlei Du, and Min Gu, "Near-field visualization of focal depth modulation by step corrugated plasmonic slits", Appl. Phys. Lett. 94, 151912 (2009).

[19]　Haofei Shi, Changtao Wang, Chunlei Du, Xiangang Luo, Xiaochun Dong, Hongtao Gao, "Beam manipulating by metallic

nano-slits with variant widths", Opt. Express 13, 6815-6820 (2005).

[20] Ting Xu, Cunlei Du, Changtao Wang, and Xiangang Luo, "Subwavelength imaging by metallic slab lens with nanoslits", Appl. Phys. Lett. 91, 201501 (2007).

[21] Changtao Wang, Chunlei Du, Yueguang Lv, Xiangang Luo, "Surface electromagnetic wave excitation and diffraction by subwavelength slit with periodically patterned metallic grooves". Opt. Express 14, 5671-5681 (2006).

[22] Yongqi Fu, Wei Zhou, Lim Enk Ng Lennie, Chunlei Du, Xiangang Luo, "Plasmonic microzone plate: superfocusing at visible regime," Appl. Phys. Lett. 91(6), 061124 (6 Aug. 2007).

[23] Yongqi Fu, Wei Zhou, Lim Enk Ng Lennie, "Propagation properties of a plasmonic micro-zone plate in near-field region". Journal of Optical Society of America A 25(1), 238-249 (2008).

[24] Zhaowei Liu, Jennifer M. Steele, Werayut Srituravanich, Yuri Pikus, Cheng Sun, and Xiang Zhang, "Focusing Surface Plasmons with a Plasmonic Lens", Nano Lett. 5, 1726-1729 (2005).

[25] Yongqi Fu, Yu Liu, Xiuli Zhou, Shaoli Zhu, Experimental demonstration of focusing and lasing of plasmonic lens with chirped circular slits, Opt. Express 18 (4), 3438–3443 (2010).

[26] Yongqi Fu, Xiuli Zhou, Yu Liu, Ultra-enhanced lasing effect of plasmonic lens structured with elliptical nano-pinholes distributed in variant period. Plasmonics, 5 (1), (2010). (in press).

[27] Yongqi Fu, Wei Zhou, Lim Enk Ng Lennie, "Nano-pinhole-based optical superlens", Research Letter in Physics 2008, 148085 (2008).

[28] Yongqi Fu, Wei Zhou, "Hybrid Au-Ag subwavelength metallic zone plate-like structures for superfocusing", Journal of Nanophotonics, Vol. 3, 033504 (22 June 2009).

[29] Yongqi Fu, Wei Zhou, "Modulation of main lobe for superfocusing using subwavelength metallic heterostructures", Plasmoncs, 4(2), 141-146 (2009).

[30] Yongqi Fu, Wei Zhou, Lennie E.N. Lim, Chunlei Du, Haofei Shi, Changtao Wang, "Geometrical characterization issues of plasmonic nanostructures with depth-tuned grooves for beam shaping". Opt. Eng. 45 (10), 108001 (2006).

[31] Yongqi Fu, Wei Zhou, Lim Enk Ng Lennie, "Influences of V-shaped plasmonic nanostructures on transmission properties". Appl. Phys. B, Vol.86, No.3, 461-466 (2007).

[32] Yongqi Fu, Wei Zhou, Lennie E N Lim, "A practical V-shaped nano-aperture flanked with surface corrugations for beam focusing". J. Comp. Theor. Nanosc. 4, 614-618 (2007).

[33] Yongqi Fu, Ngoi Kok Ann Bryan, "Investigation of physical properties of quartz via focused ion beam Bombardment". Appl. Phys. B 80, 581-585 (2005).

[34] Yongqi Fu, Ngoi Kok Ann Bryan, "Investigation of physical properties of quartz via focused ion beam Bombardment". Appl. Phys. B, 80(4), 581-585 (2005).

[35] Qiwen Zhan, "Cylindrical vector beams: from mathematical concepts to applications", Adv. Opt. Photon. 1, 1-57 (2009).

[36] D. Pohl, "Operation of a Ruby laser in the purely transverse electric mode TE01," Appl. Phys. Lett. 20, 266–267 (1972).

[37] Y. Mushiake, K. Matzumurra, and N. Nakajima, "Generation of radially polarized optical beam mode by laser oscillation," Proc. IEEE 60, 1107–1109 (1972).

[38] S. Quabis, R. Dorn, M. Eberler, O. Glöckl, and G. Leuchs, "Focusing light into a tighter spot," Opt. Commun. 179, 1–7 (2000).

[39] W. Chen and Q. Zhan, "Optimal plasmonic focusing with radial polarization," Proc. SPIE 6450, 64500D (2007).

[40] H. Kano, S. Mizuguchi, and S. Kawata, "Excitation of surface-plasmon polaritons by a focused laser beam," J. Opt. Soc. Am. B 15, 1381–1386 (1998).

[41] Q. Zhan, "Evanescent Bessel beam generation via surface plasmon resonance by radially polarized beam," Opt. Lett. 31, 1726–1728 (2006).

[42] W. Chen and Q. Zhan, "Realization of evanescent Bessel beam via surface plasmon interference excited by radially polarized beam" Opt. Lett. (to be published).

[43] A. Bouhelier, F. Ignatovich, A. Bruyant, C. Huang, G. Colas des Francs, J.-C.Weeber, A. Dereux, G. P.Wiederrecht, and L. Novotny, "Surface plasmon interference excited by tightly focused laser beams," Opt. Lett. 32, 2535–2537 (2007).

[44] A. Bouhelier, J. Renger, M. R. Beversluis, and L. Novotny, "Plasmon-coupled tip-enhanced near-field optical microscopy," J.

Microsc. 210, 220–224 (2003).

[45] W. Chen and Q. Zhan, "Numerical study of an apertureless near field scanning optical microscope probe under radial polarization illumination," Opt. Express 15, 4106–4111 (2007).

[46] W. Chen and Q. Zhan, "Field enhancement analysis of an apertureless near field scanning optical microscope probe with finite element method," Chin. Opt. Lett. 5, 709–711 (2007).

[47] Zhou Xiu-Li, Yongqi Fu, Wang Shi-Yong, Peng An-Jing, Cai Zhong-Heng, "Funnel-shaped arrays of metal nano-cylinders for nano-focusing", Chin. Phys. Lett. 25 (9), 3296-3299 (2008).

[48] Mark I. Stockman, "Nanofocusing of Optical Energy in Tapered PlasmonicWaveguides", Phys. Rev. Lett. 93, 137404 (2004).

11. METALLIC NANOPARTICLES ARRAY FOR BIOSENSING

Abstract: Design considerations for the metallic nanoparticles for biosensing were introduced firstly. Then the commonly used nanofabrication approaches were presented. After that, one important application, localized surface plasmon polritons (LSPR)-based immunoassay using the nanoparticles array was illustrated. Finally, the LSPR-related photothermal therapy was briefly described.

Metallic nanoparticles array being used as sensors for biosensing is a major application of nanophotonic devices, especially for the localized surface plasmon resonant (LSPR) effect-based nano-biosensors. It is an important approach for immunoassay. Numerous biosamples can be detected by means of the LSPR-based nano-biosensors such as biotin,[1] amyloid-derived diffusible ligands,[2] and Staphylococcus aureus enterotoxin B etc.[3] It becomes appealing for biological researchers due to its advantages of portable, cost effective, light weight, small volume, and simple system in comparison to the conventional SPR system, *e.g.*, Bicore system.

11.1 Brief introduction of design methods

11.1.1 Discrete dipole approximation algorithm-based calculation

Discrete dipole approximation (DDA) is one of the most efficient numerical methods for nanoparticles with arbitrary shapes. DDA can calculate the absorbing and scattering of the particles with arbitrary shape and dimension. This method has advantages of possessing less calculated resources, calculating the mutual action between the light and the metal nanoparticles with arbitrary shapes. It expresses the target particles as an array of point dipoles, with the fields of these dipoles determined self-consistently. Any particles can be divided into a large number of polarizable cubes (point dipoles), and the electromagnetic scattering problem can be solved essentially and exactly as long as the cubes are small enough and subject to a model for the cube polarizability.

In the present applications using an extended DDA, we modeled the rhombic structure of the particles accurately. This work aimed the problems of recent experiments that the parameters of the rhombic sliver nanoparticles arrays could not be decided only by the experiment facture. We will show that the results are in quantitative agreement to that of the experiments. The theory for determining the parameters in the fabrication of the rhombic silver nanoparticles array was provided. The DDA program DDSCAT 6.1 [4] was firstly reported by Draine and Flatau. The dielectric constants for silver were cited from reference [5]. In order to ensure that the effect of the substrate is included in the result, we considered it as the effective index of medium in calculation, which is 1.2 [4] and the wavelength values in the parameter file will be changed as well. The size of the dipole is 2 nm which can promise the perfect convergence of the calculated results and the higher calculated efficiency.

The initial DDA simulation parameters can be determined by experiments firstly. As can be seen in Fig. 11-1, the Ag nanorhombus has in-plane width of ~140 nm and out-of-plane height of ~40 nm. The angle of the arris and

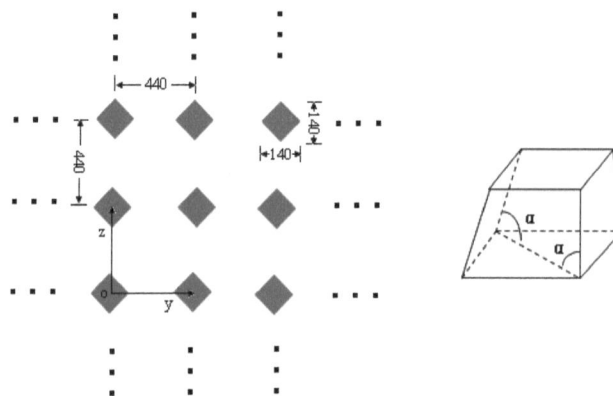

Fig.11-1 Geometrical parameters and shape of the metallic nanoparticles for DDA 3D simulation.

underside is 60^0 and the period of the Ag nanorhombus array is 440 nm. Considering that the index of air and glass substrates is a uniform index, the refraction index around the Ag nanorhombus array used in the simulation was determined. This Ag nanorhombus lies at y-z plane and the direction of the incident light is in x-axis. The polarization of the incident light is in y-axis.

In computational calculations, for the same period and in-plane widths with different out-of-plane heights, the structures can be fabricated by depositing metal with different thicknesses through the masks using thermal or electron beam evaporation. The calculations were done for the particles arranged in rhombic structures. However, there are different thicknesses for the fabrication. Thus an important issue is that the influence of this different thicknesses on the results. To study this effect, we have calculated the spectrums of different thicknesses. In these calculations, each particle is displaced relative to its position on a perfect rhombic array. Thicknesses of the rhombic structures used in the calculations are 25 nm, 30 nm, 35 nm, 40 nm, 45 nm, 50 nm, 55 nm, and 60 nm respectively.

Figure 11-2 (a) shows the results of the DDA simulation, whereas (1) presents the extinction spectrum of the DDA simulation which thickness is equivalent to 25 nm; (2) corresponds to the extinction spectrum when the thickness is changed to be 30 nm; and (3) shows the extinction spectrum for Ag nanoparticles which thickness is 35 nm and the extinction spectrums with the thicknesses of 40 nm, 45 nm, 50 nm, 55 nm and 60 nm are illustrated by (4), (5), (6), (7), and (8) respectively. The positions of the peaks are located in 579.84 nm, 589.85 nm, 599.86 nm, 609.88 nm, 619.89 nm, and 619.89 nm, respectively. The relationship between the peak wavelength and thicknesses of the rhombic silver particles was shown in Fig. 11-2 (b) which shows the effect of thicknesses to the peak positions. The relationship between the extinction efficiency and thickness of the rhombic silver particles was shown in Fig.11-2 (c). Considering FWHM of the wavelength and the usage for the biochip which mainly concerns the redshift of the peak wavelengths of nanostructure, the acceptable thickness is ranging from 35 nm to 45 nm for the experiments. The preferable thickness for fabrication is 40 nm.

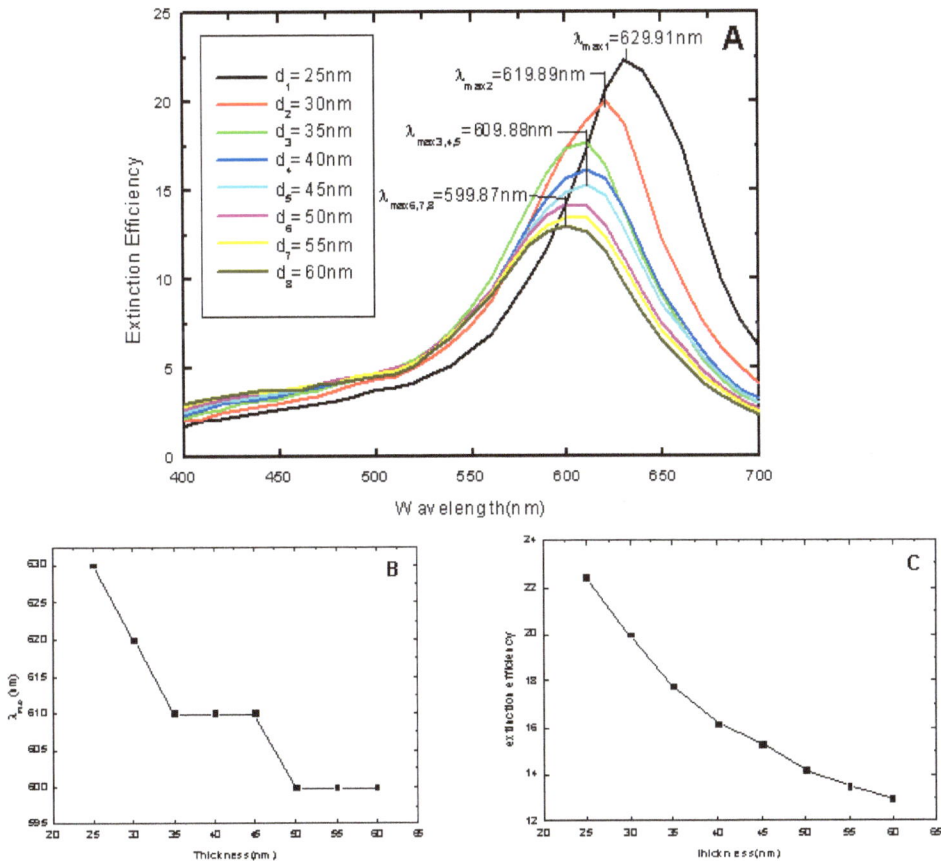

Fig.11-2 (a) Extinction spectra of rhombic silver particle array for different thickness. (b) Relationship between the peaks of wavelength and the thicknesses of rhombic silver particles. (c) Relationship between the extinction efficiency of wavelength and the thicknesses of rhombic silver particles.

The diverse diameters of the polystyrene spheres (PS) were used to adjust the period of the rhombic silver particles. The correct direction for facture of the rhombic silver particles array was offered by calculating the effect of the periods. The periods of rhombic structure are 340 nm, 380 nm, 420 nm, and 460 nm, respectively. The peak position is located in 588 nm, 597 nm, 604 nm, and 608 nm, respectively. When the periods of the rhombic silver particles are different, the peak positions are changed also. Figure 11-3 (a) shows the resulting spectra, which seem very similar to each other. However there are two significant differences among them. Firstly, the peak heights of the small periods are always higher than those of the large periods. It is easy to be understood as coherence of the dipole sums is expected to be the same area. The second effect is that the peak positions are slightly red shifted when the periods become larger. Relationship between the peak wavelengths and period of the rhombic silver particles is shown in Fig. 11-3 (b), where the first point stands for the single particle when the period is zero. The relationship between the extinction efficiency and period of the rhombic silver particles is shown in Fig. 11-3 (c). Considering the FWHM and the intensity of the extinction efficiency, the effective periods for fabrication are ranging from 350 nm to 560 nm. Here, we selected the period of 440 nm for fabrication use. The detailed experimental results of fabrication and biosensing can be seen from Ref. [6-8].

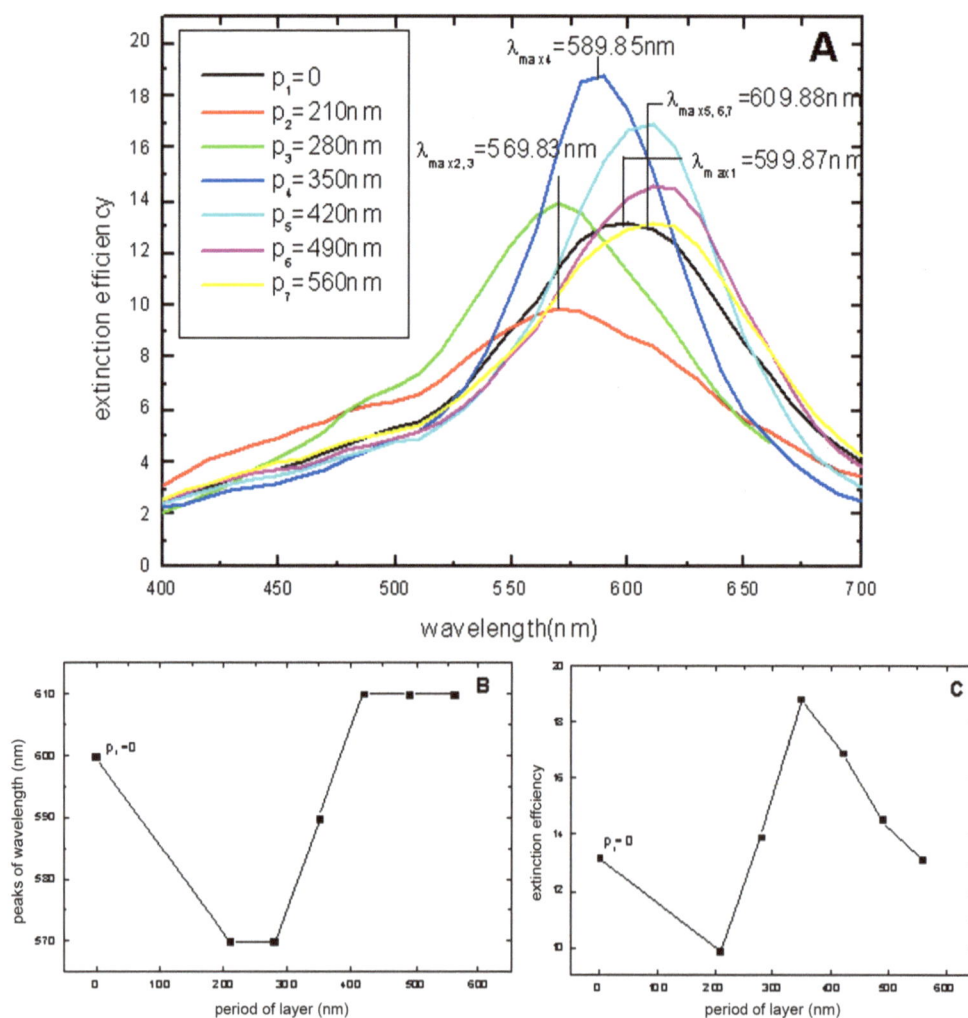

Fig.11-3 (a) Extinction spectra of rhombic silver particle array for different periods. (b) Relationship between the peaks of wavelength and the periods of rhombic silver particles. (c) Relationship between the extinction efficiency of wavelength and the periods of rhombic silver particles.

11.1.2 Finite-difference and time-domain algorithm-based design

Commercial professional software was adopted here for the computational calculations and numerical analyses. The used metal here is Ag with $\varepsilon_m = -17.24 + i0.498$, and $\varepsilon_d = 1.243$ for SiO_2 at $\lambda = 633$ nm. Broad band of the incident light is ranging from 400 nm to 750 nm with plane wave in normal incidence angle $\theta=0°$. Meshing size in x and y (two-dimensional simulation) is Δx=2 nm and Δy=2 nm, respectively. Simulation time t (theoretically, $t=\Delta x/2c$, c is the velocity of light) is set to be 125 fs. The output result is extinction spectrum.

Here we set hybrid Au-Ag nanoparticles as a design example to illustrate the FDTD-based design approach [9]. Side-effects of the pure Ag nanoparticles are oxidation and sulfuration (further study is required to reveal that which process is more significant.) which will change dielectric constant of the Ag particles, and transmission property will degrade accordingly. To solve this problem, we put forth a new structure here, a hybrid Au-Ag metallic nanoparticles array with Au thin film covered on the Ag film surface. Therefore, the Au film acts as both a protector and modulator here. On the one hand, it can protect Ag film surface from oxidation. On the other hand, it can modulate the transmission property of the nanosensors so as to achieve higher detection sensitivity. Using FDTD algorithm, we designed the extinction spectra and the corresponding extinction spectra of the hybrid nanoparticles.

The hybrid Au-Ag triangular nanoparticles were proposed as a sensitive cell of the LSPR-based nano-biosensor. [9] Using FDTD algorithm, we designed and calculated the extinction spectra as well as the corresponding electric field of the hybrid nanoparticles array. Three-dimensional geometrical model of the hybrid Au-Ag triangular nanoparticles array is shown in Fig. 11-4.

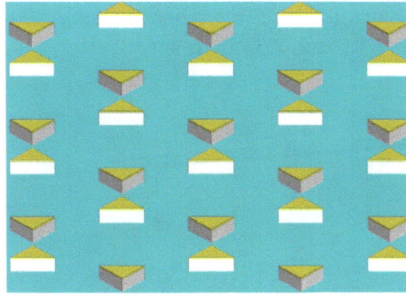

Fig. 11-4 Schematic diagram of three-dimensional geometrical model of the triangular hybrid Au-Ag nanoparticles

Out-of plane height of the silver nanoparticles is 50 nm and the upper thickness of Au nanoparticles is 5 nm, and the in-plane widths of each nanoparticles is 100 nm. The period of the nanoparticle array is 400 nm, and wavelength of the incidence white light source ranging from 400 nm to 700 nm was adopted for spectrum measurement. The incidence light beam is projected in perpendicular to the substrate. In order to investigate the transmission property of different refractive index of the media surrounding this hybrid nanoparticles, we selected the media of air (n_1=1.0) and Protein A (Protein A: PBS (0.01 M, pH 7.4) =1:100, n_2=1.3352) surrounding the nanoparticles. When the refractive indexes of the surrounding media are 1.0 and 1.3352, the computational results using FDTD method were shown in Fig. 11-5. From the result, we can obtain sensitivity of the hybrid Au-Ag triangular nanoparticles by the simple calculation as $S=(\lambda_{max1}-\lambda_{max2})/(n_2-n_1)=(551.638-484.513)/(1.3352-1.0)\approx200$ nm/RIU. Thus it is reasonable to believe that the hybrid nanoparticles can realize the higher sensitivity detection of bio-molecule.

In addition, we calculated the electric fields when wavelength of the incident light equals to the LSPR wavelength λ_{SPP}. The electric fields can be expressed as $\mathbf{E}^2=\mathbf{E}_x^2+\mathbf{E}_y^2+\mathbf{E}_z^2$. Using the FDTD algorithm, we obtained the electric field $\mathbf{E}$ when the incident light wavelength is λ_{in}=551.638 nm (refractive index of the medium surrounding this hybrid nanoparticles is n_d=1.0). The electric field is also calculated when λ_{in}=484.513 nm (refractive index of the medium surrounding this hybrid nanoparticles is n_d=1.3352). Figure 11-5 shows the calculated results of the electric field for the case of $\lambda_{in}=\lambda_{SPP}$. It can be seen from Figs. 11-6 (a) and (b) that when the refractive index of the medium surrounding the hybrid nanoparticles is n_d=1.0, the total electric field $\mathbf{E}$ from Au film in 5 nm thickness changed tremendously as well. These results indicate that the hybrid Au-Ag triangular nanoparticles can function as a platform for detecting the bio-molecule such as Protein A.

A typical protein can directly bind to the surface of Au film. Using our developed LSPR-based nano-biosensor with the hybrid Au-Ag nanoparticles, we can realize the refractive index sensitivity of 200 nm/RIU. The nano-biosensors demonstrated the potential applications in monitoring, detection and identification of biological agents, as well as characterization of intermolecular interactions. [9]

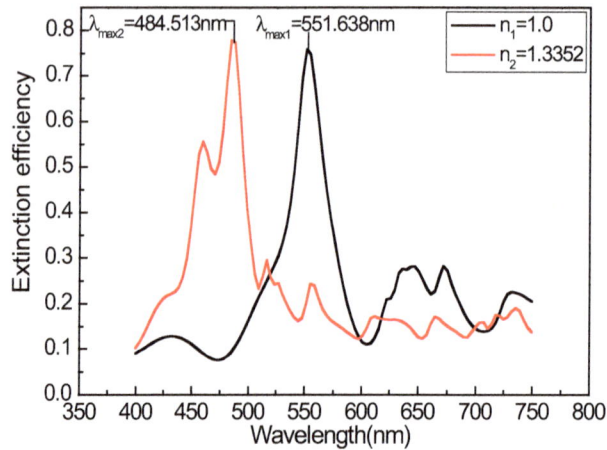

Fig. 11-5 FDTD solution calculation results when the refractive index medium surrounding this hybrid nanoparticles are 1.0 and 1.3352, respectively.

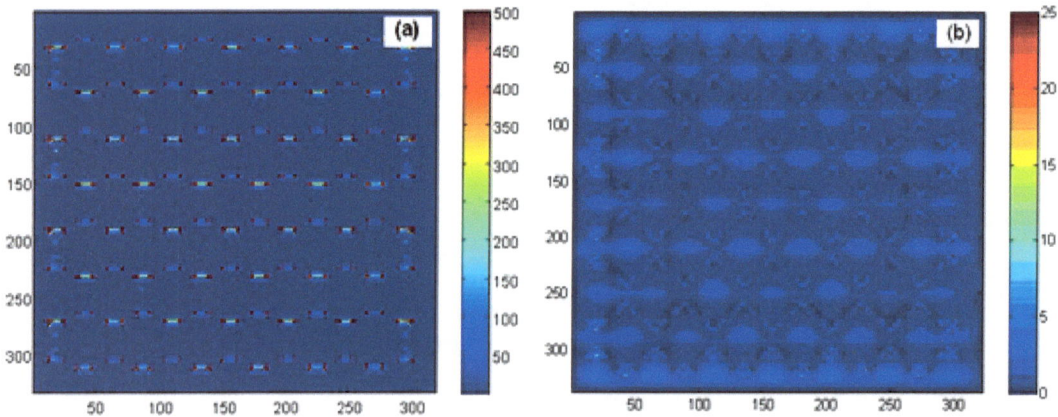

Fig. 11-6 FDTD solution calculated electric field result when the incidence light wavelength is equal to the LSPR wavelength. (a) Total electric field $\mathbf{E}^2 = \mathbf{E}_x^2 + \mathbf{E}_y^2 + \mathbf{E}_z^2$ from Au surface 5 nm; n=1.0, $\lambda_{in} = \lambda_{max} = 551.638$ nm; (b) Total electric field $\mathbf{E}^2 = \mathbf{E}_x^2 + \mathbf{E}_y^2 + \mathbf{E}_z^2$ from Au surface 5 nm; n=1.3352, $\lambda_{in} = \lambda_{max} = 484.513$ nm.

11.1.3 The multiple multipole program algorithm-based design

The multiple multipole program (MMP) is a semi-analytic method for numerical field computations that has been applied to electromagnetic fields and acoustics.[10,11] Essentially, the field is expanded by a series of basis fields. Each of the basis field is an analytic solution of the field equations within a homogeneous domain. The amplitudes of the basic fields are computed by a generalized point matching technique that is efficient, accurate, and robust. The analyses of metallic nanoparticles with different shapes such as spherical and triangle-shaped nanoparticles can be performed using the semi-analytical MMP method, which provides complex eigen-value estimations as well as proper error measurements. It was carried out especially for coupled-particle configurations with different inter-particle distances and under various illumination conditions while controlling error measures highly-accurate computations. For more information regarding how to design the metallic nanoparticles, please read the textbook in Ref. 12. Some design examples can be found in this book.

11.2 Nanofabrication

Considering integrality of this chapter, a brief introduction of commonly used three fabrication techniques: focused ion beam milling, self-assembly monolayer, and laser interference lithography, was provided in following subsections although they were illustrated in Chapter 6 already.

11.2.1 Focused ion beam technique

Focused ion beam milling (FIBM) is a point-to-point scanning technique. Material sputtering effect can be formed while high energy of accelerated Ga^+ bombs the sample surface, and thus the material removal can be realized. Therefore, it is a pure physical process. Pattern transfer from photoresist into substrate for conventional photolithography and e-beam writing techniques is not necessary for the FIBM technique. Moreover, there is no selectivity of substrate material. With development of the FIBM, some extra functions were invented such as chemical gas assistant etching and deposition. With these functions, some fine structuring processes at local region can be realized by means of the FIBM, *e.g.*, fine line joining, nanohole drilling, and nanoparticles forming *etc*. [13] It is a maskless technique. There is no shape limitation for the FIBM process because any two-dimensional patterns can be designed by users in advance. And then the users call the patterns into the defined milling windows. For current commercial FIB machines, the feature size as small as several tens nanometers can be realized. However, it has drawbacks of high expenditure, slow speed, and small fabrication area. Therefore, it is more suitable to be used for fabrication of prototype and master piece.

Fig. 11-7 FIB fabricated Ag nanoparticles array with period of 170 nm.

Figure 11-7 is a SEM micrograph of a five-star patterned Au nanoparticles array fabricated using the FIB direct milling technique. Figure 11-8 (a)-(d) is another FIB fabrication example for therapy of cancer cell. It is a hybrid plasmonic-photonic device for label-free detection of a single molecule. [14] Specifically, this novel technique combines AFM with topographic resolution below 10 nm, and chemical information with spatial resolution below 10 nm. This means that it will be possible to obtain topographic and label-free spectroscopic information directly from a living cell in the near future. Design of the nano-optical device developed by the team in Italy allowed its fabrication by an AFM cantilever. They actually had already fabricated such a prototype and demonstrated its detection capabilities on silicon oxide nanoparticles, on a single cadmium/selenium quantum dot, and a monolayer of organic compounds. Moreover, the optical detection configuration is in far field region, which means that we use a standard microscope for the purpose of detecting a Raman spectroscopic signature of the molecule. The nanoholes were drilled using the FIB milling, and then grew the sharp tip at center of the photonic crystal with material of SiN_3 using the FIB direct deposition. Deposited material is Pt. This is a typical fabrication example of the FIB deposition function.

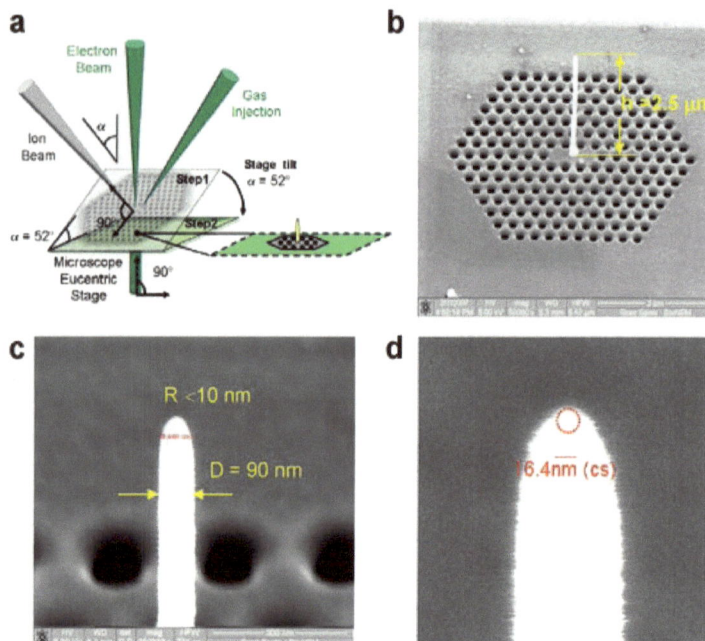

Fig.11-8 Fabrication setup and SENSe (surface-enhanced nanosensor) device. (a) Scheme of the dual beam fabrication system based on focused ion milling (step 1) and electron beam induced deposition (step 2) with a gas injection system. (b) SEM image of the whole device including the photonic cavity and, at its center, the plasmonic nanoantenna. The latter is 2.5 μm high, and its size gradually decreases from 90 nm in diameter at bottom down to 10 nm radius of curvature at the tip. (c, d) SEM details of the nanoantenna tip and its radius of curvature. Reprinted with permission from Di Fabrizio, Nano Lett. 8, 827-832 (2007) with copyright © 2007 of American Chemical Society.

11.2.2 Self-assembly monolayer

Self-assembly monolayer (SAM) also named nano-spherical lithography (NSL), is the surface consisting of a single layer of molecules on a substrate. Rather than having to use a technique such as chemical vapor deposition or molecular beam epitaxy to add molecules to a surface (often with a poor control over the thickness of the molecular layer), self-assembled monolayers can be prepared simply by adding a solution of the desired molecules onto the substrate surface and washing off the excess.

As an example, fabrication of the rhombic Ag particles using SAM was introduced here. SAM was employed to create the surface-confined rhombic Ag nanoparticles which are supported on a glass substrate.[1,4] This method was developed on the basis of the NSL technique.[14] Firstly, the glass substrate was cleaned in a piranha solution (1:3 30% H_2O_2/H_2SO_4) at 80℃ for 30 min., and then cooled by high-pressure N_2 gas. Once cooled, the glass substrates were rinsed with copious amounts of second distilled water and then sonicated for 60 min. in 5:1:1 $H_2O/NH_4OH/30\%$ H_2O_2. Next, the single-layer of size-monodispersed (the sphere size of the chemical solutions which was spin-coated as a monolayer determines the generated rhombic particle size.) polystyrene nanospheres (PS, 500 nm, 2%) and glass nanospheres (GS, 200 nm, 1%) with fluorocarbon surfactant (FSO) (100:1000:1 PS/GS /FSO) solution ~10 μl were coated onto the glass substrate so as to form a deposition mask, and then followed by hydrofluoric acid etching to remove the glass nanospheres. After that, the Ag metal thin film was deposited through the nanospheres masks using thermal evaporation or electron beam evaporation. After removal of the polystyrene nanospheres by sonication in absolute ethanol for 3 min., well ordered two-dimensional (2D) rhombic Ag nanoparticles array was finally obtained on the substrates. By modulation of the nanosphere diameter and the deposited Ag film thickness, the nanoparticles with different in-plane widths, out-of-plane heights, and interstructure spaces can be tuned. Figures 11-9 (a) and (b) are SEM images of the Ag nanoparticles array with shapes of triangular and rhombic, respectively.

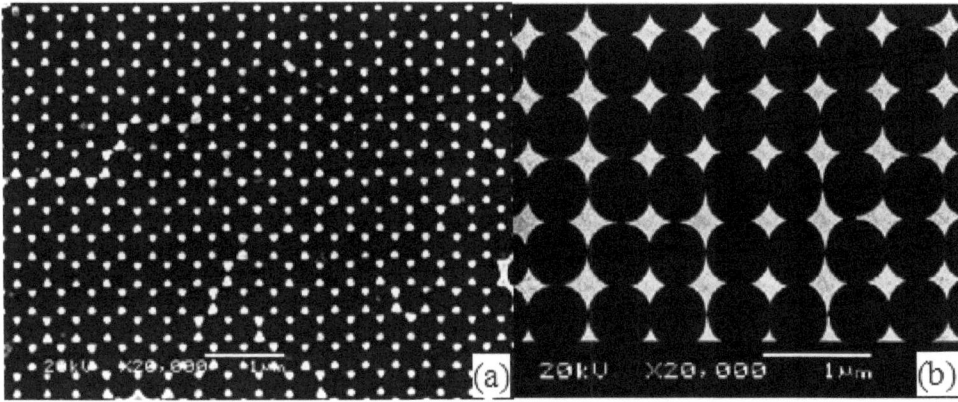

Fig.11-9 Representative SEM image of (a) Ag triangular nanoparticles array; (b) the Ag rhombic nanoparticles array fabricated using SAM technique.

11.2.3 Laser interference lithography

A Lloyd's mirror interferometer system was built, as shown in Fig. 6-8. [15] Light source is He-Cd laser with 442 nm working wavelength. The laser beam was filtered and expanded by a spatial filter which is composed of Lens 1 and a pinhole. The expanded beam is collimated by Lens 2. A part of the incident beam is reflected back by the mirror which is positioned in normal to the substrate and interferes with the other non-reflection beam to form the interference patterns. Since the beam is only split with a short path length near the substrate, this setup is very insensitive to the mechanical vibration caused instabilities. Therefore, no extra feedback control system is required to stabilize the interference fringe patterns. [16]

As a fabrication example, the formation of particles in photoresist spin-coated on quartz substrate was presented. Firstly, the quartz substrate was dipped in nitric acid solution for ~6 h. Then it was cleaned by ultrasonic vibrations and acetone to remove the dust attached on surface of the substrate. We used an oven baking the substrate for half an hour at 150 °C for the purpose of removing the solvent absorbed on the surface. After that, an etched mask layer of Cr with ~10 nm in thickness was deposited in the front side of the cleaned quartz. The Cr thin layer coated on the substrate was prepared for the purpose of enhancing adhesion between photoresist and substrate, and can also be used as a protection layer for the dry etching in next step. On top surface of this Cr layer, a layer of positive resist (AR-P3170, Allresist Co.) with a thickness of 100 nm was spin-coated, and followed by pre-baking time of 20 min. at 95 °C. The dimension of the exposed area strongly depends on the exposure dose. Because of the cosine instead of rectangular distribution of the exposure intensity, energy at wings or tails of the beam profile still has contribution to the exposure process. The wing energy causes line broadening at edge of the dots, and thus makes the dot dimensions enlarged to a certain extent. Generation of the structures with high aspect ratio will be limited due to this broadening effect accordingly. For compensation, the corrected exposure time should be slightly shorter than the normal value. The photoresist was exposed by the collimated beam from the Lloyd's mirror interferometer system. A 1D grating pattern of the photoresist layer was formed after the first time exposure and development. Exposure dose was measured to be 1.5 mW/cm^2 in normal incidence. The developer adopted in our experiments is AR 300-35 (Allresist Co.).

The photoresist was exposed by the collimated beam generated from the Lloyd's mirror interferometer system. A 1D grating pattern of the photoresist layer was formed after the first time exposure and development. Exposure dose was measured in normal incidence to be 1.5 mW/cm^2. The developer adopted in the experiments is AR 300-35 (Allresist Co.). AFM measurement results after the exposure can be seen in Fig. 11-10 (a). In order to obtain a 2D grid pattern with the dots array in different sizes on the resist, second exposure with varied exposure time can be performed. The diameter of the holes in the photoresist depends on the exposure time. The dependence gives a possible tool to tune the microhole size at a constant space. Corresponding AFM micrograph was shown in Fig. 11-10 (b) and (c), respectively. In order to analyze the pattern period, an AFM measurement was performed, which shows that the period is ~400 nm.

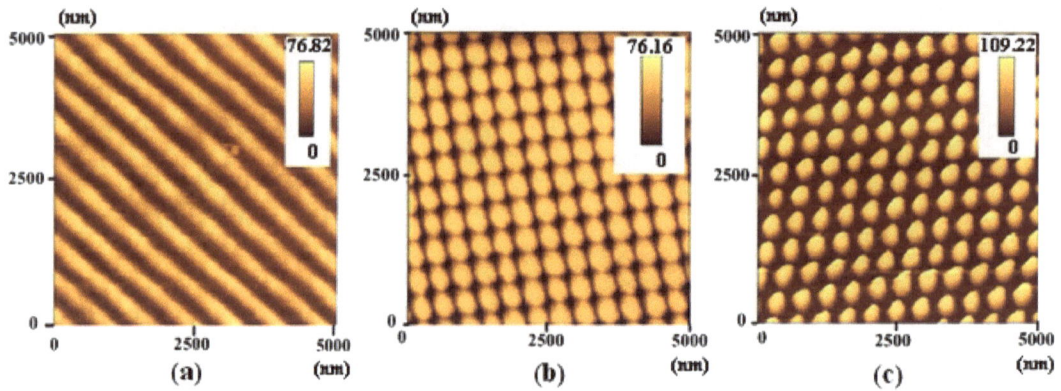

Fig. 11-10 AFM micrograph of the photoresist layer that remains after exposure and development in the laser interference setup (a) 1D grating structure after first exposure, (b) 2D grid pattern after second exposure with short exposure time of 8 s., (c) 2D dots array after second exposure with longer exposure time of 13 s.

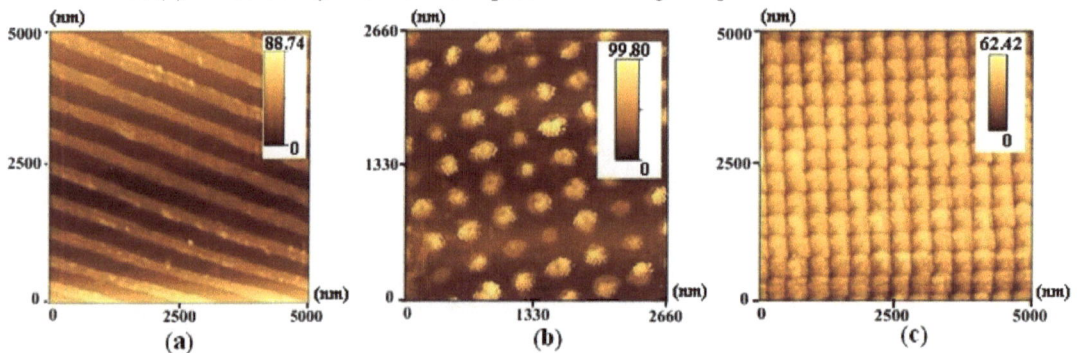

Fig. 11-11 (a) Cr gratings with wet etching method, (b) Cr dots array with wet etching method, (c) pattern of the quartz substrate with reactive ion etching technique.

The Cr patterns were obtained by the wet etching in a mixture of ceric ammonium nitrate, perchloric acid and distilled water, as shown in Fig. 11-10 (a) and (b). Subsequently, the Cr pattern was transferred to the quartz by means of CHF_3/O_2 reactive ion etching at a flow rate of 35 sccm: 1 sccm, and the etching time is 8 min. After the remained Cr layer was removed, the AFM was employed to view the samples (see Fig. 11-11(c)).

11.3 LSPR-based immunoassay using metallic nanoparticles array [17, 18]

To observe the sensor characteristics of this type of the Au nanoparticles-based biosensors, extinction spectroscopy was adopted by a Sciencetech 9055 spectrophotometer and the schematic diagram is shown in Fig. 11-12. To obtain ultraviolet-visible transmittance spectra, white light beam (400 nm~700 nm) propagates through a multimode optical fiber and firstly reaches a collimating lens, as denoted in Fig. 11-13, and then illuminates the biosensor. The transmitted light signal through the sample was collected with an identical focus lens attached to the multimode fiber. A monochromator was used to separate the light from the multimode fiber into monochromatic lights and sent them to a personal computer which was integrated with an analog photomultiplier. The transmittance spectra curves were directly shown on the screen of the computer. The measuring procedure can be divided into three steps: 1) collecting the background light without any input sources and samples; 2) measuring the incident light; and 3) collecting the transmitted light with presence of the sample placed perpendicular to the incident light.   The transmittance T can be written as follows: $T=(s-b)/(r-b)$, where s, r, and b denote the intensity of the sample, reference/light source and background, respectively. The extinction spectra E for each step were achieved by $E= -\log T$. In addition, the transmittance spectra can be directly plotted on the screen of the computer without further data treatment, which makes the test simpler.

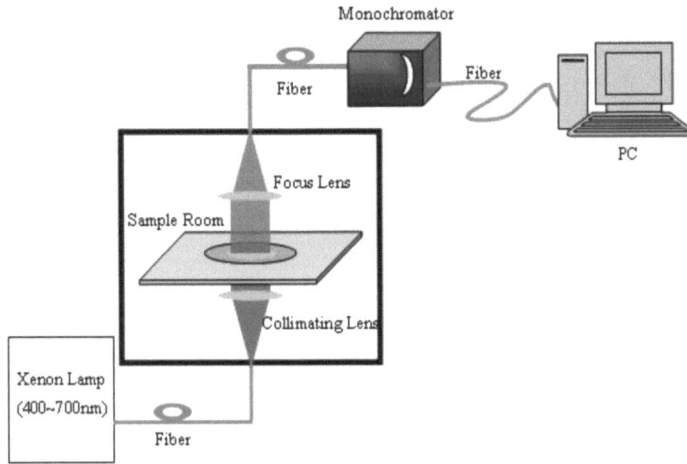

Fig. 11-12 The schematic diagram of the Sciencetech 9055 spectrophotometer used to obtain the extinction of Ag nanoparticles

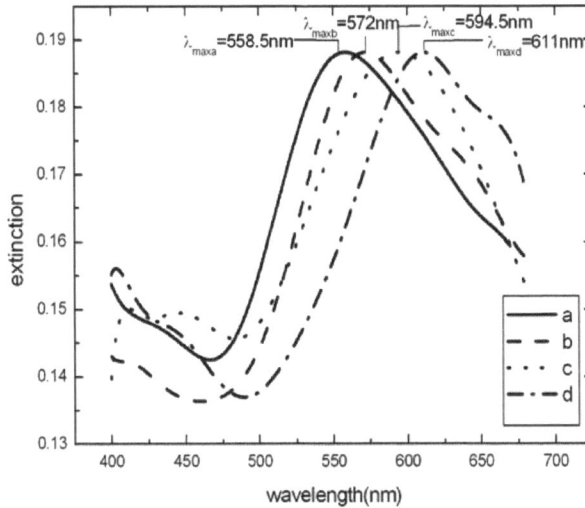

Fig. 11-13 LSPR spectra of each step in the surface modification of NSL-derived rhombic Ag nanoparticles to form a biotinylated Ag nanobiosensor and the specific binding of SA. (a) Ag nanoparticles before chemical modification, λ_{max} = 558.5 nm. (b) Ag nanoparticles modified by 1mM 1:3 11-MUA/1-OT, λ_{max} = 572 nm. (c) Ag nanoparticles after modification with 1mM biotin, λ_{max} =594.5 nm. (d) Ag nanoparticles associated with 100 nM SA, λ_{max} = 611 nm. All extinction measurements were collected in air environment.

With the incidence wavelength ranging from 400 nm to 700 nm, the resulting extinction spectrum of the bare Ag nanoparticles is depicted in Fig. 11-13-a, where the LSPR λ_{max} was measured to be 558.5 nm. Similarly, Fig. 11-13-b shows the extinction spectrum after modification of the Ag nanoparticles with 1 mM 1:3 11-MUA/1-OT. To ensure a well-ordered SAM on the Ag nanoparticles, the sample was incubated in the thiol solution for 24 h. After careful rinsing and thorough drying with N$_2$ gas, the corresponding LSPR λ_{max} was measured to be 572 nm. In comparison to the bare Ag nanoparticles, the LSPR λ_{max} in this surface functionalization step is red-shifted by approximately 13.5 nm. Next, 1 mM biotin was covalently attached via amide bond formation with a two-unit poly(ethylene glycol) linker to carboxylated surface sites, the obtained LSPR spectrum is indicated in Fig. 11-13-c, showing that the maximum occurs at 594.5 nm, corresponding to an additional 22.5 nm red-shift from the second peak (see Fig. 11-14-b). We finally plotted the extinction spectrum in Fig. 11-14-d after the reaction between 100 nM SA and 1 mM biotin. It was found that the maximum wavelength has a 16.5 nm red-shift from 594.5 nm up to 611 nm. In contrast, using the same

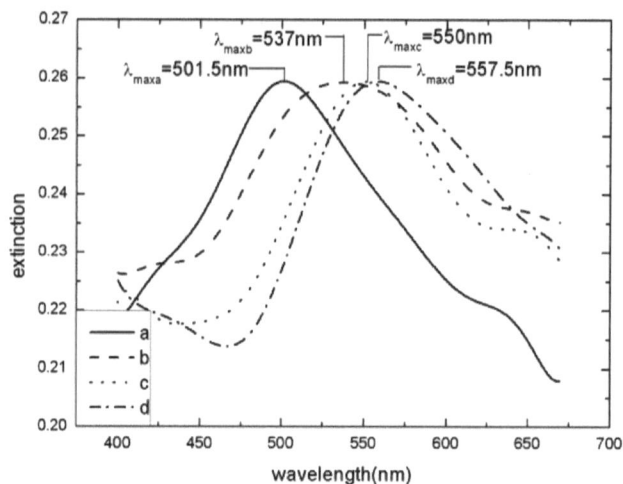

Fig. 11-14 LSPR spectra of each step in the surface modification of NSL triangular Ag nanoparticles to form a biotinylated Ag nanobiosensor and the specific binding of SA. (a) Ag nanoparticles before chemical modification, λ_{max} = 501.5 nm. (b) Ag nanoparticles modified by 1 mM 1:3 11-MUA/1-OT, λ_{max} = 537 nm. (c) Ag nanoparticles after modification with 1 mM biotin, λ_{max} =550 nm. (d) Ag nanoparticles associated with 100 nM SA, λ_{max} =558.5 nm. All extinction measurements were collected in air environment.

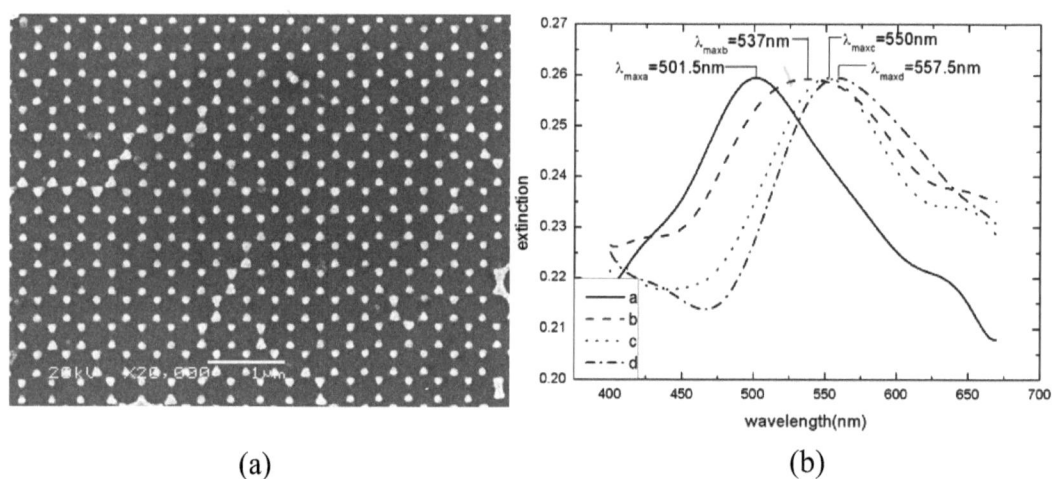

(a)   (b)

Fig. 11-15   (a) Representative scanning electron microscope(SEM) image of substrate (B)LSPR spectra of each step in the surface modification of NSL triangular Ag nanoparticles to form a biotinylated Ag nanobiosensor and the specific binding of SA. a: Ag nanoparticles before chemical modification, λ_{max} = 501.5 nm. b: Ag nanoparticles modified by 1 mM 1:3 11-MUA/1-OT, λ_{max} = 537 nm. c: Ag nanoparticles after modification with 1 mM biotin, λ_{max} =550 nm. d: Ag nanoparticles associated with 100 nM SA, λ_{max} =558.5 nm. All extinction measurements were collected in air environment.

diameter PS spheres, the uniform height of the Ag nanoparticles, under the same metrical condition and processing time, we fabricated the traditional Ag triangular nanosensor (see Fig. 11-15 (a)), and measured the extinction spectrum, as shown in Fig. 11-15 (b). However, in our experiments, for the traditional Ag triangular nanosensor, a 7.5 nm red-shift in the reaction between 100 nM SA and 1 mM biotin was observed. It is worthy to point out that the refractive index in our experiments was fixed for different bio-samples. It is LSPR-induced spectrum shifting which occurs when the biosample changed (refractive index is different accordingly). In other words, the peak shifting is caused due to the modification of the LSPR excited from the metal nanoparticles instead of the refractive index.

By analyzing the experimental results, we apparently found that the extinction spectroscopy varies with each step of surface functionalization for the Ag nanoparticles. The spectral shifts in that case can be explained by the change in the local refractive index resulting from the surface modification in each step. And the wavelength shift of the binding reaction between antigen and antibody can be improved by 2.2 folds over the whole 15×10 mm^2 detection area in comparison to that of the traditional triangular nanoparticles in the same experimental situation.

In order to evaluate our final results, we designed experiments for calibration of compound refractive index making use of a home designed fluidic chamber. Firstly, we calibrated the Ag triangle nanoparticles with hexagonal distribution which have the same material, period, and thickness with the rhombic nanoparticles with rectangle distribution. The spectrometer (USB4000 from Ocean Optics Inc.) was used in the experiment. There are four channels in the chamber, the left and right channels were designed as fluidic channels, and the other two channels located at top and bottom (perpendicular to chamber top surface) for optical fiber interconnection use. The calibrated liquid flows to the chip surface through a peristaltic pump with flowing velocity of 0.42 ml/min. An Ag triangle nanoparticles array was designed with 500 nm period and 50 nm height. Calibrating process is described as follows: 1) detecting spectrum of the extinction efficiency in the case of without any liquid, *i.e.* the chip is surrounded by air, $n_{air}=1$, as shown in Fig. 11-16 (black curve); 2) replacing air to be water evaporated in triple times, after the spectrum being stable, the derived extinction spectrum is shown in Fig. 11-15 with red curve; 3) refractive index value of the evaporation water measured

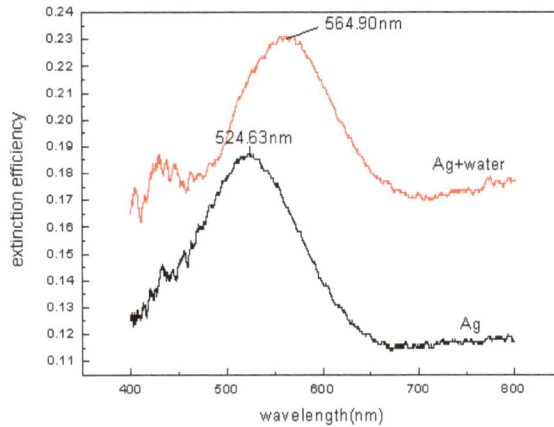

Fig. 11-16 Extinction spectra of the triangle metallic nanoparticles-based biochip with air and water on its surface.

using an apparatus with model of ZWA-J is 1.3325 (experiment condition: 20℃±0.1 ambient temperature, ±0.0002 sensitivity of the apparatus). Then sensitivity of the refractive index can be calculated as $\Delta\lambda_{max}/\Delta n =$ (564.90-524.63)/(1.3325-1.0)=121 nm/RIU. We mentioned above that spectrum shifting of the rhombic nanoparticles is 2.2 times of the triangle nanoparticels for detection of the target molecules with the same concentration. Therefore, we can conclude that the sensitivity of refractive index for the Ag rhombic nanoparticles is 266.2 nm/RIU. Moreover, the refractive indexes of the biotin and the SA were fitted by using the discrete dipole approximation (DDA) [16~19] numerical fitting method. Then the change for the refractive index was divided by the experimental maximum wavelength shift. We indicated that the refractive index sensitivity of the Ag rhombic nanosensor is 330 nm per refractive index unit (330 nm/RIU) for the Ag rhombic nanoparticles with an in-plane width of ~140 nm and out-of-plane height of 47 nm. Difference between the theoretical calculation and experimental detection may attributes to fabrication defects of the triangle nanoparticles and bubbles existing in the flowing liquid through the chamber. It can be further improved by modifying the fabrication process and detecting method.

It should be noted that all the measurements regarding the extinction spectrum were collected in atmosphere environment, where the samples were exposed in ordinary condition with dust and vapor which is suitable for ambient applications. It can be expected that the detection sensitivity will be further increased if the sample is measured in a bio-sample chamber filled with N$_2$.

The pure Ag nanoparticles have a problem of corrosion which will cause invalidity of the biosensing function. To solve this problem, a hybrid Au-Ag nanoparticle was put forth.[18] The hybrid Au-Ag triangular nanoparticles were proposed as a sensitive cell of the LSPR-based nano-biosensor. Using the FDTD algorithm, we designed and calculated the extinction spectra as well as the corresponding electric field of the hybrid nanoparticles array. A cross-section of a single particle labeling the materials of the substrate and particle as well as their thicknesses is shown in Fig. 11-17. Glass substrate was considered in the simulation while we created three-dimensional model. This model is almost close to the actual experimental sample. To simulate the Protein A binding to surface of the gold nanoparticles, we change the refractive index around the gold nanoparticles including the effect of the glass substrate only. A global index change was not assumed when protein A is bound.

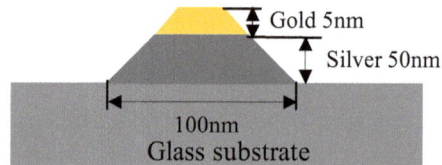

Fig. 11-17 Schematic diagram of the model of the hybrid Au-Ag triangular nanoparticles array.

Out-of plane height of the silver nanoparticles is 50 nm and the upper Au nanoparticles are 5 nm in height, and the in-plane width of each nanoparticle is 100 nm. The period of the nanoparticle array is 400 nm, and wavelength of the incidence white light source is defined ranging from 400 nm to 700 nm. The incidence light beam is projected in perpendicular to the substrate.

The LSPR-based nano-biosensors are extremely sensitive to variation of effective refractive index (ERI) within a few hundred nanometers on gold surface. Capture of the target analyte (Protein A) by the specific reaction between the metal Au and the Protein A. Protein A bound to the sensing face changes the apparent RI due to solution displacement by the analytes of higher refractive index. To test the detection capability of the hybrid Au-Ag nano-biosensors, experiments were performed using solutions of Protein A in PBS buffer (1:100), and the refractive index of Protein A

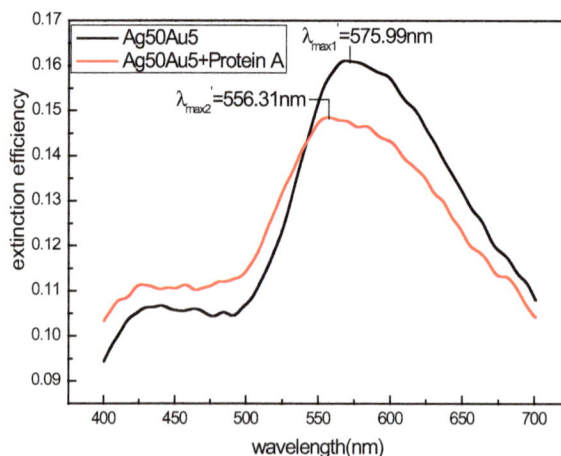

Fig. 11-18 Measured spectra for both pure hybrid Au-Ag (thickness of the Ag and Au is 50 nm and 5 nm, respectively) nanoparticles array and binding with protein A.

(1.3352) was detected by Abbe refractometer ZWA-J (at temperature 20 □, and Δn=$\pm$0.0002). In this study, the LSPR spectra of the specific binding signal between the hybrid Au-Ag nano-biosensor and Protein A were monitored by the integrated LSPR sensor (see Fig. 11-18).

Resonant wavelength λ_{max} of the bare hybrid Au-Ag nanoparticles (see Fig. 11-18, black line) was measured to be 575.99 nm. Exposure to 1:100 Protein A resolution, λ_{max}=556.31 nm was produced (see Fig. 11-18, red line), corresponding to a -19.68 nm shift. It should be noted that $\Delta \lambda_{max}$=-19.68 nm is smaller than the calculated result of the FDTD. It attributes the experimental defects caused by the NSL fabrication technique. Further study and modification will be performed in our future relevant research project. All the extinction measurements were collected in atmosphere environment.

11.4 Photothermal therapy

Metal nanoshells are core/shell nanoparticles that can be designed to either strongly absorb or scatter within the near-infrared (NIR) wavelength region (650-950 nm). Nanoshells were designed that possesses both absorption and scattering properties in the NIR to provide optical contrast for improved diagnostic imaging at higher light intensity, and rapid heating for photothermal therapy. Imaging results can be displayed using optical coherence tomography (OCT). The use of contrast agents in OCT offers the promise of enhanced diagnostic power, similar to the results with contrast agents in computed tomography (CT) and magnetic resonance imaging (MRI). To increase contrast in OCT, imaging agents including microbubbles 25 and microspheres with gold nanoparticles were employed.[26, 27] Optimal contrast agents for OCT would both further enhance contrast in vivo over ~5% improvement obtained with microbubbles and offer a smaller alternative to engineered microspheres (~2-15 µm) so as to improve distribution into the microcirculation. Because of their biocompatibility, tunable optical properties, and small scale size, nanoshells are well suited to OCT, providing easily detected backscattering of NIR light. [19]

11.4.1 Nanoshell fabrication

The basic fabrication strategy for metal nanoshells proceeds as shown in Fig. 11-19. Firstly, the nanoparticles were chemically modified so that gold colloid will readily attach to their surfaces. Then, the nanoparticles were coated with

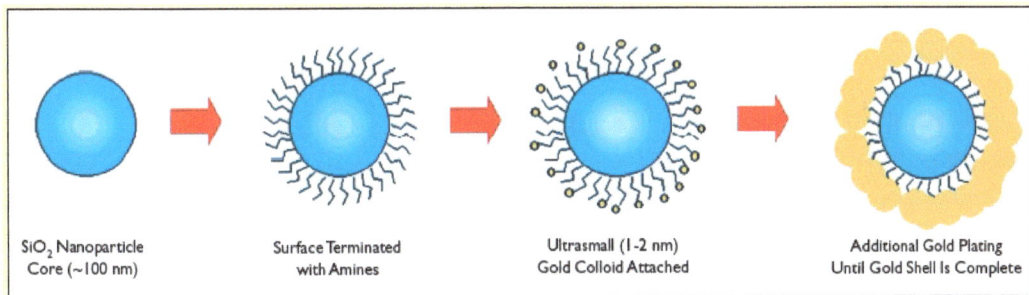

| SiO₂ Nanoparticle Core (~100 nm) | Surface Terminated with Amines | Ultrasmall (1-2 nm) Gold Colloid Attached | Additional Gold Plating Until Gold Shell Is Complete |

Fig.11-19 Schematic of nanoshell fabrication method, in which a nanoscale metallic overlayer is grown on top of a silica nanoparticle core. Reprinted with permission from "Naomi Halas, Optics & Photonics News, August 2002, 26-30" with copyright ©2002 of Optical Society of America.

extremely small gold colloid, which typically binds to the nanoparticle surface with coverage of approximately 25%. This "dressed" nanoparticle was then immersed in an electroless plating solution in which the total amount of metal to be plated corresponds to the desired thickness of gold caped on the nanoparticle surface and the total surface area of the nanoparticles. By this approach, the achievable minimal metal thickness is ~5 nm. The close, predictive correspondence between Mie scattering theory and nanoparticle core and shell layer thicknesses allows us to design or "nanoengineer" the optical response quite accurately. This strategy for planar fabrication on a nanoparticle surface has recently been modified to fabricate silica core-Ag shells, which also show quantitative agreement between their core and shell dimensions as well as Mie scattering theory. The nanoshells are similar to all optical materials. One important issue is thermal stability. Melting temperatures depressed significantly below the bulk melting point of the corresponding metal, and are characteristic of nanoscale metallic structures. Since the metal layer will seek to minimize its free energy and segregate into a solid metal nanoparticle at elevated temperatures, a depressed melting temperature will be accompanied by a drastic change in nanoshell morphology, and thus in nanoshell spectral response. To counteract this temperature sensitivity and essentially stabilize the nanoparticle with respect to intense heating, either by thermal or optical sources, we grew silica encapsulation layers around nanoshells. It was found that addition of a silica encapsulation layer of 60-70 nm raises the effective melting temperature (the temperature of observable structural and optical deformation) from 325° C to be greater than 600° C. [19]

11.4.2 Antibody conjugation [20]

Either anti-HER$_2$ or a nonspecific antibody (anti-IgG) was firstly attached to a PEG linker (OPSSPEG-NHS, MW) 2 K) through an amidohydroxysuccinimide group (NHS). The antibody-PEG linker complex was then attached to the nanoshell surface through a sulfur-containing group located at the distal end of the PEG linker. The antibody-PEG complex was reacted with nanoshells for 1 h. Additional PEG-thiol (MW 5 K, 50 nM) was later added to the nanoshell suspension in order to block nonspecific adsorption sites. Nanoshells were centrifuged to remove excess PEG and antibody and resuspended in water at a final concentration of 3×10^9 nanoshells per milliliter. HER$_2$-positive SKBr$_3$ breast adenocarcinoma cells were grown in McCoy's 5A growth media containing 10% FBS and 1% antibiotics at 37 °C and 5% CO$_2$. Concentrated 10× McCoy's media (free of FBS and antibiotics to eliminate nonspecific interactions with nanoshells) was quickly added to the nanoshells at a volumetric ratio of 1:9. Next, 500 µL of this McCoy's nanoshell suspension was placed on cells, followed by 1 h. incubation. McCoy's media containing FBS and antibiotics was added following rinsing of unbound nanoshells.

11.4.3 In Vitro Photothermal Therapy

Cells were imaged under a dark-field microscope which is sensitive to scattered light. Images were taken with a Zeiss Axioscope microscope equipped with a black/ white CCD camera. All images were taken at the same magnification under the same lighting conditions. Immediately following imaging, cells were exposed to NIR irradiation (820 nm, 0.008 W/m^2 for 7 min). The overlap of peak nanoshell absorbance with the emission wavelength of the laser source promoted optimal laser-induced nanoshell heating. Cells were stained for viability using calcein AM. Stained cells were examined under fluorescence and phase contrast microscopy with a Zeiss Axiovert 135 microscope. Silver staining was then performed to assess the presence of nanoshell binding on cell surfaces. Figure 11-20 presents results from combined imaging and therapy of SKBr$_3$ breast cancer cells using nanoshells targeted against HER$_2$ (right column).[21] In addition, control images of cells taken without nanoshells (left column) and of cells incubated with nonspecifically labeled nanoshells (middle column) were presented. Significantly increased scatter-based optical contrast due to nanoshell binding was observed in cells incubated with anti-HER$_2$ nanoshells (top row, right column) as compared to the two control cell groups (top row, left and middle columns). After photothermal therapy, cell death was observed only in cells treated with NIR laser following exposure to anti-HER$_2$ nanoshells (middle row, right column). This effect was not observed in cells treated with either nanoshells conjugated to a nonspecific antibody or NIR light alone (middle row, left and middle columns). Greater silver staining intensity was seen in cells exposed to anti-HER$_2$ nanoshells (bottom row, right column) compared to controls (bottom row, left and middle columns), suggesting enhanced nanoshell binding to cell surfaces over-expressing HER$_2$. To establish that anti-HER$_2$ nanoshells alone do not induce cytotoxicity, researchers incubated SKBr$_3$ cells with anti-HER$_2$ nanoshells over a range of concentrations and incubation times. Figure 11-21 shows calcein fluorescence of SKBr$_3$ cells that were exposed to HER$_2$-targeted nanoshells (3×10^9 nanoshells/mL). In statistical analysis comparing cells incubated with nanoshells and control cells not exposed to nanoshells, no differences in viability were observed.

The nanoshell was firstly put forth by Halas [19]. In 2004, working with her Rice colleague Jennifer West, she injected plasmonic nanoshells into the bloodstream of mice with cancerous tumors and found that the particles were nontoxic. What is more, the nanoshells tended to embed themselves in the rodents' cancerous tissues rather than the

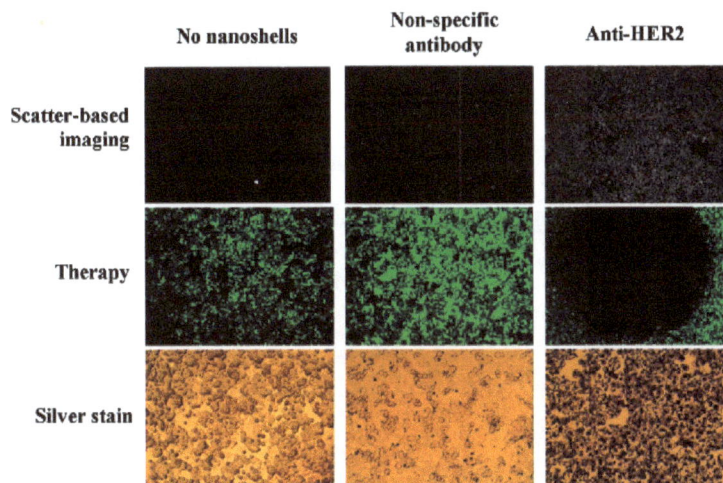

Fig.11-20 Combined imaging and therapy of SKBr₃ breast cancer cells using HER₂-targeted nanoshells. Scatter-based dark-field imaging of HER₂ expression (top row), cell viability assessed via calcein staining (middle row), and silver stain assessment of nanoshell binding (bottom row). Cytotoxicity was observed in cells treated with a NIR-emitting laser following exposure and imaging of cells targeted with anti-HER₂ nanoshells only. Note increased contrast (top row, right column) and cytotoxicity (dark spot) in cells treated with a NIRemitting laser following nanoshell exposure (middle row, right column) compared to controls (left and middle columns). Reprinted with permission from "Christopher Loo, Amanda Lowery, Naomi Halas, Jennifer West, and Rebekah Drezek, Nano Lett. 5, 709-711 (2005)" with copyright © 2005 of American Chemical Society.

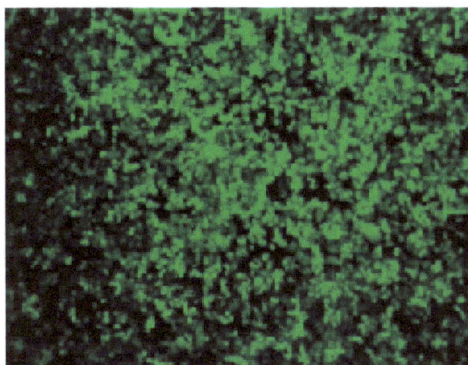

Fig.11-21 Calcein fluorescence of HER₂-positive SKBr₃ cells incubated with anti-HER₂ nanoshells without NIR photothermal therapy. No cytotoxicity was observed. Reprinted with permission from "Christopher Loo, Amanda Lowery, Naomi Halas, Jennifer West, and Rebekah Drezek, Nano Lett. 5, 709-711 (2005)" with copyright © 2005 of American Chemical Society.

healthy ones because more blood was circulated into the fast-growing tumors. (The nanoshells can also be attached to antibodies to ensure that they target cancer cells.) Fortunately, human and animal tissues are transparent to radiation at certain infrared wavelengths. When the researchers directed near-infrared laser light through the mice's skin and at the tumors, the resonant absorption of energy in the embedded nanoshells raised the temperature of the cancerous tissues from about 37 °C to about 45 °C. The photothermal heating kills the cancer cells while leave the surrounding healthy tissue unharmed. In the mice treated with nanoshells, all signs of cancer disappeared within 10 days; in the control groups, the tumors continued to grow rapidly. Houstonbased Nanospectra Biosciences is currently seeking permission from the Food and Drug Administration so as to conduct clinical trials of nanoshell therapy in patients with breast and neck cancer, as shown in Fig. 11-22. [22]

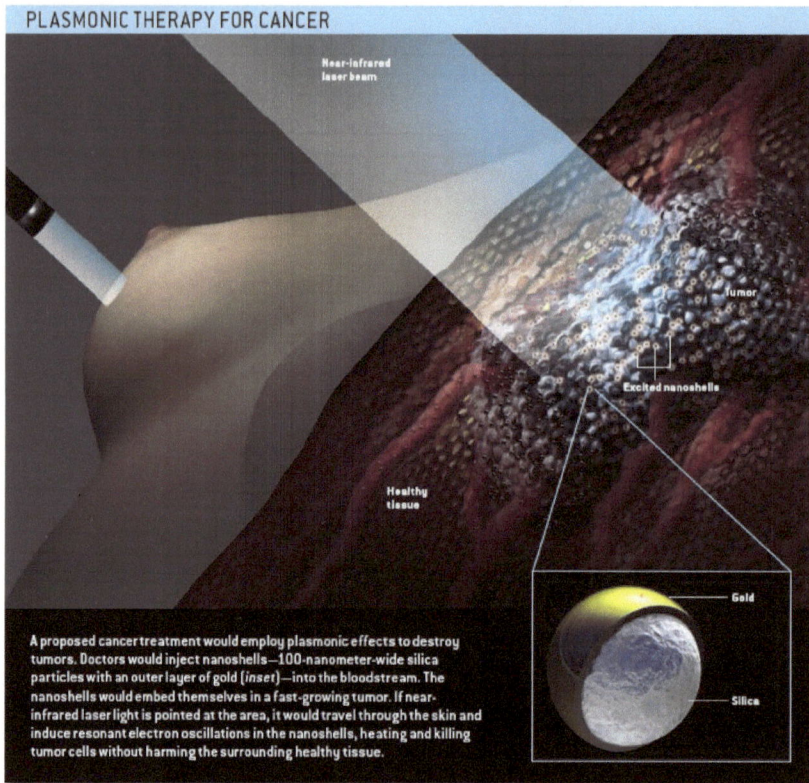

Fig. 11-22 Schematic diagram shows the plasmonic nanoshells for therapy of cancer. Reprinted with permission from "Harry A. Atwater, A promise of plasmonic, Scientific America, April 2007, 55-63" with copyright © 2007 of Scientific American, Inc.

Another example, SKBr3 breast cancer cells were cultured in 24-well plates until fully confluent. Cells were then divided into two treatment groups: nanoshells + NIR-laser and NIR-laser alone. Cells exposed to nanoshells alone or cells receiving neither nanoshells nor laser were used as controls. Nanoshells were prepared in FBS-free medium (2 × 109 nanoshells per milliliter). Cells were then irradiated under a laser emitting light at 820 nm at a power density of ~35 W/cm^2 for 7 minutes with or without nanoshells. After NIR-light exposure, cells were replenished with FBS-containing media and were incubated for an additional hour at 37 ºC. Cells were then exposed to the calcein-AM live stain for 45 min. in order to measure cell viability. The calcein dye causes viable cells to fluoresce green. Fluorescence was visualized with a Zeiss Axiovert 135 fluorescence microscope equipped with a filter set specific for excitation and emission wavelengths at 480 nm and 535 nm, respectively. Membrane damage was assessed using an aldehyde-fixable fluorescein dextran dye. Cells were incubated for 30 min. with the fluorescent dextran, rinsed, and immediately fixed with 5% glutaraldehyde. Photothermal destruction of cells was attributed to hyperthermia induced via nanoshell absorption of NIR light. The relevant OCT imaging results can be seen in Ref. [23] and [24].

Not only the circular shape, but other shapes such as elliptical nanoshells was reported recently.[25] A new hybrid nanoparticle that combines the intense local fields of nanorods with the highly tunable plasmon resonances of nanoshells were designed and fabricated. This dielectric core-metallic shell prolate spheroid nanoparticle bears a remarkable resemblance to a grain of rice, inspiring the name "nanorice", as shown in Fig. 11-23. This geometry possesses far greater structural tunability than either a nanorod or a nanoshell, along with much larger local field intensity enhancements and far greater sensitivity as a surface plasmon resonance (SPR) nanosensor than any dielectric-metal nanostructures reported previously. Invoking the plasmon hybridization picture allows the readers to understand the plasmon resonances of this geometry, as arising from a hybridization of the primitive plasmons of a solid spheroid and an ellipsoidal cavity inside a continuous metal. The fabrication of nanorice involves seeded metallization of spindle-shaped hematite nanoparticle cores. Small Au nanoparticles (~2 nm in diameter) are immobilized onto the

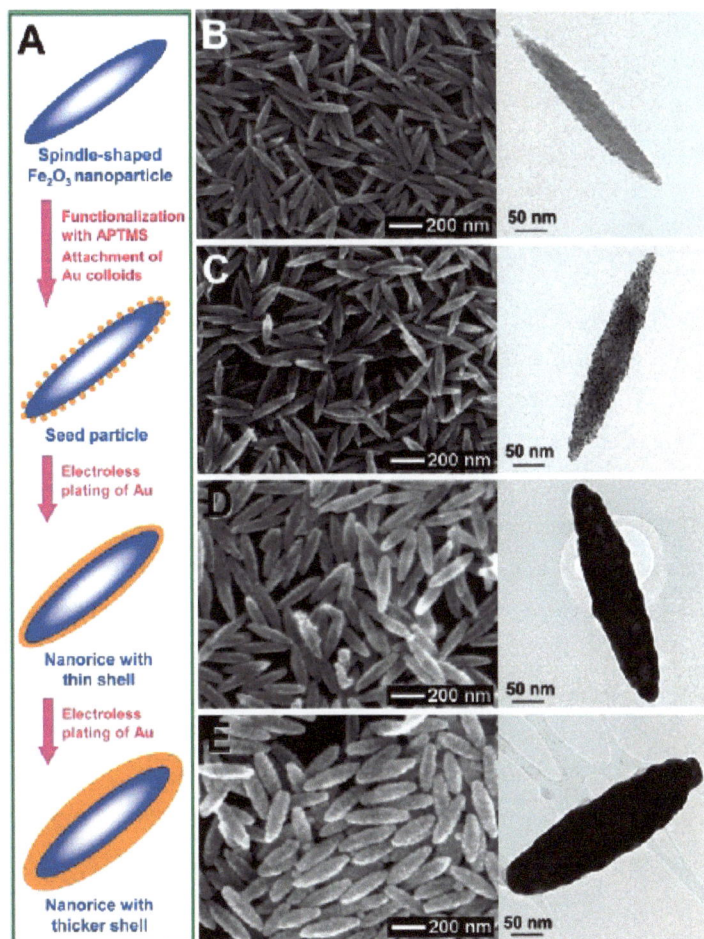

Fig. 11-23 (A) Schematics of the fabrication of hematite-Au core-shell nanorice particles. SEM (left) and TEM (right) images of (B) hematite core (longitudinal diameter of 340 (20 nm, and transverse diameter of 54 (4 nm), (C) seed particles, (D) nanorice particles with thin shells (13.1 (1.1 nm), and (E) nanorice particles with thick shells (27.5 (1.7 nm). Reprinted with permission from "Hui Wang, Daniel W. Brandl, Fei Le, Peter Nordlander, and Naomi J. Halas, Nano Lett. 6, 827-832 (2006)" with copyright © 2006 of American Chemical Society.

surface of (3-aminopropyl) trimethoxysilane (APTMS) functionalized cores at a nominal coverage of ~30%. The immobilized Au colloids act as nucleation sites for electroless Au plating onto the surface of core particles, leading to the gradual formation of a continuous and complete Au shell layer.

11.5 Nanoparticles for imaging

A number of molecular imaging techniques, such as optical imaging (OI), magnetic resonance imaging (MRI), ultrasound imaging (USI), positron emission tomography (PET), and others were reported for imaging of *in vitro* and *in vivo* biological specimens.[26, 27] Most nanoparticle-based optical imaging agents can be subdivided into two categories: quantum dots (QDs) [28] and dye-doped nanoparticle QDs. The requirement of external excitation for QDs sometimes limits their *in vivo* applications due to tissue opacity. QD–peptide conjugates were applied to bioimaging systems.[29] A QD-peptide luminescent probe with inherent signal amplification upon interaction with a proteolytic enzyme was reported by Chang *et al.*[30] In this system, QDs were bound to gold nanoparticles via a proteolytically degradable peptide sequence. Upon proteolysis of the peptide linker, the QD becomes highly luminescent owing to the separation from the quenching gold NP. Other examples of bioimaging using QDs were summarized in many recent reviews,[31, 32] demonstrating their utility in imaging lymph nodes and blood vessels. The QD-based nanoparticles are a large category in bioimaging. Considering complication of the QD theory, the application concept was put forth only in this chapter so as to keep contents integrality of this book. Numerous research papers reported the QD-based bioimaging in recent years. For more detailed information, please see relevant reference papers and books.

11.6 Summary

A rhombic Ag nanoparticle array as the biosensors' substrates was proposed in this chapter. A nanoscale biosensor experiment for detecting the binding signal between biotin and SA was demonstrated. The nanosensors designed on the basis of the rhombic nanoparticles with more hot spots have higher sensitivity than the sensors developed on the basis of the traditional Ag triangular nanoparticles. The low concentration and high sensitivity detection of the biology molecules has been realized by this LSPR-based nanosensor. Reproducibility was achieved by removing the silver nanoparticles on the substrate. The advantage of this method is that the substrate can be thoroughly cleaned so as to improve the fabrication quality of the chips for future experimental use. In comparison to conventional SPR method, the rhombic Ag nanoparticles-based LSPR method has advantages of higher spatial resolution, simple configuration of system, and cost effective. It has higher sensitivity in comparison to that of the reported triangular Ag nanoparticles. This type of nanosensor will have potentially applications in many fields such as medical science and biological technology. The hybrid Au-Ag nanoparticles can effectively prevent the Ag surface from corrosion, and is possible to be used as a new type of biosensor.

This new LSPR-based nano-biosensor with rhombic Ag nanoparticles array presented a detection approach with higher sensitivity compared to the previously reported triangle Ag particles. In addition, the LSPR-based sensing system is simple and cost effective in comparison to the SPR-based systems. Therefore, it is reasonable to believe that the reported modified LSPR-based systems will have a potential market in the next 5-10 years. Moreover, it may have significant advantages for some specific bio-samples immunoassay such as Staphylococcus aureus enterotoxin B, and amyloid-derived diffusible ligands.

Nanoparticles present a highly attractive platform for a diverse array of biological applications. The surface and core properties of these systems can be engineered for individual and multimodal applications, including biomolecular recognition, therapeutic delivery, biosensing, and bioimaging. Nanoparticles have already been used for a wide range of applications for both in *vitro* and in *vivo*. Full realization of their potential, however, requires addressing a number of open issues, including acute and long-term health effects of nanomaterials as well as scalable, reproducible manufacturing methods and reliable metrics for characterization of these materials.[33] Not only the immunoassay, imaging, and photothermal therapy, but the metallic nanoparticles can be transported by drug delivery to the diseased area, *e.g.*, breast with cancer cells distributed underneath the skin through injection. Then the cancer cells can be killed or stopped growing by means of illuminating the area to produce SPP-induced localized heating effect.

In addition, optical cloaking is another important phenomenon. It may have potential applications of camouflage in some areas such as military and life science. It is another story and not a topic being involved in this chapter because it is beyond biosensing scope. For more information regarding this topic, interested readers can see Ref. [34-36].

Reference

[1]   Shaoli Zhu, Fei Li, Xiangang Luo, Chunlei Du, Yongqi Fu, "A Localized Surface Plasmon Resonance Nanosensor Based on Rhombic Ag Nanoparticle Array", Sens. Actuat. B Chemical 134, 193–198 (2008).

[2]   Amanda J. Haes, W. Paige Hall, Lei Chang, William L. Klein, and Richard P. Van Duyne, "A Localized Surface Plasmon Resonance Biosensor: First Steps toward an Assay for Alzheimer's Disease", Nano Lett. 4, 1029-1034 (2004).

[3]   Shaoli Zhu, ChunLei Du, Yongqi Fu, Aiqun Liu, "Detection of Staphylococcus aureus enterotoxin B using localized surface plasmon resonance-based hybrid Au-Ag nanoparticles", Opt. Mater. (in press).

[4]   Draine B T, Flatau P J, 2004 User Guide for the Discrete Dipole Approximation Code DDSCAT.6.1 with the following website: http://arxiv.org/abs/astro-ph/0409262.

[5]   Patric C Chaumet, Adel Rahmani, Garnett W Bryant, "Generalization of the coupled dipole method to periodic structures", Phys. Rev. B 67, 165404 (2003).

[6]   Shaoli Zhu, Xiangang Luo, Chunlei Du, Fei Li, Shaoyun Ying, Qilin Deng, Yongqi Fu, "Hybrid metallic nanoparticles for excitation of surface plasmon resonance", J. Appl. Phys. 101(6), 064701 (2007).

[7]   Shaoli Zhu, Fei Li, Xiangang Luo, Chunlei Du, Yongqi Fu, "Fabrication and characterization of rhombic silver nanostructure array used as nanosensor chip". Opt. Mater. (in press)

[8]   Shaoli Zhu, Fei Li, Xiangang Luo, Chunlei Du, Yongqi Fu, "A novel bio-nanochip based on localized surface plasmon resonance spectroscopy of rhombic nanoparticles", Nanomedicine 3(5), 669-677 (2008).

[9]   Shaoli Zhu, Aiqun Liu, and Yongqi Fu, "Hybridization of localized surface plasmon resonance-based Au-Ag nanoparticles", Biomedical Microdevices 11(3), (2009). (published online on 6 April 2009)

[10] C. Hafner, "Multiple multipole program computation of periodic structures," J. Opt. Soc. Am. A 12, 1057-1067 (1995).

[11] Christian Hafner, Xudong Cui, Andre Bertolace, and Rüdiger Vahldieck, "Multiple multipole program analysis of metallic optical waveguides", Proc. SPIE, Vol. 6617, 66170C (2007).

[12] Lukas Novotny, Bert Hecht edt. *Principles of nano-optics*, Cambridge University Press, 2006.

[13] Yongqi Fu, Ngoi Kok Ann Bryan, "Fabrication and characterization of slanted nanopillars array". Journal of Vacuum Sciences and Technology B, Vol.23, No.3, pp. 984-989, May/June 2005.

[14] Di Fabrizio, "A Hybrid Plasmonic-Photonic Nanodevice for Label-Free Detection of a Few Molecules". Nano Lett. 8(8), 2321-2327 (2008).

[15] J. C. Hulteen, and R. P. Van Duyne, "Nanosphere lithography: A materials general fabrication process for periodic particle array surfaces", J. Vac. Sci. Technol. A 13, 1553-1558 (1995).

[16] Haiying Li, Xiangang Luo,   Chunlei Du, Xunan Chen, and Yongqi Fu, "Analysis of sensing performance of Gaussian-shaped metallic nano-gratings", Journal of Nanophotonics 2, 023508 (1 August 2008).

[17] Cees J. M. van Rijn, "Laser interference as a lithographic nanopatterning tool", J. Microlith., Microfab., Microsyst. 5(1), 0110121-0110126 (2006).

[18]   Lukas Novotny, "Effective Wavelength Scaling for Optical Antennas", Phys. Rev. Lett. 98, 266802 (2007).

[19] Naomi Halas, "Optical properties of nanoshells". Optics & Photonics News, August 2002, 26-30.

[20] Andre´ M. Gobin, Min Ho Lee, Naomi J. Halas, William D. James, Rebekah A. Drezek, and Jennifer L. West, "Near-Infrared Resonant Nanoshells for Combined Optical Imaging and Photothermal Cancer Therapy", Nano Lett. 7, 1929-1934 (2007).

[21] Harry A. Atwater, "A promise of plasmonic", Scientific America, April 2007, 55-63.

[22] Christopher Loo, Alex Lin, Leon Hirsch, Min-Ho Lee, Jennifer Barton, Naomi Halas, JenniferWest, Rebekah Drezek, "Nanoshell-Enabled Photonics-Based Imaging and Therapy of Cancer", Technology in Cancer Research & Treatment, Volume 3, Number 1, February 2004.

[23]   Christopher Loo, Amanda Lowery, Naomi Halas, Jennifer West, and Rebekah Drezek, "Immunotargeted Nanoshells for Integrated Cancer Imaging and Therapy", Nano Lett. 5, 709-711 (2005).

[24]   Hui Wang, Daniel W. Brandl, Fei Le, Peter Nordlander, and Naomi J. Halas, "Nanorice: A Hybrid Plasmonic Nanostructure", Nano Lett. 6, 827-832 (2007).

[25]   D. J. A. Margolis, J. M. Hoffman, R. J. Herfkens, R. B. Jeffrey, A. Quon, S. S. Gambhir, "Molecular Imaging Techniques in Body Imaging", Radiology 245, 333（2007）.

[26]   R. Weissleder, "Scaling down imaging: molecular mapping of cancer in mice". Nat. Rev. Cancer 2, 11 (2002).

[27]   A. P. Alivisatos, "Semiconductor Clusters", Nanocrystals, and Quantum Dots. Science 271, 933 (1996).

[28]   M. Zhou, I. Ghosh, "Quantum dots and peptides: A bright future together". Biopolymers 88, 325 (2007).

[29]   E. Chang, J. S. Miller, J. T. Sun, W. W. Yu, V. L. Colvin, R. Drezek, J. L. West, "Protease-activated quantum dot probes", Biochem. Biophys. Res. Commun. 2005, 334, 1317.

[30]   T. Jamieson, R. Bakhshi, D. Petrova, R. Pocock, M. Imani, A. M.Seifalian, "Biological applications of quantum dots", Biomaterials 28, 4717 (2007).

[31]   J. H. Rao, A. Dragulescu-Andrasi, H. Q. Yao, H. Q. Yao, Curr. Opin. "Fluorescence imaging in vivo: recent advances", Biotechnol. 18, 17 (2007).

[32] By Mrinmoy De, Partha S. Ghosh, and Vincent M. Rotello, "Applications of Nanoparticles in Biology", Adv. Mat. 20, 1-17 (2008).

[33] J. B. Pendry, D. Schurig, and D. R. Smith，"Controlling Electromagnetic Fields", Science 312, 1780 (2006).

[34] Ulf Leonhardt, "Optical Conformal Mapping". Science 312 (5781), 1777 (2006).

[35] Andrea Alù and Nader Engheta, "Achieving transparency with plasmonic and metamaterial coatings", Phys. Rev. E 72, 016623 (2005).

[36] Graeme Milton, Nicolae-Alexandru Nicorovici, Ross McPhedran, Viktor Podolskiy, "A proof of superlensing in the quasistatic regime, and limitations of superlenses in this regime due to anomalous localized resonance" Proc. R. Soc. A (2005) 461, 3999–4034.

12 PLASMONIC LASERS

Abstract: Surface plasmon polaritons (SPP)-based subwavelength metallic structures for laser amplification, emission, and collimation were introduced in this chapter. The plasmonic structures here act as a type of microlens which is integrated together with the laser for beam shaping or collimation. Then one application of the plasmonic laser for ablation of Silicon substrate was described.

12.1 Introduction

With rapid development of plasmonic nanostructures, a plasmon-based nanocavity for lasing of semiconductor lasers was proposed recently. [1] It can not only reduce the laser dimension, but enable ultrawide bandwidths for direct current modulation. Focusing semiconductor lasers usually requires individual bulky optical lenses acting as a "collimator". The micro-cylindrical lens and diffractive lens directly integrated with diode laser on emitting facet were reported several years ago.[2, 3] The microlenses were directly fabricated on the facet of both vertical cavity surface emitting laser (VCSEL) and edge emitting lasers using focused ion beam technology. The separated lens for beam shaping and collimation is not required accordingly. Following the similar idea, researchers demonstrated a plasmonic collimator that utilizes grooves etched directly into emitting facet of the semiconductor laser. Then semiconductor lasers can be downsized to a bare die without a lens. Some plasmonic structures such as bow-tie and nanorods were adopted. They will be illustrated in the following sections in this chapter.

12.2 Semiconductor Plasmon Lasers

Recently, Christina *et al.* [1] reported their nanoscale cavities for electrically-pumped surface-emitting semiconductor lasers that used surface plasmons to provide optical mode confinement in cavities which have dimensions in the 100-300 nm range. The proposed laser cavities are in many ways nanoscale optical versions of micropatch antennas that are commonly used at microwave/RF frequencies. Surface plasmons are not only used for mode confinement but also for output beam shaping to realize single-lobe far-field radiation patterns with narrow beam waists from subwavelength size cavities. We identify the cavity modes with the largest quality factors and modal gain, and show that in the near-IR wavelength range (1.0-1.6 µm) cavity losses (including surface plasmon losses) can be compensated by the strong mode confinement in the gain region provided by the surface plasmons themselves and the required material threshold gain values can be smaller than 700 cm^{-1}.

Three examples of nanopatch laser structures are shown in Fig. 12-1. The basic nanopatch laser structure consists of a bulk semiconductor gain medium in the form of a p-i-n heterostructure sandwiched on both sides by metal layers that confine the lasing optical mode via surface plasmons and also serve as electrical contacts for current injection. The light is radiated out from the sides of the cavity. The bottom metal layer acts like an antenna reflector and directs the light radiated out from the sides of the cavity in the upward direction thereby contributing to the surface emission characteristics of the nanopatch lasers. Ground planes in micropatch antennas used at microwave/RF frequencies perform similar functions [4]. The radii of the nanopatch lasers are between 100 nm and 300 nm and the heights of the

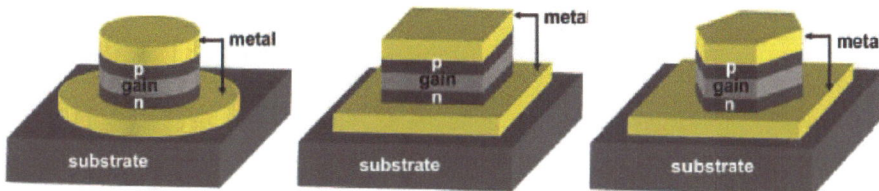

Fig. 12-1 Examples of nanopatch laser structures: (LEFT) a circular nanopatch laser, (MIDDLE) a rectangular nanopatch laser, and (RIGHT) a hexagonal nanopatch laser. The structures are not drawn to scale. Reprinted with permission from "Manolatou, C.; Rana, F., IEEE Journal of Quantum Electronics 44, 435-447 (2008)" with copyright © 2008 of IEEE.

dielectric part of the cavity are between 100 nm and 250 nm. The metal layers were assumed to be thick enough to prohibit light transmission. Figure 12-2 shows the far-field radiation pattern for the TM_{111} mode of a 160 nm radius and 220 nm cavity height circular nanopatch cavity. The angular dependence of the z-component of the field is assumed to be proportional to $\cos(\phi)$ rather than $\exp(\pm i\phi)$ to show the ϕ-dependence of the pattern. The radiation is emitted out from the sides of the cavity but interferes constructively in the upward direction. A weak surface plasmon wave is also emitted along the bottom metal layer. The surface wave decays as it propagates and its power would have made no contribution to the radiation pattern if the radiation pattern were computed over a hemispherical surface of much larger radius.

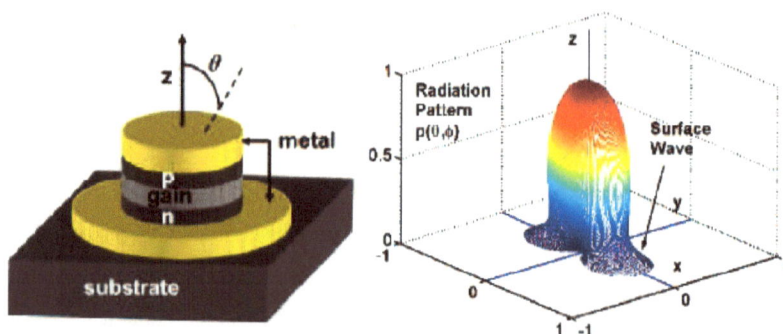

Fig. 12-2 A circular nanopatch laser (LEFT) and the computed far-field radiation pattern (RIGHT) for the TM_{111} mode of a 160 nm radius and 220 nm cavity height laser are shown. The angular dependence of the z-component of the field is assumed to be proportional to $\cos(\phi)$ rather than $\exp(\pm i\phi)$. A weak surface plasmon wave emitted along the bottom metal layer is also visible in the radiation pattern. Reprinted with permission from "Manolatou, C., Rana, F., IEEE Journal of Quantum Electronics 44, 435-447 (2008)" with copyright © 2008 of IEEE.

Challenges associated with the fabrication of nanopatch lasers are not expected to be impossible to meet. Substrate removal techniques used in the fabrication of dual-metal-waveguide semiconductor far-IR lasers can be employed for the realization of the dual-metal nanopatch laser structures. [5] However, in the case of nanopatch lasers, the semiconductor films obtained after substrate removal would be much thinner and would require more care. The metal used for the top and bottom layers must make good ohmic contacts to the contact/cladding layers, not react with or diffuse into the semiconductor, and also have small surface plasmon losses. Silver seems to be the ideal choice. Electrical contact leads to the top metal layer would need to be microfabricated since the laser structure is too small to be contacted directly with electrical probes. Finally, the exposed side walls of the cavity would need to be passivated to reduce surface recombination. Quantum well gain media can also be used in place of bulk gain media in the nanopatch lasers. Quantum wells can generally provide larger material gain compared to bulk for the same carrier density due to the reduced density of states. Benefits of the nanopatch laser structures are not expected to be limited to the just reduced dimensions. The small photon lifetimes in nanopatch lasers compared to conventional semiconductor lasers could enable ultrawide bandwidths (> 100 GHz) [1] for direct current modulation. The dual-metal layer structure of nanopatch lasers could enable them to be used as functional elements and placed on desired substrates, such as Silicon microchips, to provide lasers on demand capabilities for multifunctional micro- and nano-systems.

It is well known that microscopic lasers can reach the diffraction limit, based on photonic crystals, metal-clad cavities and nanowires. However, such lasers are restricted, both in optical mode size and physical device dimension, to being larger than half the wavelength of the optical field, and it remains a key fundamental challenge to realize ultra-compact lasers that can directly generate coherent optical fields at the nanometer scale, far beyond the diffraction limit. A way of addressing this issue is to make use of surface plasmons. This approach is capable of tightly localizing light, but so far ohmic losses at optical frequencies have inhibited the realization of truly nanometer scale lasers based on such approaches.

Most recently, a "Hybrid plasmonics" approach was reported and showed experimentally the laser action of surface plasmon polaritons with mode as small as $\lambda^2/400$ [6]. The truly nanometre-scale plasmonic laser devices consist of cadmium sulphide (CdS) nanowires on a silver film, where the gap layer is magnesium fluoride (MgF_2, as shown in Fig.12-3 a). The close proximity of the semiconductor and metal interfaces concentrates light into an extremely small area as much as a hundred times smaller than a diffraction-limited spot (see Fig.12-3 b). To show the unique properties of the hybridized plasmon modes, we compare the plasmonic lasers directly with CdS nanowire lasers on a quartz substrate, similar to typical nanowire lasers reported before. Here, we will refer to these two devices as plasmonic and photonic lasers, respectively. The plasmonic lasers can maintain strong confinement and optical mode-gain overlap for the diameters as small as 52 nm, a diameter for which a photonic mode does not even exist. The relatively small difference in the thresholds of the plasmonic and photonic lasers can be attributed to high cavity mirror losses, which are of the same order of magnitude as plasmonic losses.

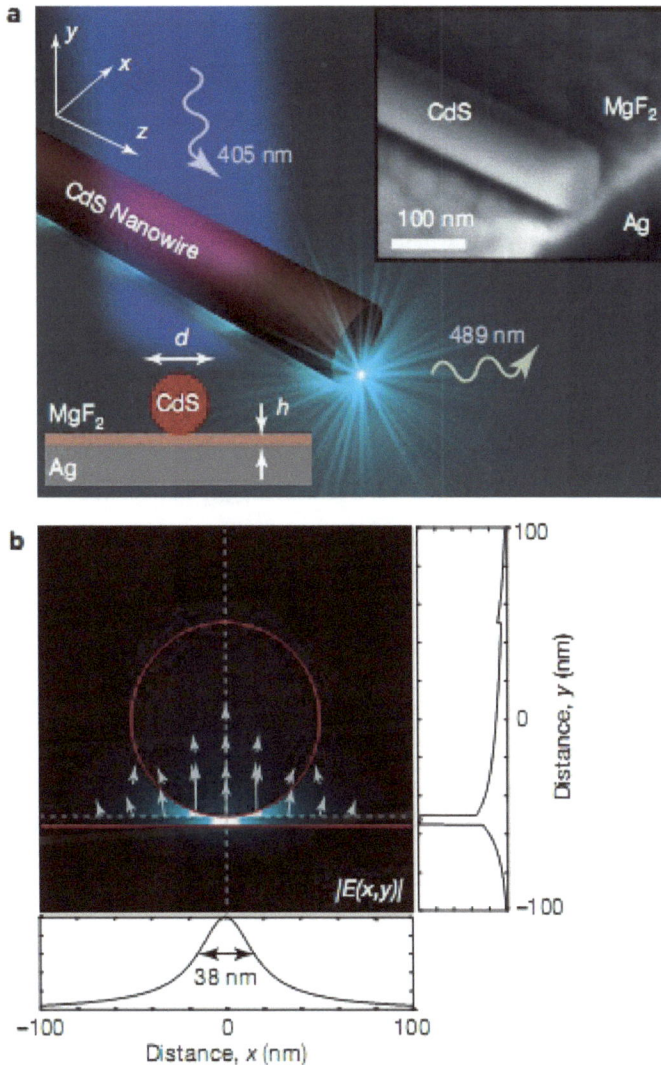

Fig.12-3 The deep subwavelength plasmonic laser. a, The plasmonic laser consists of a CdS semiconductor nanowire on top of a silver substrate, separated by a nanometre-scale MgF_2 layer of thickness h. This structure supports a new type of plasmonic mode the mode size of which can be a hundred times smaller than a diffraction-limited spot. The inset shows a scanning gelectron microscope image of a typical plasmonic laser, which has been sliced perpendicular to the nanowires axis to show the underlying layers. b, The stimulated electric field distribution and direction $|E(x,y)|$ of a hybrid plasmonic mode at a wavelength of $\lambda=489$ nm, corresponding to the $CdSI_2$ exciton line. The cross-sectional field plots (along the broken lines in the field map) illustrate the strong overall confinement in the gap region between the nanowire and metal surface with sufficient modal overlap in the semiconductor to facilitate gain. Reprinted with permission from "Rupert F. Oulton, Volker J. Sorger, Thomas Zentgraf, Ren-Min Ma, Christopher Gladden, Lun Dai, Guy Bartal, Xiang Zhang, Nature 461, 629 (2009)" with copyright ©2009 of Macmillan Publishers Limited.

This demonstration of deep subwavelength plasmonic laser acting at visible frequencies suggests new sources that may produce coherent light far below the diffraction limit. Extremely strong mode confinement, which is evident from the up to six fold increase of the spontaneous emission rate, and the resulting high β-factor, are key aspects of operation of the deep subwavelength lasers. They have also shown that the advantage of plasmonic lasers is their ability to downscale the physical size of devices, as well as the optical modes they contain, unlike diffraction-limited lasers.

Furthermore, the use of metals in plasmonics can provide a natural route towards electrical injection schemes that do not interfere with mode confinement. The impact of plasmonic lasers on optoelectronics integration is potentially significant because the optical fields of these devices rival the smallest commercial transistor gate sizes and thereby reconcile the length scales of electronics and optics.

Apart from the CdS semiconductor nanowire-based plasmonic laser, a room-temperature single nanoribbon lasers were reported by Zapien *et. al.* earlier in 2004 [7]. Using a single nano-object measurement methodology that enables the correlation between size/morphology/structure and photoluminescence (PL) characteristics, they showed that nanoribbons are an excellent model system to study single nano-objects. In particular, they measured the PL characteristics of optically pumped individual single-crystal zinc-sulfide nanoribbons. Small collection angle measurements showed that nanoribbons form excellent optical cavities and gain medium with high (full width at half maximum <0.1 nm) lasing modes free of PL background even for a low pumping density of 9 kW/cm². Large collection angles add abroad PL component and obscure the power correct high-quality lasing of the nanowires/nanoribbons. The low threshold and easy manipulation of the nanoribbons suggest it is worthwhile to process them to diodes and achieve electrical pumping. Nanoribbons, in addition to their possible use in nanodevices, offer an easy model system to study how different properties are affected by the structure, morphology, and size of the nanoribbons as demonstrated in the present work.

The research reported in Ref. [6, 7] is just a beginning for exploration of the plasmonic lasers. With rapid development of plasmonics and nanophotonic devices, it is reasonable to believe that the hybrid plasmonics plus nanowire-based waveguide will be one of tendency for the research topic of plasmonic lasers in the future.

12.3 SPP subwavelength metallic structures for amplification

A new type of nanoscale optical device called a lasing spaser combines metamaterials and spasers to create a versatile planar source of coherent radiation. The spaser (surface-plasmon laser) is a quantum amplifier of surface-plasmon emission with potential in a wide range of applications from nanoscale lithography and ultra-microscopy to ultra-sensitive surface-enhanced Raman scattering. Introduced in 2003, [8] the spaser is the equivalent of a laser but it amplifies plasmon resonances rather than photons, with plasmonic resonators instead of a resonant cavity. In the lasing spacer, a two-dimensional array exploits a slightly perturbed symmetry within its structure to emit spatially and temporally coherent radiation normal to its surface.[9] The structure consists of a regular array of metallic asymmetric split-ring (ASR) resonators, which generates spatially coherent and high-quality-factor (high-Q) current oscillations. Here the high-quality factor needed for lasing and amplification is a feature of the collective "coherent" response of the whole array—an individual ARS does not show such a high-quality response. On top of this array sits a thin substrate of amplifying dielectric material such as a semiconductor that can be optically or electrically pumped. A minor asymmetry designed into the ASR resonators breaks the trapped-mode oscillation in the amplifying substrate, causing a portion of the energy to be emitted into free space (see Fig.12-4).[10] If the array of resonators is in contact with a gain medium, for example when it is supported by a thin slab of gain material, then radiation losses and Joule losses in the metal can be overcome. Various gain media such as optically and electrically pumped semiconductor structures or quantum-dot-doped dielectrics may be suitable for this purpose. We show below that on reaching the threshold value of gain, the intensity of the resonant wave is reflected and transmitted through the structure increases dramatically. By combining a thin layer of a gain medium with a high Q-factor ASR array, it is possible to achieve orders of magnitude enhancement of single-pass amplification in comparison to the amplification of the bare gain medium layer. In an infinite array, the width and magnitude of the amplification peak are only limited by radiation losses, and are controlled by the asymmetry of the split-ring resonators. In a realistic case the finite size of the array and fabrication tolerances will limit amplification.

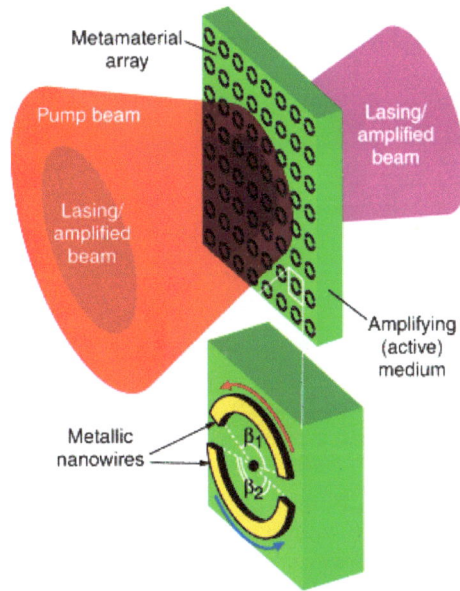

Fig.12-4 A lasing spaser consists of a thin slab of a dielectric gain medium (center) underneath an array of asymmetric split-ring resonators. In a single translation cell of the array, asymmetric currents of plasmonic oscillations in phase with the other cells produce emission of spatially and temporally coherent light that propagates in a direction normal to the array. The arc angles of the nanowire segments, β_1 and β_2, determine the Q-factor and coupling intensity. Reprinted with permission from "N. I. Zheludev1, S. L. Prosvirnin, N. Papasimakis, and V. A. Fedotov, Nature Photonics 2, 351 (May 2008)" with copyright © 2008 of Nature Publishing Group.

Zheludev and his group considered two cases of amplification derived using the ASR arrays on the dielectric substrate. The first case assumed negligible Joule losses in the array but accounted for losses in the substrate. A gain exceeding the threshold value of 70 cm^{-1} is enough to overcome substrate losses, resulting in resonant amplification in the bare substrate of approximately 5% in the mid-infrared (mid-IR) range of the wavelengths near 8.4 µm (35 THz). The second case considered Joule losses in the metallic wires and the substrate: a threshold gain of 1800 cm^{-1} is required to overcome the losses. At a peak amplification of α=2550 cm^{-1}, the correlated amplification in the bare substrate is 5.5% at 1.65 µm (181 THz), as shown in Fig. 12-5, and the spectral width of the peak resonance narrowed from 3 THz to 500 GHz. The final attraction of the scheme is that manipulation of the ring size determines the amplification/lasing frequency and is therefore tunable to various luminescence frequencies in numerous types of gain media. The researchers believe that the approach should eventually enable high gain amplifications and lasing with a nanoscale or micron-scale layer of material, having potential in numerous highly integrated devices, particularly in those devices with sensitivity to heat management.

The lasing spaser allows high amplification and lasing in a very thin layer of material with a more modest gain level, making it a very practical proposition. The thin-layer geometry is a desirable feature for some highly integrated devices and from the point of view of heat management and integration. Here the amplification/lasing frequency is determined by the size of the ring and may be tuned to match luminescence resonances in a large variety of gain media. This therefore makes the lasing spaser a generic concept for many applications. Finally, the ring currents in the metamaterial array can be seen as classical analogues of magnon quasi-particles, and the striking similarity between the coherent regime of the lasing spaser and Bose–Einstein condensation of magnons under pumping should be considered. [9]

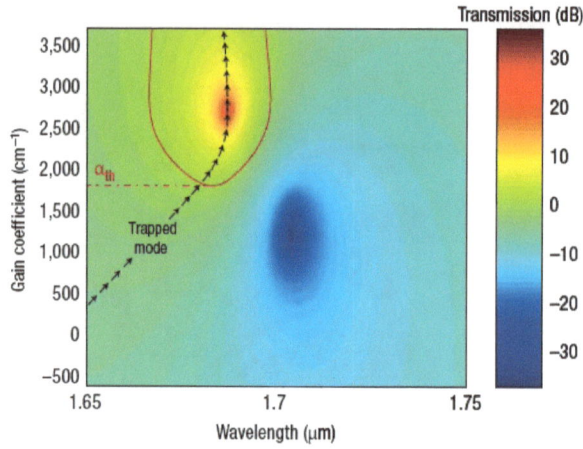

Fig.12-5 Transmission spectra of the near-IR resonator array. Spectra of the planar ASR metamaterial are provided in the vicinity of the trapped-mode transmission resonance corresponding to different values of bare substrate gain a. Solid contour: region of unity transmission. The line of arrows shows the evolution of the trapped-mode resonance frequency with increase of a. Reprinted with permission from "N. I. Zheludev1, S. L. Prosvirnin, N. Papasimakis, and V. A. Fedotov, Nature Photonics 2, 351 (May 2008)" with copyright © 2008 of Nature Publishing Group.

Most recently, Banerjee *et. al.* reported their surface plasmons lasers: threshold, gain, spectral line narrowing and feedback in the visible range.[11] Their structures were made of surface waveguides: in one construction, a perforated 50 nm thick oxide layer (aluminum oxide) separates the metal (aluminum) from air; in the other, the perforated thin oxide layer is bound by two-layer graphene, as well as the supporting metal substrate. Gain is provided by highly fluorescing dye chromophores (fluorescein), which are imbedded within the hole-array in the perforated oxide. Another fluorescence line from oxide defect states served as a reference spectral line. The feedback to the SP waves is provided by the same sub-wavelength, periodic hole-array, similarly to the construction of photonic crystal for semiconductor lasers [12-13] with the exception that the hole-array pitch is 1/6 of SP wavelength. Their construction is also unlike previously suggested scheme [14], which relied on V-shape metallic waveguides with a quantum dots serving as gain media. The surface metallic waveguides are consisted of a nanoscale hole-array in a 50 nm thick layer of aluminum oxide on top of aluminum substrate (anodized aluminum oxide, or abbreviated as AAO). In some cases, two-layer graphene was added on top of the perforated oxide layer, as well. The sub-wavelength array of holes enable coupling to and from the waveguides as well as, providing feedback to the surface modes. The gain media molecules (fluorescein) are imbedded in the structure's pores. Threshold and spectral line narrowing of 30% were clearly demonstrated when pumped with a pulsed laser.

12.4 Plasmonic laser antenna

12.4.1 A bowtie plasmonic quantum cascade laser antenna

A bowtie plasmonic quantum cascade laser antenna was reported that can confine coherent mid-infrared radiation well below the diffraction limit.[15] The antenna was fabricated on the facet of a mid-infrared quantum cascade laser and consists of a pair of gold fan-like segments, whose narrow ends were separated by a nanometric gap. The bowtie antenna efficiently suppresses the field enhancement at the outer ends of the structure, making it more suitable for spatially-resolved high-resolution chemical and biological imaging and spectroscopy.

The laser antenna was designed by aid of the finite-difference time-domain (FDTD) algorithm. Schematics of the simulated antenna structures are shown in Fig. 12-6 (a), with their geometric parameters indicated. The antennas have the same gap width and QCL material which is used as the substrate in the model. The excitation source consists of a mid-IR plane wave polarized along the antenna axis. The plane wave is launched from the inside of the laser material at

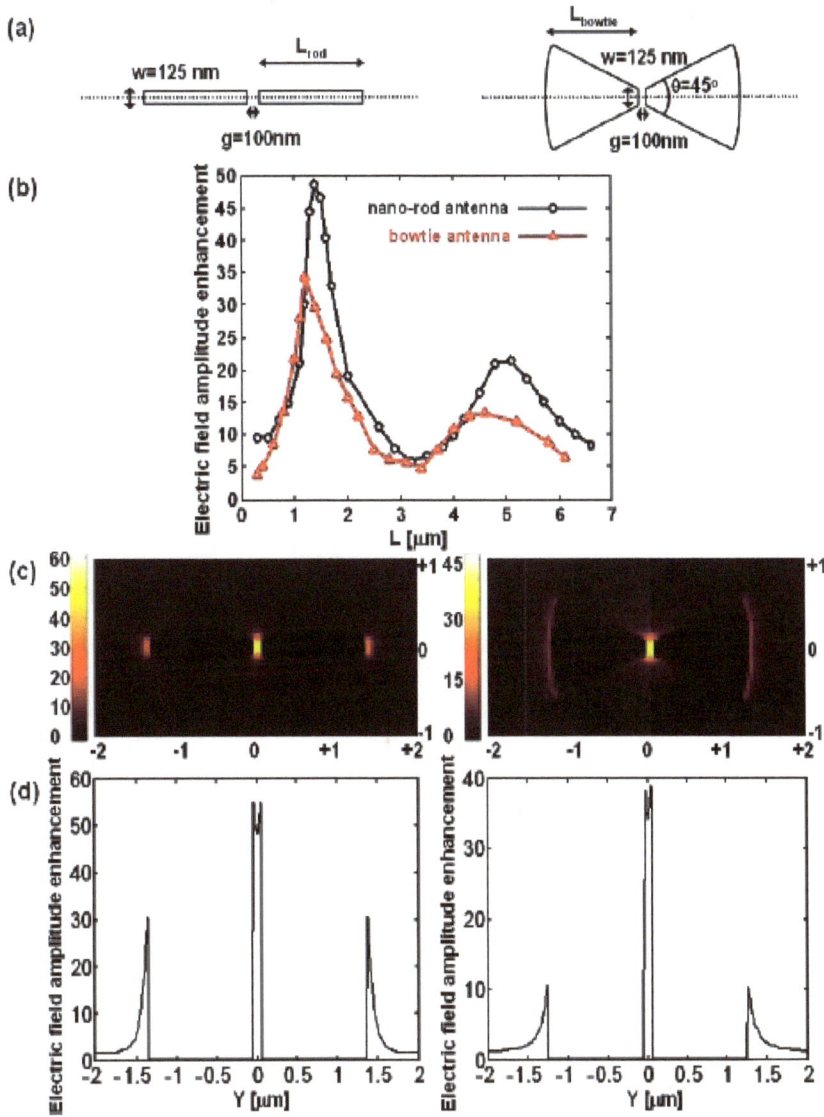

Fig. 12-6 A comparative study of the mid-infrared nano-rod antenna and the bowtie antenna by FDTD simulations. The free space wavelength of the incident plane wave normal to the plane is assumed to be 7 μm. (a) Simulated antenna structures. Left panel: a schematic of the nano-rod antenna. It is composed of two gold nano-rods separated by a nanometric gap. The size of the gap, the length and the width of the nano-rods, are indicated in the figure as g, L_{rod}, and w, respectively. Right panel: a schematic of the bowtie antenna. It consists of two gold fan-like segments separated by a nanometric gap. The geometries of the antenna are indicated in the figure. Both antennas have a thickness of 70 nm and they are defined on the output facets of quantum cascade lasers (QCLs). The facets are assumed to be coated with a 70 nm-thick electrically insulating layer of alumina. The antenna axes (dotted lines) are aligned with the polarization of the incident electric field. (b) Electric field amplitude enhancement vs. the antenna length L for the nano-rod antenna and the bowtie antenna. The field is calculated in the middle of the antenna gap at the level of the antenna top surface and is normalized to the amplitude of the incident field. The antenna length L is varied; other geometric parameters are kept unchanged and have the values as indicated in (a). (c) FDTD simulation results showing the electric field amplitude enhancement distribution of the two antennas at the first resonance (L_{rod}=1.4 μm and L_{bowtie}=1.2 μm). The enhancement is calculated on the plane that is at the same level as the antenna top surface. (d) Line scans of (c) along the antenna axes. Reprinted with permission from "N. Yu, L. Diehl, E. Cubukcu, C. Pflügl, D. Bour, S. Corzine, J. Zhu, G. Höfler, K. B. Crozier, and F. Capasso, Opt. Express 15, 13272-13281 (2007)" with copyright © 2007 of Optical Society of America.

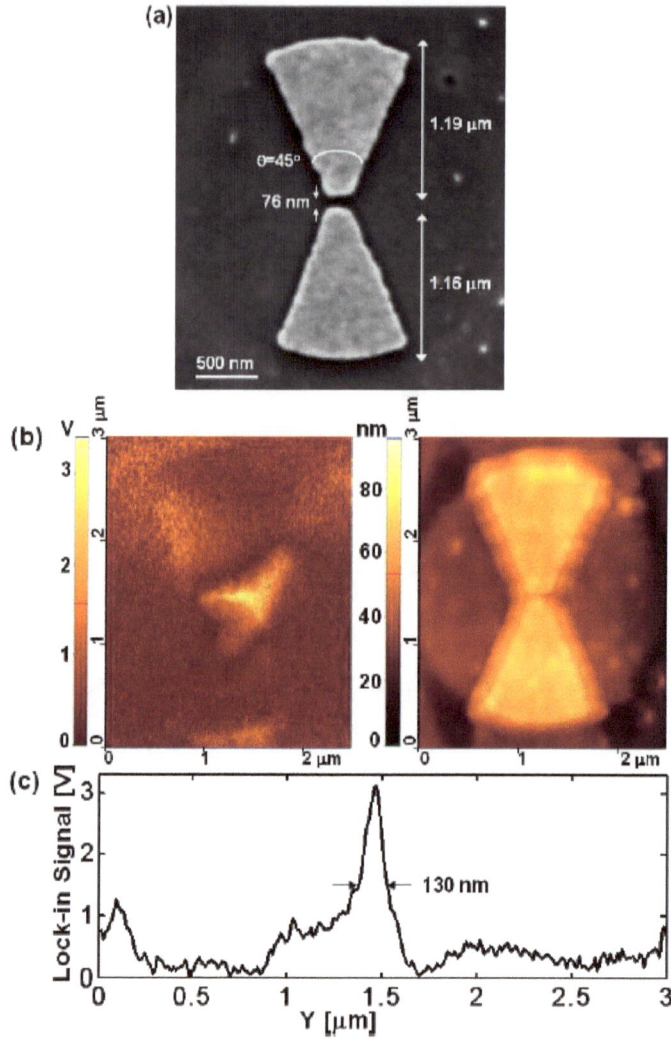

Fig. 12-7 Characterizations of another bowtie plasmonic laser antenna. (a) A scanning electron microscope image of the antenna. It has a smaller gap size compared with the antenna shown in Figs. 3 and 4. (b) Simultaneous NSOM image and AFM topography for the antenna. It is designed to be working at the first dipolar resonance with λ=7.0 μm. (c) Line scan of the NSOM image along the antenna axis. The center peak has a full-width at half-maximum of about 130 nm. Reprinted with permission from "N. Yu, L. Diehl, E. Cubukcu, C. Pflügl, D. Bour, S. Corzine, J. Zhu, G. Höfler, K. B. Crozier, and F. Capasso, Opt. Express 15, 13272-13281 (2007)" with copyright ©2007 of Optical Society of America.

normal incidence to the substrate. The refractive index of the laser medium is taken as 3.15, which is the weighted average of the refractive indices of the two constituents of the laser active region: InGaAs and AlInAs. This is a good approximation because the wavelength in the laser medium (about 2.2 μm at λ=7.0 μm) is much larger than the thickness of each individual quantum well layer in the active region. Adaptive meshing is used in all the FDTD simulations [16]: the area around the antenna gap has a grid size of 5 nm and other parts of the simulation region have a grid size of 20 nm. This helps to reduce the amount of calculations without sacrificing the resolution around the antenna gap. The field enhancement in the antenna gap is maximized when the antenna is illuminated at one of its resonant wavelengths, as shown in Fig. 12-6 (b). It can be seen that the bowtie antenna has smaller maximum field amplitude enhancement compared to that of the nano-rod antenna. This is most likely due to a smaller Q-factor for the former because of the larger antenna area as well as correspondingly greater losses. This was also predicted by Crozier *et. al.* for triangular optical antennas. [17] In Fig. 12-6 (c), FDTD simulations of the 2D distributions of the field amplitude enhancement for the two antennas at the first resonance are presented. For both structures, the strong electric field is confined to the antenna gap with a spatial extent mostly determined by the gap size. In comparison to the nano-rod structure, weaker field enhancement appears at the outer ends of the bowtie structure. This is more apparent in Fig. 12-6 (d), in which line scans along the antenna axes of the data presented in Fig. 12-6 (c) are given. In comparison to the nanorod antenna, the bowtie antenna has a larger ratio of the near field amplitude in the gap to the near field amplitude at the antenna outer ends. The ratio is about 3.5 for the latter and about 1.8 for the former. This demonstrates the benefit of the bowtie structure, which is more efficient in suppressing the side spots. Simulations indicate that the suppression ratio is even larger for bowtie antennas with larger tip angle θ.

To optically characterize the designed device, a near-field scanning optical microscope (NSOM) based on a commercial atomic force microscope (AFM) (PSIA XE-120) working at mid-IR is utilized to characterize the spatial profiles of the laser transverse modes [18] and the near field distribution of the bowtie antennas. The characterization of the transverse modes enables us to define the antennas at the laser mode maxima. For these measurements, they used commercial silicon AFM tips (Veeco Probes and Budget Sensors) that have resonance frequencies ranging from about 40 kHz to about 80 kHz. The tips were coated with 70 nm gold and the tip apex curvature is about 30 nm. The QCLs are driven by 125 ns current pulses with a repetition rate of about 500 kHz to reduce the average power, thereby preventing melting of the antennas and the metallic coatings of the AFM tips. In order to enhance the detection sensitivity, the interference between the near field scattered into the far field by the apex of the AFM tip and a reflected laser output is measured [16]. Therefore, the detected NSOM signal is proportional to the near-field magnitude rather than the near-field intensity, assuming that the magnitude of the reflected laser field is a constant [19, 20]. In Fig. 12-7 (a), the SEM image of another bowtie antenna working at the first dipolar resonance is presented. The antenna gap width is about 76 nm. Figure 12-6 (b) shows the NSOM image and the AFM topography taken simultaneously for this device. The line scan through its axis [Fig. 12-7 (c)] shows correspondingly a smaller FWHM of about 130 nm.

A plasmonic laser was theoretically constructed by considering plasmonic nanoantennas immersed in active host medium. Specifically shaped metal nanoantennas can exhibit strong magnetic properties in the optical spectral range due to the excitation of magnetic plasmon resonance. A case when a metamaterial comprising such nanoantennas can demonstrate both "left handiness" and negative permeability in the optical range is considered herein.[21] In Ref.16, theoretical computational results show that high losses predicted for optical "left-handed" materials can be compensated in the gain medium. Gains required to achieve local generation in such magnetic active metamaterials were calculated for real metals. We proposed a plasmonic nanolaser, where the metal nanoantenna operates similar to a resonator. The size of the proposed plasmonic laser is much smaller than the wavelength. The devices can provide confinement of the electric field in the antenna gap to dimensions ranging from $\lambda/50$ to $\lambda/30$ at mid-IR wavelengths. Therefore, it can serve as a very compact source of electromagnetic radiation.

12.4.2 Nanorode-based plasmonic laser antenna

The plasmonic laser antenna consists of a resonant optical antenna integrated on the facet of a laser diode.[22] Such a compact laser source with subwavelength spatial resolution provides distinct advantages in a number of applications including microscopy, spectroscopy, optical data storage, lithography, and laser processing. A schematic diagram of the proposed device is shown in Fig. 12-8. A pair of coupled triangle-like particles (for example, a bow tie antenna) [23, 24] or of nanorods [25] separated by a very small gap, which is less than 30 nm in this work, is defined on one of the facets of a diode laser. Gold nanorods can generate enhanced near fields through the excitation of SPs.[26] The large field enhancement in these structures, compared to gold nanospheres, is due to the substantially reduced plasmon dephasing rate, an effect caused by suppression of interband damping at near-infrared wavelengths.[27]

a-NSOM was used to study the optical near field of an aperture fabricated on a laser diode. The NSOM characterization results show that half maximum of the central peak of the near-field intensity distribution is 40 nm in the x direction and 100 nm in the y direction, as shown in Figs. 12-8 (b) and (c). This intense optical spot is localized within an area that is 50 times smaller than what one would obtain with conventional optics such as lenses (Rayleigh limit) in addition to the large intensity enhancement, as calculated from our simulations. The fabricated antenna exhibits also field enhancement on the far ends of the nanorods (see Figs. 12-8 (b) and (c)) as predicted by the simulations. For some applications such as optical data storage where a single optical spot is desired, these multiple spatial peaks around the structure could be problematic. This can be overcome by employing a bow tie design (see Fig. 12-9 (b)) since the tapering in the latter produces a "lighting rod" effect so that the fields will be mostly confined in the gap region.

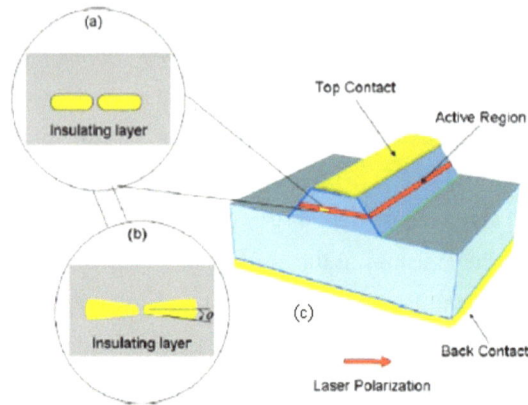

Fig. 12-8 (a) AFM topography and (b) a-NSOM image of resonant optical antenna fabricated on one of the facets of a commercial diode laser operating at ~0.83 μm wavelength. (c) Line scan of the near-field distribution along the antenna axis. Reprinted with permission from "Ertugrul Cubukcu, Eric A. Kort, Kenneth B. Crozier, and Federico Capasso, Plasmonic laser antenna, Appl. Phys. Lett. 89, 093120 (2006)" with copyright © 2006 of American Institute of Physics.

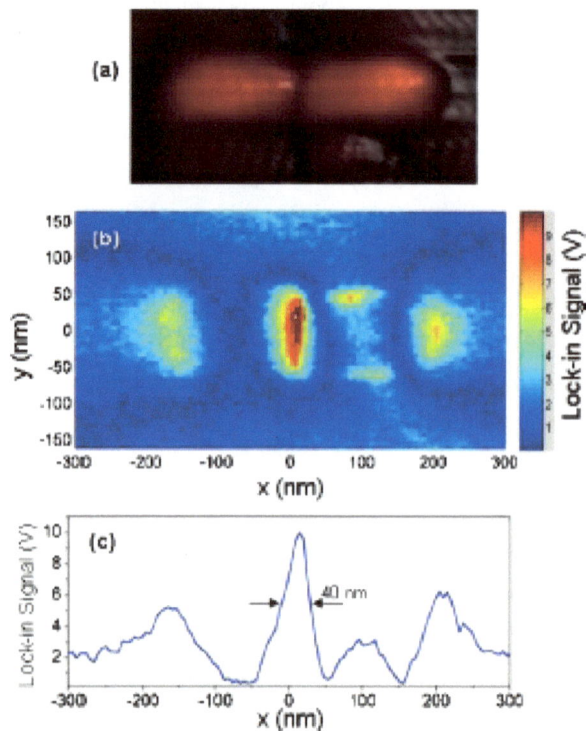

Fig. 12-9 (a) AFM topography and (b) a-NSOM image of resonant optical antenna fabricated on one of the facets of a commercial diode laser operating at ~0.83 μm wavelength. (c) Line scan of the near-field distribution along the antenna axis. Reprinted with permission from "Ertugrul Cubukcu, Eric A. Kort, Kenneth B. Crozier, and Federico Capasso, Plasmonic laser antenna, Appl. Phys. Lett. 89, 093120 (2006)" with copyright © 2006 of American Institute of Physics.

The small bump seen in the AFM image of Fig. 12-9 (a), below the right-hand side nanorod, is most likely due to nano-masking during FIB patterning, which causes stray near fields, as shown in Figs. 12-9 (b) and (c). It is likely that the composition of this bump has changed due to doping by Ga ions used in the FIB system and this may be the reason why the near field is perturbed strongly as if this bump acts like a metallic nanoparticle. Since we can only measure the relative near-field intensity but not its magnitude, we can provide an estimate of the actual near-field intensity only. On the basis of the simulated intensity enhancement of 800, an average optical power of 0.3 mW, as used in the experiments of Fig. 12-9, leads to peak intensity ~1 GW/cm^2 in the gap. This corresponds to an electric field of 105 V/cm at 10 nm above the surface, where the total intensity drops to 20% of its value right on the antenna surface. Such large localized optical intensities are very important for single molecule SERS.

12.5 Plasmonic lasers for ablation of Si

A fabrication of nanostructures ablated on silicon (100) by the plasmonic scattering of 780 nm, 220 fs laser pulses in the near-field of gold nanospheres was reported recently. [28] The enhanced plasmonic scattering of ultrashort laser light in the particle near-field was adopted to ablate well-defined nanocraters. Gold nanospheres of 150 nm diameter were deposited onto a silicon surface and irradiated with a single laser pulse.

Figure 12-10 illustrates the experimental setup. A femtosecond Ti: sapphire laser system (Spitfire, Spectra Physics, Mountain View, CA) delivered laser pulses of 220 fs temporal width and 780 nm wavelength at a 1 kHz repetition rate. An attenuator consisting of a half-wave plate and a polarizing cube beam-splitter was used to control the delivered laser power. Pulse energies were measured before the focusing system using an energy meter (Ophir PL10). The measured pulse energy transmission through the focusing system was estimated to be 64% by finding the ratio of energy before and after the objective. The sample was mounted onto an *x-y* translational stage. The setup provided simultaneous optical imaging and ablation of the sample through the same objective lens. A silicon (100) wafer (band gap energy of 1.14 eV) having a 21 Å native oxide layer as measured by ellipsometry was cut into 5×5 mm pieces and washed using a four step processes: sulfuric acid, distilled water, acetic acid, and methanol. Before and after irradiation, samples were characterized using SEM and AFM (Dimension 3100; Digitial Instruments). The AFM was operated in tapping mode.

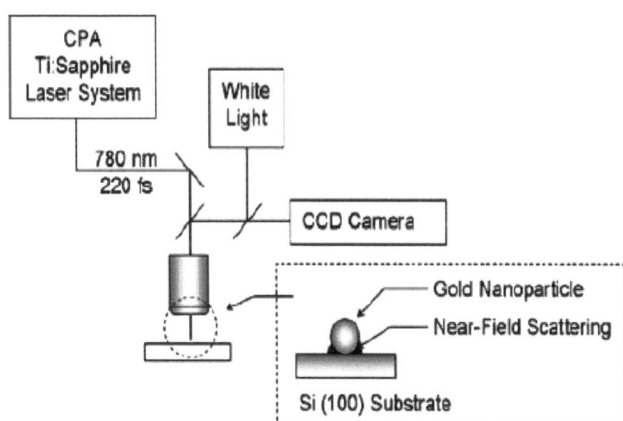

Fig 12-10 Schematic of the femtosecond laser ablation system. Reprinted with permission from "D. Eversole1, Luk'Yanchuk, A. Ben-Yakar, Appl. Phys. A 89, 283–291 (2007)" with copyright © 2007 of Springer-Verlag.

Figure 12-11 (a)~(f) shows AFM images and cross-sectional profiles of three nanocraters obtained using different cases of laser polarizations. In each case, the scattering pattern in the particle near-field was directly imprinted into the silicon surface. When the incident radiation E-field is directed at a 45° angle into the silicon substrate surface (*p*-polarized laser light), only one lobe of the dipolar scattering region interacts with the underlying substrate, and leads to the formation of a single circular crater. In Fig. 12-11 (a), the relative fluence interacting with the particle is 88 mJ/cm^2. A crater having a 100 nm diameter and 53 nm maximum depth was generated. Images of craters produced by laser polarizations parallel to the substrate surface are shown in Fig. 12-11 (b–c) when the substrate is angled 0° and 45° to the incident irradiation (*s*-polarized laser light), respectively. In each of the cases, when the laser polarization is parallel to the substrate surface, the generated craters have a double lobed crater structure, which follows the dipolar scattering pattern of the nanoparticle. Eversole *et. al.* found re-solidified gold at the ends of each dipolar crater structure after ablation. In Fig. 12-11, the gold can be clearly visualized and is represented as the peaks surrounding the craters in the cross-sectional profiles.

With further advances in nanoparticle deposition techniques, this type of ablative technology has the potential to directly impact the fields of laser-assisted nanomachining and nanolithography, and provides a patterning method which is not limited by diffraction and achieving high throughput with reduced production cost. Additionally, the use of antibody specific gold bio-conjugates presents the possibility for plasmonic laser-based nanosurgery, which could be used to manipulate molecules and subcellular structures.

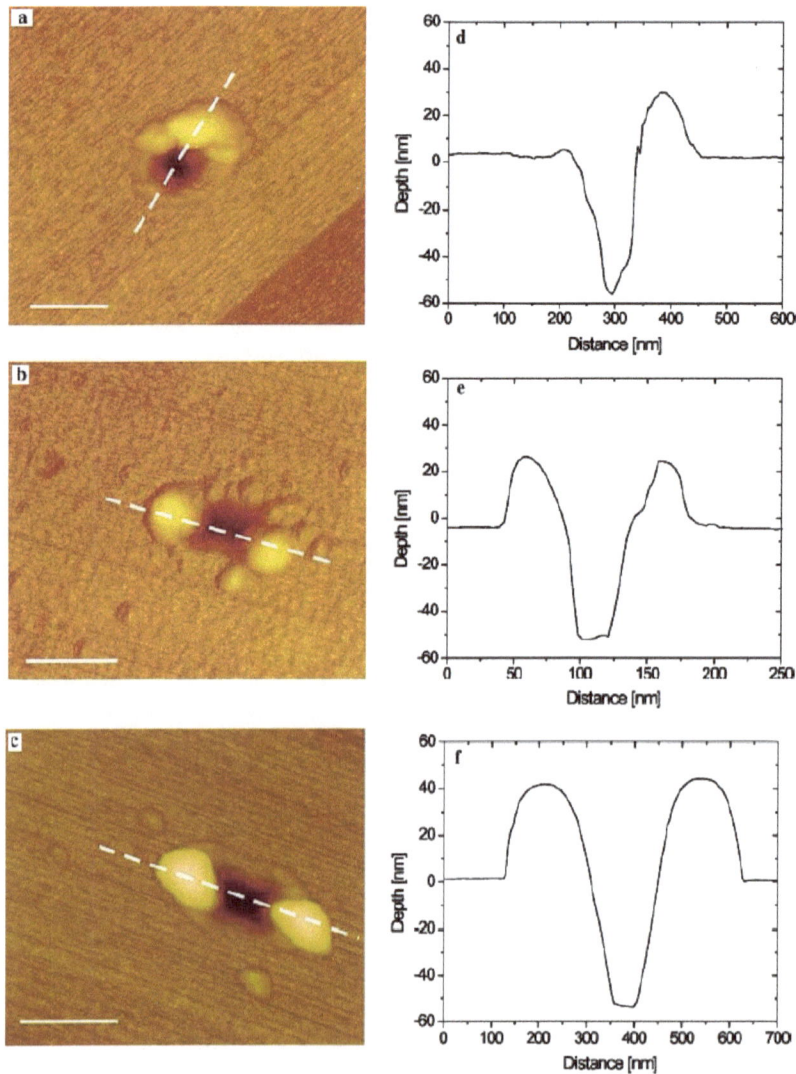

Fig. 12-11 AFM images of nano-craters ablated by 150 nm gold nanoparticles on silicon (100) and corresponding cross sections as found along the white dotted line: (a) and (d) 88mJ/cm^2 pulse fluence having *p*-polarization at 45° incident angle, (b) and (e) 58mJ/cm^2 pulse fluence having normal incidence, and (c) and (f) 128 mJ/cm^2 pulse fluence having *s*-polarization at 45° incident angle. The scale bars indicate 200 nm. Reprinted with permission from "D. Eversole1, Luk'Yanchuk, A. Ben-Yakar, Appl. Phys. A 89, 283–291 (2007)" with copyright ©2007 of Springer-Verlag.

12.6 Summary

Plasmonic lasers for amplification, beam confinement, and ablation were illustrated in this chapter. Essentially, they are applications of plasmonic structures such as bow-tie and nanorods combining with semiconductor lasers. These combinations can play important roles for the purpose of improving output quality of laser beam, and provide new performance of the laser applications such as ablation and fabrication. The research results presented here are just in the phases of initial and theoretical with a proof-of-concept. There are still many technical challenges from the experimental results to practical engineering applications. Hopefully, with rapid development of the plasmonic research, practical plasmonic lasers can be seen in industry in the near future.

Reference

[1] Manolatou, C., Rana, F., "Subwavelength Nanopatch Cavities for Semiconductor Plasmon Lasers", IEEE Journal of Quantum Electronics 44, 435 - 447 (2008).

[2] Yongqi Fu, Ngoi Kok Ann Bryan, Ong Nan Shing, "Integrated micro-cylindrical lens with laser diode for single-mode fiber coupling". IEEE Photonics Technology Letters. 12, pp. 1213-1215 (2000).

[3] Yongqi Fu, Ngoi Kok Ann Bryan, "Investigation of micro- diffractive lens with continuous relief with vertical-cavity surface emitting lasers using focused ion beam direct milling". IEEE Photonics Technology Letters. 13, pp.424-426, (2001).

[4] C. A. Balanis, *Antenna Theory*, John Wiley, NY (1997).

[5] B. S. Williams, S. Kumar, H. Callebaut, and Q. Hu, "Terahertz quantum-cascade laser at λ=100 *μ*m using metal waveguide for mode confinement", Appl. Phys. Letts., 83, 2124 (2003).

[6] Rupert F. Oulton, Volker J. Sorger, Thomas Zentgraf, Ren-Min Ma, Christopher Gladden, Lun Dai, Guy Bartal, Xiang Zhang, "Plasmon lasers at deep subwavelength scale", Nature 461, 629-632 (2009).

[7] J.A. Zapien,Y. Jiang, X.M. Meng, W. Chen, F.C.K.Au, Y.Lifshitz, and S.T.Lee, "Room-temperature single nanoribbon lasers". Appl. Phys. Lett. 84, 1189-1191 (2004).

[8] David J. Bergman and M.I. Stockman, "Surface Plasmon Amplification by Stimulated Emission of Radiation: Quantum Generation of Coherent Surface Plasmons in Nanosystems," Phys. Rev. Lett. 90, 027402 (2003).

[9] N. I. Zheludev1, S. L. Prosvirnin, N. Papasimakis, and V. A. Fedotov, "Lasing spacer," Nature Photonics 2, 351-354 (2008).

[10] V. A. Fedotov, M. Rose, S. L. Prosvirnin, N. Papasimakis, and N. I. Zheludev, "Sharp Trapped-Mode Resonances in Planar Metamaterials with a Broken Structural Symmetry," Phys. Rev. Lett. 99, 147401 (2007).

[11] R. Li, A. Banerjee and H. Grebel, "The possibility for surface plasmons lasers", Opt. Express 17 (3), 1622-1627 (2009).

[12] S. Noda, M. Yokoyama, M. Imada, A. Chutinan, and M. Mochizuki, "Polarization Mode Control of Two-Dimensional Photonic Crystal Laser by Unit Cell Structure Design," Science 293, 1123-1125 (2001).

[13] X. Wu, A. Yamilov, X. Liu, S. Li, V. P. Dravid, R. P. H. Chang, and H. Cao, "Ultraviolet photonic crystal laser," Appl. Phys. Lett. 85, 3657-3659 (2004).

[14] D. J. Bergman and M. I. Stockman, "Surface Plasmon Amplification by Stimulated Emission of Radiation: Quantum Generation of Coherent Surface Plasmons in Nanosystems," Phys. Rev. Lett. 027402 (2003).

[15] Nanfang Yu, Ertugrul Cubukcu, Laurent Dieh, David Bour, Scott Corzine, Jintian Zhu, Gloria Höfler, Kenneth B. Crozier, and Federico Capasso, "Bowtie plasmonic quantum cascade laser Antenna," Opt. Express 15, 13272-13281 (2007).

[16] FDTD simulations were performed using a commercial software XFDTD (Remcom Inc.): http://www.remcom.com/

[17] K. B. Crozier, A. Sundaramurthy, G. S. Kino, and C. F. Quate, "Optical antennas: resonators for local field enhancement," J. Appl. Phys. 94, 4632-4642 (2003).

[18] N. Yu, L. Diehl, E. Cubukcu, C. Pflügl, D. Bour, S. Corzine, J. Zhu, G. Höfler, K. B. Crozier, and F. Capasso, "Near-field imaging of quantum cascade laser transverse modes," Opt. Express 15, 13227-13235 (2007).

[19] R. Hillenbrand, B. Knoll, and F. Keilmann, "Pure optical contrast in scattering-type scanning near-field microscopy," J. Microsc. 202, 77-83 (2001).

[20] B. Knoll, and F. Keilmann, "Enhanced dielectric contrast in scattering-type scanning near-field optical microscopy," Opt. Commun. 182, 321-328 (2000).

[21] 10. Andrey K. Sarychev1 and Gennady Tartakovsky2, "Magnetic plasmonic metamaterials in actively pumped host medium and plasmonic nanolaser," Phys. Rev. B 75, 085436 (2007).

[22] Ertugrul Cubukcu, Eric A. Kort, Kenneth B. Crozier, and Federico Capasso, "Plasmonic laser antenna", Appl. Phys. Lett. 89, 093120 (2006).

[23] R. D. Grober, R. J. Schoelkopf, and D. E. Prober, "Optical antenna: Towards a unity efficiency near-field optical probe," Appl. Phys. Lett. 70, 1354 (1997).

[24] P. J. Schuck, D. P. Fromm, A. Sundaramurthy, G. S. Kino, and W. E. Moerner, "Improving the Mismatch between Light and Nanoscale Objects with Gold Bowtie Nanoantennas," Phys. Rev. Lett. 94, 017402 (2005).

[25] P. Muhlschlegel, H. J. Eisler, O. J. F. Martin, B. Hecht, and D. W. Pohl, "Resonant Optical Antennas", Science 08, 1607 (2005).

[26] A. Bouhelier, R. Bachelot, G. Lerondel, S. Kostcheev, P. Royer, and G. P. Wiederrecht, "Band-Selective Measurements of Electron Dynamics in VO_2 Using Femtosecond Near-Edge X-Ray Absorption," Phys. Rev. Lett. 95, 067405 (2005).

[27] C. Sönnichsen, T. Franzl, T. Wilk, G. von Plessen, J. Feldmann, O. Wilson, and P. Mulvaney, "Drastic Reduction of Plasmon Damping in Gold Nanorods", Phys. Rev. Lett. 88, 077402 (2002).

[28] D. Eversole1, Luk'Yanchuk, A. Ben-Yakar, "Plasmonic laser nanoablation of silicon by the scattering of femtosecond pulses near gold nanospheres", Appl. Phys. A 89, 283–291 (2007).

13 METAMATERIALS-BASED ANTENNAS

Abstract: As one of important applications of metamaterials, the metamaterials-based antennas were introduced in this chapter. Firstly, theoretical background regarding the antenna was given. Then two types plasmonic structures being used as antennas: electromagnetic band-gap (EBG) structures-based antenna and waveguide slit array antenna were presented. Some experimental results were demonstrated also. Some unique properties of the antennas were shown in terms of these results.

13.1 Introduction

Metamaterials are artificial periodic media with unusual electromagnetic properties. The size of their unit cells is much smaller than the incident electromagnetic wavelength, thus allowing the definition of effective material parameters such as a permittivity ε, a permeability μ, and a refractive index n. A brief review of metamaterials was given already in Chapter 5. In this chapter, we mainly targeted one of its principal applications, microwave antennas. Here "left-handed" metamaterials (LHM) were considered for which their permittivity and permeability are both negative. These metamaterials exhibit characteristics of backward-wave propagation and therefore a negative index of refraction, as was predicted in the pioneering work of Victor Veselago [1]. LHM has the following characteristics: (1) simultaneous negative permittivity (-ε) and permeability (-μ); (2) reversal of Snell's Law (negative index of refraction), Doppler effect, and Cerenkov Effect; (3) electric field, magnetic field, and wavevector of electromagnetic wave in a LHM form a left-handed triad; (4) LHM supports backward waves: anti-parallel group and phase velocity; and (5) artificial effectively homogenous structure: metamaterial. Currently, naturally occurring LHM has not been discovered yet.

Such "left-handed" or "Negative-Refractive-Index" (NRI) media were firstly implemented using bulk periodic arrays of thin wires to synthesize negative permittivity, meanwhile, split-ring resonators to synthesize a negative permeability [2]. A different approach for synthesizing 2-dimensional (2D) NRI media in planar form was proposed in Ref. [3] and [4] which was called transmission-line approach. The other analyzing method of the LHM is a resonant approach. Apart from the LHM, there is composite right/left-handed metamaterial (CRLH). Physical realization of CRLH metamaterials, distributed microstrip implementation based on Sievenpiper mushroom structure was reported in 1999.[5] Currently, the metamaterial-based microwave devices have four types: (1) dominant leaky-wave antenna, (2) small metamaterial antenna, (3) dual-band hybrid coupler, and (4) negative refractive index flat lens.

LHM being realized on the basis of split ring resonators (SRR) was named "Resonant approach" towards LHM, as shown in Fig. 13-1.[6] The SRR has the features of μ<0 at resonance ring and ε<0 at metal wire. The SRR-based metamaterials only exhibit LH properties at resonance (inherently narrow-band) and lossy. In addition, the SRR-based LHM is bulky. Thus it is not practical for microwave engineering applications. In contrast, backward wave transmission line can form a non-resonant LHM.[7, 8] Transmission line approach is put forth on the basis of the dual of a conventional transmission line. It is a perfect left-handed (LH) transmission line and not resonant dependent, and thus low-loss and broad-band performance. However, perfect LH transmission line is not possible due to unavoidable parasitic right-handed (RH) effects occurring with physical realization.

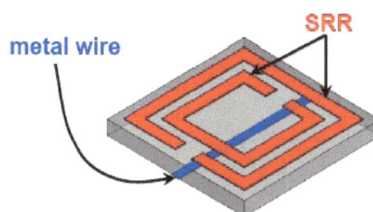

Fig.13-1 Schematic diagram of SRR-based LHM

CRLH leaky-wave antenna operated at dominant mode was reported in Ref. 9. It has the following characteristics: (1) operating in leaky regions which propagates in backward for $\beta < 0$ and forward for $\beta > 0$; (2) broadside radiation ($\beta = 0$) balanced case: $v_g(\beta=0) \neq 0$; and (3) working in fundamental mode. Small metamaterial antenna was designed using resonant antenna theory. Typical example is CRLH patch antenna which has two CRLH unit cells. Compared with conventional right-handed patch antenna (treat as periodic, consisting of 2 right-handed "unit-cells"), CRLH can have the same half wavelength field distribution, but at much lower frequency. In case of $n=0$, as the unit number N increases, gain of the CRLH antenna increases also, but resonant frequency does not change much.[10] Typical example of the dual-band hybrid coupler is CRLH / CRLH hybrid structures.[11] It has characteristics of dual-band functionality for an arbitrary pair of frequencies f_1, f_2; working principle that transition frequency (f_0) provides DC offset an additional degree of freedom with respect to the phase slope; and applications in multi-band systems.

In this chapter, we paid more attention to the negative refractive index flat lens. Theoretical design background, prototype design, fabrication and performance testing were illustrated in detail in the following sections. In addition, artificial LHM-based magnetic conductors and electrically small radiating and scattering systems were emphasized. Single negative, double negative, and zero-index metamaterial systems were discussed as a means to manipulate their size, efficiency, bandwidth, and directivity characteristics.

13.2 Theoretical design background

Composite electromagnetic materials can be regarded as a homogenous medium under the following conditions: (1) lattice unit size is far smaller than wavelength of incident electromagnetic (EM) wave; and (2) in case that radiation cannot recognize internal structure of the materials, interaction effect between the materials and electromagnetic wave can be described by corresponding effective permittivity ε_{eff} and effective permeability μ_{eff}.

Figure 13-2 shows four cases of permittivity ε and permeability μ. For the case of size of the lattice unit cell of LHM far smaller than working wavelength λ. Theory of effective homogeneous medium is available for the EM wave analysis. The metallic periodic opened surface resonant ring (SRR) and wire, as shown in Fig. 13-2 (c), which were put forth by J. B. Pendry *et. al.* can realize negative permeability and negative permittivity, respectively. This report greatly inspired researchers for studying the artificial metamaterials in depth. This section focuses on study of electromagnetic response of metallic wire for an incident EM wave for the purpose of further realizing zero-approaching effective dielectric constants $\varepsilon_{eff} \approx 0$. For the non-magnetic materials, relationship between dielectric and refractive index is

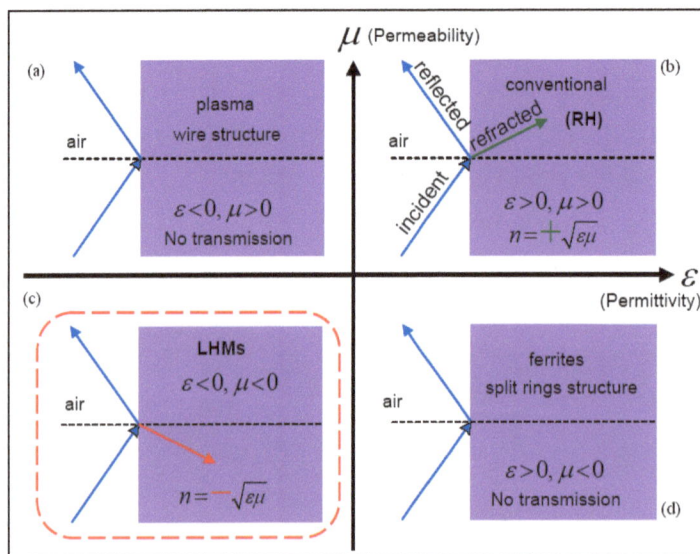

Fig.13-2 Chart explaining the four cases of variations of the dielectric constants ε and permeability μ: (a) $\varepsilon > 0$, $\mu > 0$; (b) $\varepsilon < 0$, $\mu > 0$; (c) $\varepsilon < 0$, $\mu < 0$; and (d) $\varepsilon > 0$, $\mu < 0$. LHMs is available for case (c).

$n_{eff}^2 = \varepsilon_{eff}$. The low refractive index materials discussed in this chapter indicates those artificial materials with dielectric ε

ranging from 0 to 1 and approaching zero. The materials are constructed to be periodic lined metallic wires or periodic thin planar metallic grid multilayers with rectangular cross section in which is filled with low dielectric polymers. Abbreviation name "MTMs" is used to represent the artificial medium, and named Metamterials hereinafter. Considering fabrication issue, we use metallic grid structures to realize the low refractive index medium. Retrieval method of dispersion parameter (S) can be adopted to refine ε_{eff} and μ_{eff} of the structures.[12-14]

13.2.1 Model of metallic free electron gas

Surface plasmon is composed of a mass of free movement charges on metal surface. The entity of the charges is neutralization. Electromagnetic interaction between the two adjacent charges presents plasmon oscillation behavior. Similar to the charged particles of gas plasmon, the metallic conductivity can be described using Drude model. In this model, the metal is simply regarded as covalent electron gas which behaves oscillation as well if external electric field is applied on the metal. Like plasmon frequency of ionized gas, it can be written as:

$$\omega_p^2 = \frac{Ne^2}{\varepsilon_0 m} \tag{13.1}$$

where N is the electron density (electron amount of unit volume), e is the charge of electrons, m is the electron mass, ε_0 is the dielectric constant in vacuum. ε_0 and e/m are constant parameters, and hence ω_p depends on electron mass only. Therefore, the plasmon frequency ω_p can be tuned by adjusting electron density N. According to Drude model, effective relative dielectric constant is expressed as

$$\varepsilon_{eff}(\omega) = 1 - \frac{\omega_p^2}{\omega^2 + j\gamma\omega} \tag{13.2}$$

where, ω_p is the plasmon frequency which is discussed in Eq. (13.1), γ is the damping factor.

It can be seen from Eq. (13-2) that ε_{eff} is a complex number, and can be rewritten as

$$\varepsilon_{eff}(\omega) = \varepsilon_1 + j\varepsilon_2 = \left(1 - \frac{\omega_p^2}{\omega^2 + \gamma^2}\right) + j\frac{\gamma\omega_p^2}{\omega(\omega^2 + \gamma^2)} \tag{13.3}$$

where real part of $\varepsilon_{eff}(\omega)$ is $\varepsilon_1 = 1 - \dfrac{\omega_p^2}{\omega^2 + \gamma^2}$. It determines dispersion relationship of EM wave propagation. Imaginary

part of $\varepsilon_{eff}(\omega)$ is $\varepsilon_2 = \dfrac{\gamma\omega_p^2}{\omega(\omega^2 + \gamma^2)}$. It determines absorption and lossy of the EM wave propagation. For the metal

materials, $\omega_p \approx 10^{16}$ Hz which is equivalent to violet frequency, $\gamma \approx 10^{13} \sim 10^{14}$ Hz which is equivalent to infrared frequency. For the non-ferromagnetism substances $\mu_{eff}=1$, their complex refractive index is

$$n = \sqrt{\varepsilon_{eff}(\omega)} = \sqrt{\varepsilon_1 + j\varepsilon_2} \tag{13.4}$$

For the definition $n=n_1+jn_2$, have

$$\left(n_1 + jn_2\right)^2 = \varepsilon_1 + j\varepsilon_2 \tag{13.5}$$

then

$$n_1^2 - n_2^2 = \varepsilon_1 \tag{13.6}$$

$$2n_1 n_2 = \varepsilon_2 \tag{13.7}$$

Real and imaginary parts of refractive index n is

$$n_1 = \sqrt{\frac{1}{2}\left(\sqrt{\varepsilon_1^2 + \varepsilon_2^2} + \varepsilon_1\right)}$$
(13.8)

and

$$n_2 = \sqrt{\frac{1}{2}\left(\sqrt{\varepsilon_1^2 + \varepsilon_2^2} - \varepsilon_1\right)}$$
(13.9)

, respectively. They are all function of frequency. In respect to above expressions, it can be foreseen that different characteristics of the metals can be produced for the different frequency.

For EM wave at microwave band (3×10^8 Hz$\sim$$3\times10^{12}$ Hz), $\omega\approx\gamma$, from Eq.(13.3) we have $\varepsilon_1 = 1 - \omega_p^2/(\omega^2+\gamma^2) \approx -\omega_p^2/\gamma^2 < 0$, and $\varepsilon_2 \approx \omega_p^2/\omega\gamma$. Substituting ε_1 and ε_2 into Eq. (13.8) and (13.9), we have

$$n_1 = \sqrt{\frac{1}{2}\left(\sqrt{\left(-\frac{\omega_p^2}{\gamma^2}\right)^2 + \left(\frac{\omega_p^2}{\omega\gamma}\right)^2} - \frac{\omega_p^2}{\gamma^2}\right)} = \sqrt{\frac{1}{2}\left(\frac{\omega_p^2}{\gamma^2}\sqrt{1+\left(\frac{\gamma}{\omega}\right)^2} - 1\right)} \approx \sqrt{\frac{\omega_p^2}{2\omega\gamma}}$$
(13-10)

$$n_2 = \sqrt{\frac{1}{2}\left(\sqrt{\left(-\frac{\omega_p^2}{\gamma^2}\right)^2 + \left(\frac{\omega_p^2}{\omega\gamma}\right)^2} + \frac{\omega_p^2}{\gamma^2}\right)} = \sqrt{\frac{1}{2}\left(\frac{\omega_p^2}{\gamma^2}\sqrt{1+\left(\frac{\gamma}{\omega}\right)^2} + 1\right)} \approx \sqrt{\frac{\omega_p^2}{2\omega\gamma}}$$
(13-11)

It can be seen that $n_1 \approx n_2$ in this case. The EM wave will rapidly decay because of the large n_2, and is too difficult to propagate further. The inherent negative dielectric response of the metals is covered accordingly. For the frequency as high as $\gamma < \omega < \omega_p$, Eq. (13.3) can be well approximated as

$$\varepsilon_{eff}(\omega) = \varepsilon_1 \approx 1 - \frac{\omega_p^2}{\omega^2} < 0$$
(13.12)

Thus expression of the metal refractive index can be written as

$$n = \sqrt{1 - \frac{\omega_p^2}{\omega^2}}$$
(13.13)

Then n becomes a complex, and $n_1 = 0$, so the EM wave decays. However, if $\omega > \omega_p$, then $\varepsilon_{eff}(\omega) = 1 - \frac{\omega_p^2}{\omega^2} > 0$, and n is a real number with $n_2 = 0$. The metal behaves like a transparent medium. If $\omega \approx \omega_p$, $n \approx 0$, the low refractive index (also called near-zero refractive index) can be realized accordingly in this case.

13.2.2 Model of low refractive index materials

As the information reflected by Eq. (13.12) and (13.13), for obtaining the metals behaves as a transparent medium, must satisfy the condition of $\omega > \omega_p$. But it is impossible to be realized at microwave band unless the ω_p for metals can be degraded to the one at the microwave band. As well known, for metals $\omega_p^2 = Ne^2/\varepsilon_0 m$, and ε_0 and e are all constant parameters. ω_p is determined by electron density N and mass m. If we can adjust N and m, then ω_p can be tuned accordingly. Inspired by this point, J. B. Pendry put forth an idea that using an effective model of the metallic periodic wires to realize ω_p at the microwave band. Its basic mind is that limits electrons in the metallic wires so as to reduce the effective electron density and increases the effective mass, as shown in Fig.13-3.[15] In this figure, r is the radius of the metallic wires and a the lattice constant. When incident wavelength $\lambda \gg r$ and a, EM wave cannot recognize its internal structures. Structures of the metallic wires array can be regarded as a bulk compound electromagnetic material. Interaction response between the array and EM wave can be described using ε_{eff} and μ_{eff}. Here $\mu_{eff}=1$, thus ε_{eff} is relevant to ω_p only.

According to filling factor of the wires array, effective electron density of the metallic array is reduced and can be expressed as

$$N_{eff} = N\frac{\pi r^2}{a^2} \tag{13.14}$$

where N is electron density of the metallic wires.

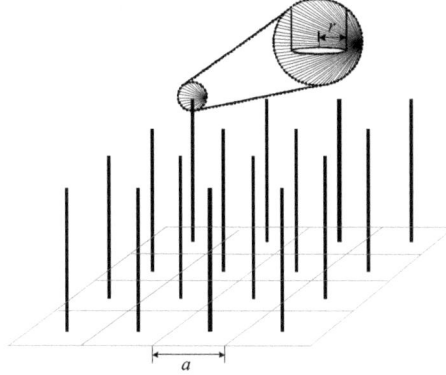

Fig. 13-3 Wire structure for realizing modulation of plasmon frequency ω_p at microwave. Reprinted with permission from "J B Pendry, A J Holdenz, D J Robbinsz and W J Stewartz, J. Phys.: Condens. Matter 10, 4785–4809 (1998)" with copyright of © 1998, IOP Publishing Ltd.

When electric field direction is the same as the wires axis, inducting current will be generated. Meanwhile, electric current flows along the wires will induce surrounding magnetic field. Then the wires are equivalent to an inductance which will resist current in the metal wires, *i.e.*, electron movement in the wires will be slow down due to the resistance from the inductance. It can be equivalently understood that the effective mass increases. In terms of classical electrodynamics, there is momentum conservation in the metal wires, hence we have

$$e\mathbf{A}(R) = m_{eff}v \tag{13.15}$$

where $A(R)$ is the vector of magnetic field, m_{eff} is the effective mass of free electrons, R is the distance between arbitrary point of the structure and the metal wires, and v is the mean velocity of the free electrons movement. $A(R)$ can be written as

$$\mathbf{A}(R) = \frac{\mu_0 \pi r^2 nve}{2\pi}\ln(a/R) \tag{13.16}$$

Put Eq. (13.6) into Eq. (13.15) and (13.16), and obtain the effective mass expressed as

$$m_{eff} = \frac{\mu_0 e^2 \pi r^2 N}{2\pi}\ln(a/r) \tag{13.17}$$

At this moment, ω_p can be written as

$$\omega_p = \sqrt{\frac{N_{eff}e^2}{\varepsilon_0 m_{eff}}} = \sqrt{\frac{N\frac{\pi r^2}{a^2}e^2}{\varepsilon_0 \frac{\mu_0 e^2 \pi r^2 N}{2\pi}\ln(a/r)}} = \sqrt{\frac{2\pi c_0^2}{a^2 \ln(a/r)}} \tag{13.18}$$

where c_0 is the light speed in free space.

It can be seen from Eq. (13.18) that ω_p is none of relations to the physical parameters in microscale such as N and v

in the metal wires, and depends on R and r only. The metal wires radius r should be small enough so as to degrade its resonant frequency. Then the wavelength at ω_p is far larger than lattice constant a, and Bragg diffraction is avoided accordingly. The effective medium theory is applicable at this moment. If r becomes large, lets $\ln(a/r) \approx 1$, calculation result shows that the wavelength at ω_p is two times of a. Plasmon behavior will be influenced because interference will occur between the Bragg diffraction effect and the plasmon in this case. Therefore, sufficiently thin metal wires are required for the structure design.

If conductivity of the metal wires σ is considered, effective dielectric of the metal wires array is

$$\varepsilon_{\mathit{eff}}(\omega) = 1 - \frac{\omega_p^2}{\omega\left(\omega + j\varepsilon_0 a^2 \omega_p^2/\pi r^2 \sigma\right)} = 1 - \frac{\omega_p^2}{\omega\left(\omega + j\left(a/r\right)^2 \varepsilon_0 \omega_p^2/\pi\sigma\right)} \qquad (13.19)$$

and the damping factor is

$$\gamma = \left(a/r\right)^2 \varepsilon_0 \omega_p^2/\pi\sigma = \omega_p\left(a/r\right)^2 \varepsilon_0 \sqrt{\frac{2\pi c_0^2}{a^2 \ln(a/r)}}/\pi\sigma \qquad (13.20)$$

Normally, conductivity of metal is large, *e.g.*, for Cu, $\sigma = 5.8\times10^7$ s/m. For the metal wires with $\alpha = 5$ mm, $r = 0.001$ mm, by Eq. (13.20) obtain $\gamma \approx 0.01\omega_p$. As can be seen, influence of γ on $\varepsilon(\omega)$ is much less than that of ω_p. In the case of $\omega \ll \gamma$, the wires array strongly attenuates EM wave; and for the case of $\omega \gg \gamma$, the influence of damping factor can be ignored here. Hence effective dielectric of the metal wires array is

$$\varepsilon_{\mathit{eff}}(\omega) = 1 - \frac{\omega_p^2}{\omega^2} \qquad (13.21)$$

If $\omega_p > \omega$, then $\varepsilon_{\mathit{eff}}(\omega) < 0$; and if $\omega_p < \omega$, then $0 < \varepsilon_{\mathit{eff}}(\omega) < 1$.

If we ignore the metal lossy, *i.e.*, $\sigma \to \infty$, at any cases, $\varepsilon_{\mathit{eff}}$ becomes

$$\varepsilon_{\mathit{eff}}(\omega) = 1 - \frac{2\pi c_0^2}{\omega^2 a^2 \ln(a/r)} \qquad (13.22)$$

Synthesizing Eq. (13-18) to (13-22), we can conclude that ω_p and effective dielectric $\varepsilon_{\mathit{eff}}$ of the metal wire model can be expressed universally by Eq. (13-18) and (13-21), respectively. Thus refractive index of the metal wire array can be deduced from Eq. (13-13) as

$$n_{\mathit{wire}} = \sqrt{1 - \frac{\omega_p^2}{\omega^2}} \qquad (13.23)$$

For the case of $\omega_p < \omega$, we have $0 < n_{\mathit{wire}} < 1$. At this condition, if $\omega \to \omega_p$, then $n_{\mathit{wire}} \to 0$, and $n_{\mathit{wire}} \approx 0$ can be realized accordingly.

It is worthy to point out that the modulation method realizing near zero $\varepsilon_{\mathit{eff}}$ or n_{wire} greatly upper extended the plasmon frequency to far infrared and even microwave region. However, the circular cross section of the metal wires is adopted in order to simplify the theoretical analysis. Practically, because of limitation of fabrication, rectangular cross

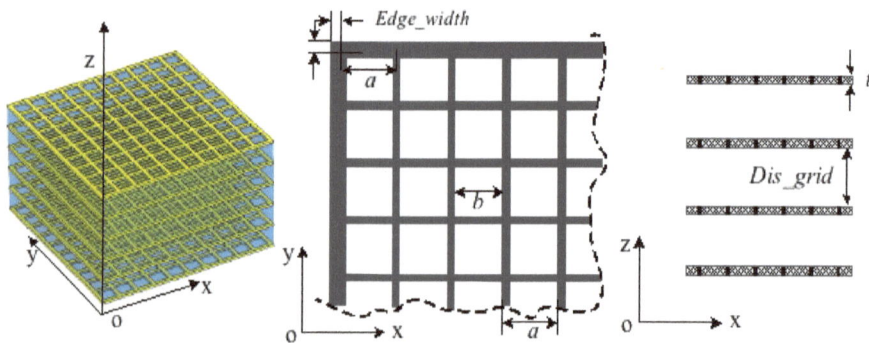

Fig. 13-4 Schematic diagram of metal grid-lattice structures.

section of the metal grid layer is employed to replace the circular metal wires because the rectangular metal strips and square is easier to be fabricated by means of the technique of printing circuit board (PCB). Considering this, we paid more attention to the rectangular metal grid layer (as shown in Fig. 13-4) for modeling ω_p of the low refractive index of metamaterials. For the metal grid structures with superimposed layers, thickness of the square metal grid is t, and period is a. For the central empted square holes with lateral length b, width of the metal strip is $(a-b)$, gap between the two adjacent layers is Dis_grid. As t is very thin, flatness of the grid plane becomes worse due to gravity effect. It is not what we desired. To solve this problem, the polymers with strong mechanical strength and low refractive index is filled into the gap for the purpose of keeping the flatness of the grid plane.

Now let us to consider such a problem, can realize ω_p at microwave band by changing the cross section of metal wire/strip from circular to be rectangular? If yes, can ω_p still be expressed by Eq. (13.18)? Actually, the expression will differ slightly even if the answer is positive. We discussed this problem below.

Starting from Eq. (13.18), we cited the concept of effective radius. Assuming width of the metal grid is $(a-b)$, thickness t, then area of the cross section is

$$S = (a-b) \cdot t \tag{13.24}$$

If the circular instead of the rectangle cross section of the metal wires is considered, the corresponding effective radius becomes

$$r_{eff} = \sqrt{\frac{S}{\pi}} = \sqrt{\frac{(a-b) \cdot t}{\pi}} \tag{13.25}$$

Putting Eq. (13-25) into Eq. (13-18), we obtain

$$\omega_p = \sqrt{\frac{2\pi c_0^2}{a^2 \ln\left(a / \sqrt{\frac{(a-b) \cdot t}{\pi}}\right)}} \tag{13.26}$$

By the relationship $\omega_p = 2\pi f_p$, ω_p (GHz) can be derived as

$$f_p = \frac{c_0}{a \sqrt{2\pi \ln\left(a / \sqrt{\frac{(a-b) \cdot t}{\pi}}\right)}} \tag{13.27}$$

If the filling polymer with dielectric constant ε_r is employer in the gap, then

$$f_p = \frac{c_0}{a \sqrt{2\pi \varepsilon_r \ln\left(a / \sqrt{\frac{(a-b) \cdot t}{\pi}}\right)}} \tag{13.28}$$

It is worthy to point out that for the metal wire model with lattice constant a, and period a of grid units at x-y plane of the metal grid model, the parameter a may be different from Dis_grid at x-z and y-z planes. What Eq. (13.28) expressed is the case of $Dis\_grid=a$; for the case of $Dis\_grid \neq a$, the equation is an approximate expression only. To derive a more accurate expression, a correction of Eq. (13-27) and (13-29) is required.

It can be seen that once we know f_p and ω_p, the refractive index of the metal grid layered structures can be easily calculated by Eq. (13-23). Therefore, influence study of the metal grid layered structures on f_p becomes more important. With respect to Eq. (13-28), the parameters affecting f_p are units period a, lateral size b, thickness t, and dielectric constant of the filling medium ε_r. Initial parameters of the structures are $a=0.3\lambda_{14.5GHz}$, $b=0.25\lambda_{14.5GHz}$, $t=0.01\lambda_{14.5GHz}$, and $\varepsilon_r=1.0$, where $\lambda_{14.5GHz}$ is the wavelength of EM wave at frequency 14.5 GHz (20.67 mm).

For the metal wires with arbitrary cross section, as shown in Fig. 13-5, n_{eff} and m_{eff} can be deduced as the following equations: [16]

$$n_{eff} = \frac{l\delta n_0}{a^2} \tag{13.29}$$

where n_0 is the electron density for unit length of the metal wires, δ is the skin depth, l is the peripheral length of the metal wires, and a is the period of the wires.

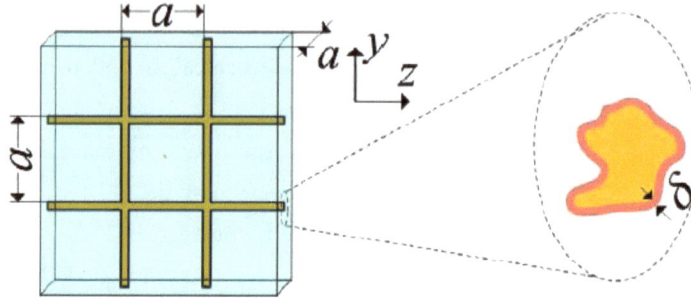

Fig. 13-5 Two-dimensional metal grid with peripheral length l of the metal wires. The wires were designed with arbitrary cross section.

$$m_{eff} = \frac{\mu_0 \cdot l \cdot \delta \cdot n_0 \cdot e^2}{2 \cdot \pi} \left[\ln(\frac{l}{2 \cdot \sqrt{\pi} \cdot a}) - \frac{l^2}{8 \cdot \pi \cdot a^2} + \frac{1}{2} \right] \tag{13.30}$$

and the corresponding plasmon frequency can be expressed as

$$\omega_p^2 = \frac{2 \cdot \pi \cdot c_0^2}{a^2 \left[\ln(\frac{l}{2 \cdot \sqrt{\pi} \cdot a}) - \frac{l^2}{8 \cdot \pi \cdot a^2} + \frac{1}{2} \right]} \tag{13.31}$$

where c_0 is the light speed in vacuum.

It can be seen from Eq. (13.31) that plasmon frequency depends on period a and peripheral length l of the cross section of the wires only. It can be explained that most of the effective free electrons concentrate on the metal surface when skin depth δ is small. The number of effective free electrons N is in proportional to l. For a fixed a, the smaller value of the l, the lower of the corresponding ω_p will be. Similarly, for a fixed l, the larger of the a, the lower of the corresponding ω_p will be also. Therefore, for the two structures with the same value of l and a, the corresponding plasmon frequency will be the same, and irrelevant to the concrete internal structures. In the case of extreme small area of the cross section of the metal wires, we can use area S of the cross section to replace the peripheral length l for the purpose of deducing the equations regarding n_{eff}, m_{eff}, and ω_p.

In order to obtain effective features of the artificial electromagnetic materials, *e.g.*, permittivity, permeability, and refractive index, we can make use of S parameter method to restore the corresponding effective dielectric constants by means of the calculated or measured (*e.g.*, using network analyzer) transmission of the artificial structures. The so-called S parameter method here was put forth by D.R. Smith firstly in 2002.[17] Its basic idea is similar to the conventional optical method that calculating optical constants using measured reflection or transmission of thin film layers. Nowadays, it becomes an important approach for characterization of the artificial electromagnetic materials. According to basic theory of electromagnetic, for a homogeneous planar material with a certain thickness and medium of air at input and output ports ($n_{air}=1$, $z_{air}=1$), transmission parameter S_{21} and reflection parameter S_{11} (number 1 denotes input port, and 2 output port) can be written as [18]

$$\frac{1}{S_{21}} = \left[\cos(nkd) - \frac{i}{2}(z + \frac{1}{z})\sin(nkd) \right] \tag{13.32}$$

$$S_{11} = \frac{i}{2}\left(\frac{1}{z} - z\right)\sin(nkd) \tag{13.33}$$

where $k=2\pi/\lambda$, d is the material thickness (for the artificial periodic structures, minimum value is one unit length), n and z is refractive index and impedance of the materials, respectively. n and z can be deduced from above two equations accordingly as

$$z = \sqrt{\frac{(1+S_{11})^2 - S_{21}^2}{(1-S_{11})^2 - S_{21}^2}} \tag{13.34}$$

$$n = \pm \frac{1}{kd}\cos^{-1}\left[\frac{1}{2S_{21}}(1 - S_{11}^2 + S_{21}^2)\right] \tag{13.35}$$

where n and z are complex numbers. Real part and imaginary part of n can be expressed respectively as follows

$$\operatorname{Im}(n) = \pm \operatorname{Im}\left\{\frac{1}{kd}\cos^{-1}\left[\frac{1}{2S_{21}}(1 - S_{11}^2 + S_{21}^2)\right]\right\} \tag{13.36}$$

$$\operatorname{Re}(n) = \pm \operatorname{Re}\left\{\frac{1}{kd}\cos^{-1}\left[\frac{1}{2S_{21}}(1 - S_{11}^2 + S_{21}^2)\right]\right\} + \frac{2\pi m}{kd} \tag{13.37}$$

where m is the arbitrary integral. ε_{eff} and μ_{eff} can be derived according to the following relationship once n and z are defined.

$$\varepsilon = n/z, \quad \mu = nz \tag{13.38}$$

One of difficulty using Eq. (13.36) and (13.37) for the S parameters reversal is that how to correctly determine m value in the equations. Solved n may have multi-values because anti-cosine function is involved in the two equations. Normally, it is necessary that wavelength of electromagnetic wave in the medium is far larger than the medium thickness, i.e., $d \leq \lambda$ so as to obtain a correct branch solution of the anti-cosine function. In addition, reversed dispersion relationship must satisfy electromagnetic characteristics of the source free materials, i.e., $\operatorname{Re}(z)>0$, and $\operatorname{Im}(n)>0$. For determining $\operatorname{Re}(n)$, the correct branch solution can be derived by means of simulating the mediums with different thicknesses. A region of resonant frequency exists because LHM is a resonant material. Multi-solutions of the anti-cosine function are close each other in the region. Thus it is difficult to determine a suitable sole solution, and leads to difficulty of determination of ε and μ. Through we can roughly estimate m using wavelength-based scaling of medium thickness, the problem is that the wavelength here cannot be foreseen in advance. This makes m becomes an extra unknown variant. Furthermore, effective wavelength is less than the medium thickness at resonant frequency. It is on the contrary to the definition of the effective medium. Therefore, the S parameter reversal results are not completely reliable. Fortunately, the definition of effective wavelength and effective medium are identical because the frequency range we considered here is far larger than the resonant frequency. It is demonstrated that the S parameter reversal is relevant to the number of medium layers. It was reported that two or three layers are enough for the purpose of getting an accurate result because the corresponding calculation error is less than 1% already. More layers will waste memory of computer and is time consuming for the calculation.[19]

The deduction above was carried out on the basis of the assumption of homogeneous artificial materials only. For the inhomogeneous materials, the deduction is not applicable.

13.3 Plasmonic structures for antenna

Compact directive antennas with a single feeding point are highly attractive in practice. Parabola has the property of high directivity. However, parabola is quite bulky to restrict its applications in some special applications. Conventional patch antenna has simple feeding mechanisms, whereas its radiated pattern is affected by the surface wave and has low gain. On the other hand, patch array antenna can offer the directive feature, but the complex feeding mechanism and the radiation efficiency limit its application range. Therefore, high directive antenna with more compact structure and simple feeding is of great interest in recent years. For computational numerical calculation, the following boundary conditions need to be satisfied:

1) Radiation boundary applied on top and side of air box.

2) Finite conductivity (copper) applied on bottom of air-box, PPWG trace, and mushroom patches.

3) Symmetry boundary (perfect-H) applied to side to reduce problem size.

Two typical examples, EBG structure-based antenna and waveguide slit array antenna were illustrated in the following sub-sections.

13.3.1 Electromagnetic band-gap structures-based antennas

Recently, the electromagnetic band-gap (EBG) structures have been widely studied in the electromagnetic and communication antenna applications.[20–25] EBG structures are artificial units composed of metallic patches arrays, which are periodically printed on a dielectric substrate and connected to the metallic ground plane with vias. The structures have frequency band-gap feature, which is revealed in two important ways: the suppression of surface-wave propagation and the in-phase reflection coefficient. The feature of surface-wave suppression can be applied to patch antenna designs to improve antenna's radiation performance and reduce the mutual coupling of the array elements.[25, 26] Meanwhile, the feature of in-phase reflection coefficient can be lead to low profile antenna designs.[27–29] A patch antenna and microstrip method (PAMM) was reported for the purpose of identifying the surface-wave band-gap range.[30, 31] The key idea of PAMM is that a patch antenna, fed by a microstrip line, is served for a radiator and an open-end microstrip line with characteristic impedance of 50 Ω, is employed as a detector for the electromagnetic field along the substrate, as shown in Fig. 13-6. The surface-wave is excited by a microstrip patch antenna, which is different from that of open-ended microstrip line (being used in Ref. 25). The patch antenna is a square metal patch, which has the size of 6.7 mm × 6.7 mm and the resonant frequency is 13.55 GHz. Several rows of EBG structures are inserted between the detector and the radiator. The band-gap range of surface-wave can be identified by measuring the transmission through the detector and the radiator. Simulation results show that the transmission coefficient with EBG

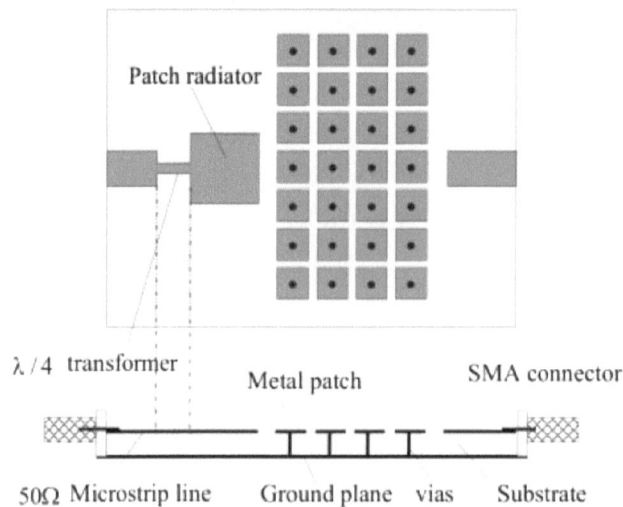

Fig.13-6 Sketch of the experiment setup of PAMM. SMA is utilized to be connected easily to the network analyzer. Reprinted with permission from "G. Cakir and L. Sevgi, Microwave and Optical Technology Letters 49(11), 2668-2672 (2007)" with copyright © 2007 of Wiley Periodicals, Inc.

presence is lower than that of the case with EBG absence above 10 dB. On the other hand, the gain of the antenna increases 1.9 dB when four rows of EBG are present. The results verify that the propagation of surface-wave is suppressed and the method is effective.

The other type of antenna with high directivity was introduced for the Ku-band (12–18 GHz). [32] The device has a simple feeding source made from a co-plane EBG structure patch antenna. The metamaterial, which consists of metallic grids and foam slices, was used to simulate a low refractive index homogeneous medium and placed in front of the feeding source. The simulation results show that the gain of the metamaterial antenna is improved to be ~ 21.6 dB at 14.6 GHz and the antenna directivity is enhanced obviously. A schematic view of the antenna is shown in Fig. 13-7. Metamaterial is made of two-layer metallic grids and foam slices (ε_r =1.07 at 10 GHz). Metamaterial superstrate is placed above the EBG substrate with a perfect electronic conductor (PEC) backing plate. A patch antenna surrounded with several rows of mushroom-like EBG structures is served as the radiation source, which can enhance the patch radiation efficiency and reduce the antenna height. The mushroom-like EBG structure consists of four parts, which are: 1) a ground plane, 2) a dielectric substrate, 3) square metal patches, and 4) connecting vias, as shown in Fig. 13-8. The frequency band-gap is determined by the patch width w, the gap size g, the substrate thickness t, and the substrate dielectric constant ε_r. Theoretically, the maximum directivity of an aperture antenna is $D_{max}=4\pi A/\lambda_0^2$, and the maximum gain $G_{max}=kD_{max}$, whereas k is the efficiency. Given $k=1$, the theoretical value of the maximum gain is $G_{max}(dB)=10\log(4\pi A/\lambda_0^2)$. Here the area of the aperture is A=116 mm×116 mm, and $\lambda_0=c_0/f_0$=20.55 mm (14.6 GHz), therefore the maximum gain is $G_{max}(dB)$=25.9 dB. The simulated result is G=21.6 dB, which approaches the theoretical limitation. If the aperture size of the metamaterial superstrate is larger than that of our simulation, the higher directivity can be obtained.

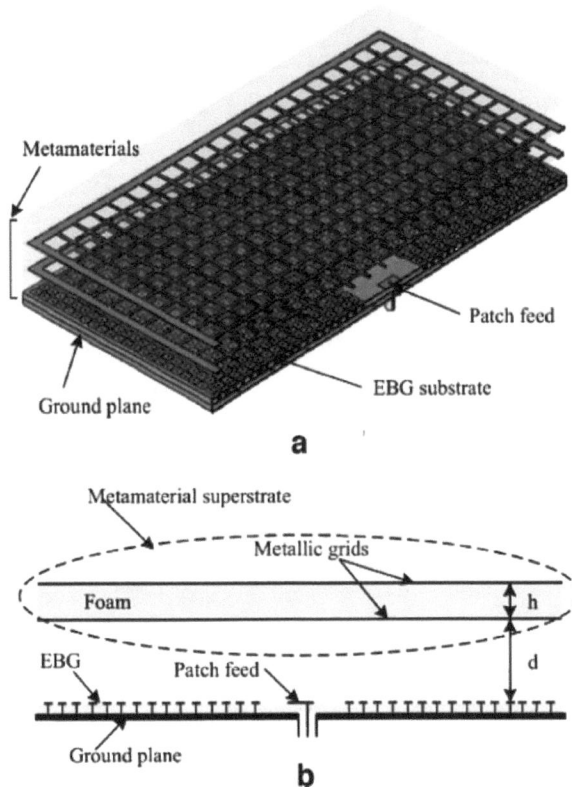

Fig.13-7 Schematic diagram of a metamaterial superstrate antenna. Metamaterial is made of two-layer metallic grids and foam slices (ε_r =1.07 at 10 GHz).

The computed E- and H plane radiation patterns of the EBG structure in Fig. 13-8 are shown in Fig. 13-9. It can be seen that the front peak radiation of EBG case is over 1.9 dB which is higher than that of conventional case. The beam-width in E-plane of the EBG case is much narrower than that of conventional case denoting an effective suppression of surface waves, while the beam-width is similar to the two cases in H-plane. It is illuminated that the radiation performance obtains obviously improvement. Herein it is worth stressing that the improvement is obtained by surrounding the antenna with only one side of the EBG structures other than completely surrounding it. In addition, we

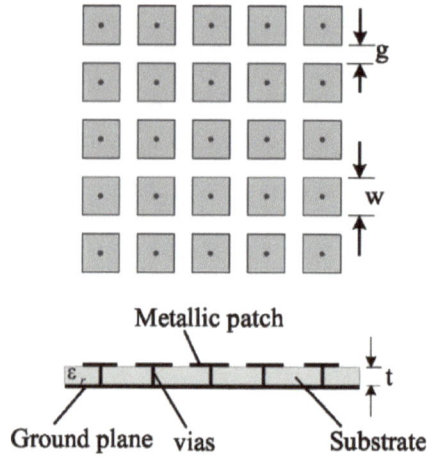

Fig.13-8 Geometry of the mushroom-like electromagnetic band-gap structures.

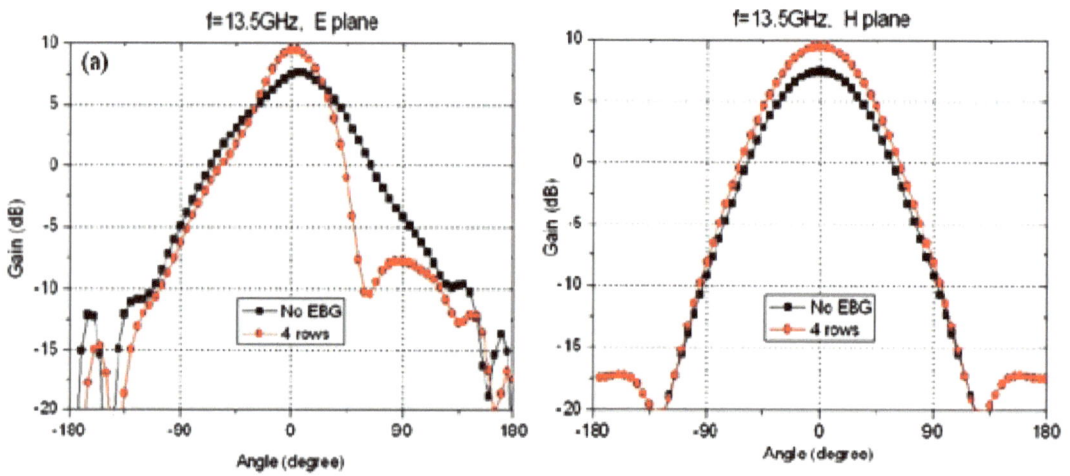

Fig.13-9 Comparison of the radiation pattern of the patch antenna: (a) E-plane radiation pattern; (b) H-plane radiation pattern. Reprinted with permission from "G. Cakir and L. Sevgi, Microwave and Optical Technology Letters 49(11), 2668-2672 (2007)" with copyright © 2007 of Wiley Periodicals, Inc.

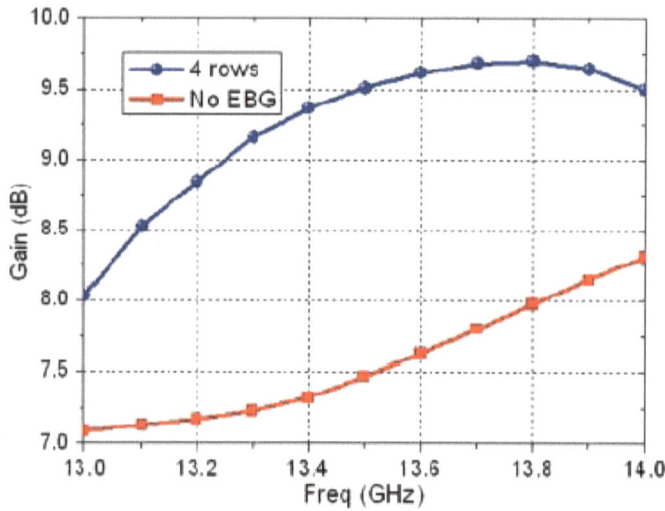

Fig.13-10 The gain comparison of the patch antennas as the function of the frequency. Reprinted with permission from "G. Cakir and L. Sevgi, Microwave and Optical Technology Letters 49 (11), 2668-2672 (2007)" with copyright © 2007 of Wiley Periodicals, Inc.

Fig.13-11 Photograph of the manufactured and assembled patched metamaterial-based antenna.

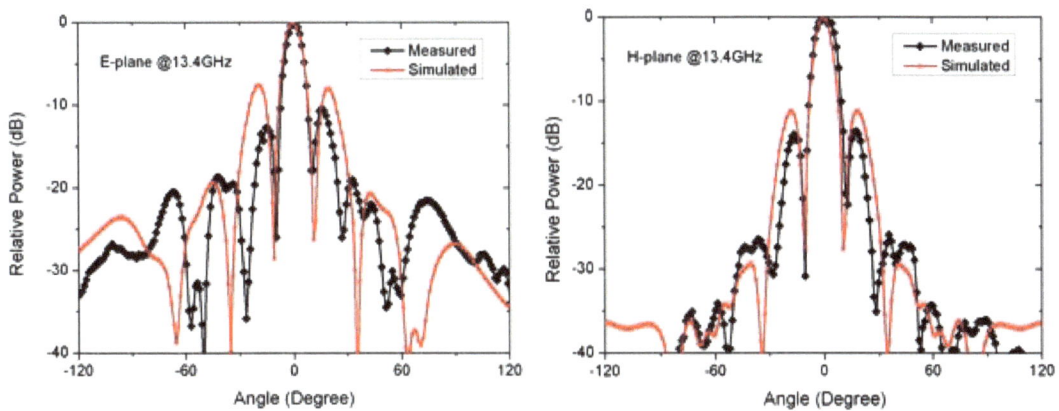

Fig.13-12 Comparison of the designed and measured results at f=13.4GHz (substrate dielectric ε_r=3.5). Directivity plots at (a) E-plane; and (b) H-plane.

compare the gain of four rows of EBG presented and EBG absented as the function of the frequency. As shown in Figure 13-10, the gain increases with the frequency varying from 13 to 14 GHz. At the frequency 13.55 GHz, the gain is about 9.2 dB. Especially at the frequency 13.8 GHz, the peak gain of 9.6 dB is achieved. In order to verify the theoretical designed antenna, a prototype was manufactured, as shown in Fig. 13-11. Figure 13-12 is the comparison between theoretical designed and experimentally measured results. It can be seen that they are fitted well each other. The measurement result is in good agreement with that of the designed.

13.3.2 Waveguide slit array antenna

Most recently, a waveguide slit array-based antenna with high gain and narrow beam width was reported by Huang *et. al.*[33] This antenna was formed by subwavelength slits surrounded with periodic grooves, as shown in Fig. 13-13, in which a narrow slit replaces a long slit, and a rectangle waveguide replaces the corresponding excitation of the plane wave. The far field radiation characteristics of the improved antenna were investigated. The physical mechanism for the performance improvement is governed by the resonance excitation of surface electromagnetic waves and can be well described by the coherent superposition of power radiated from the grooves and central slits. The finite grooves are symmetrically distributed at both sides of the narrow slit, and their detailed parameters are determined as follows: period p=20 mm, depth w=3.6 mm, width a=2.8 mm, and number N=6. This antenna was designed to work at 13 GHz. The gain and angular width of the half-power beam of this proposed antenna are compared to the conventional slit array antenna using the FDTD algorithm. It was demonstrated that the gain of the proposed antenna surrounded with periodic grooves can be improved by 11 dB, and the beam angle is confined to the normal direction. The mechanism for the

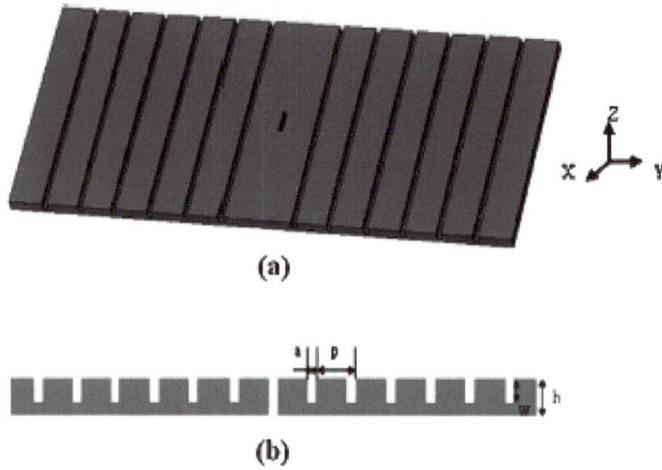

Fig. 13-13 (a) The model of the waveguide slit antenna integrated with periodic grooves and (b) a cross section of the periodic corrugated structure. Reprinted with permission from "Cheng Huang, Chunlei Du, and Xiangang Luo, A waveguide slit array antenna fabricated with subwavelength periodic grooves, Appl. Phys. Lett. 91, 143512 (2007)" with copyright © 2007 of American Institute of Physics.

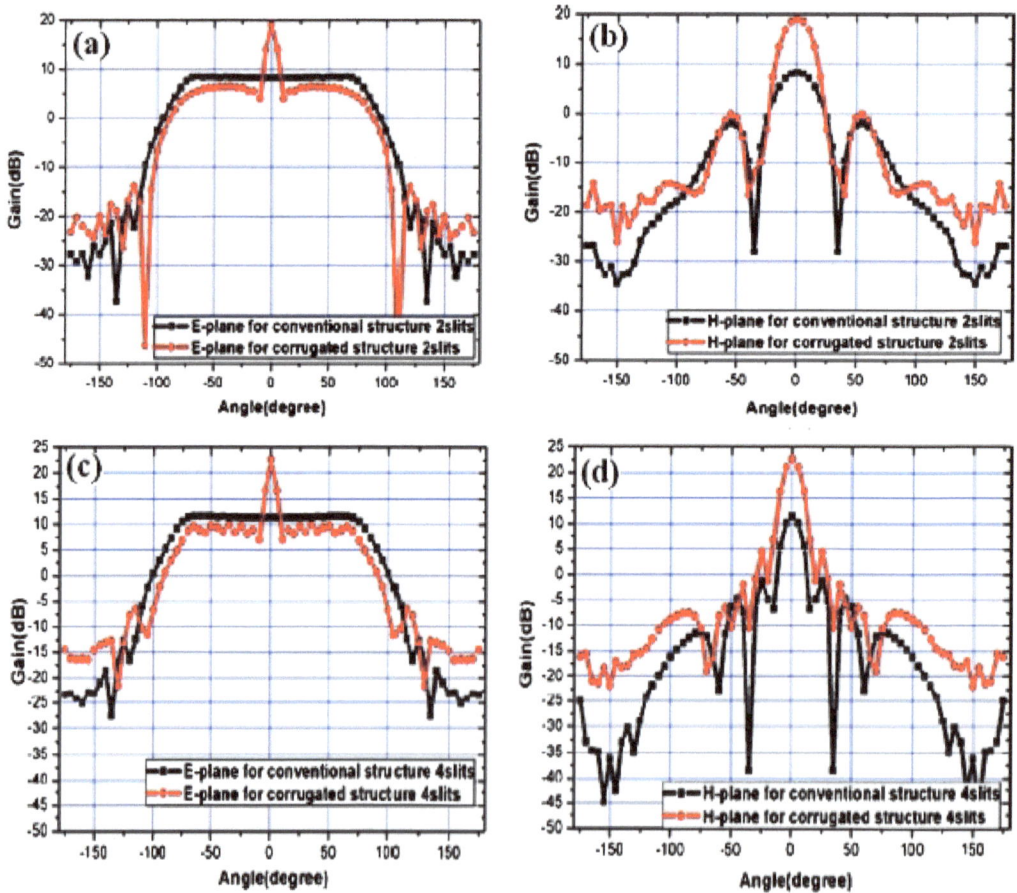

Fig. 13-14 Comparison of radiation patterns between waveguide slit array antennas with and without periodic grooves for f =13 GHz: (a) E plane for two slits, (b) H plane for two slits, (c) E plane fo four slits, and (d) H plane for four slits. Reprinted with permission from "Cheng Huang, Chunlei Du, and Xiangang Luo, A waveguide slit array antenna fabricated with subwavelength periodic grooves, Appl. Phys. Lett. 91, 143512 (2007)" with copyright © 2007 of American Institute of Physics.

improvement of the radiation pattern can be explained by the resonance excitation of the surface EM wave and coherent superposition of power radiated from the grooves and central slits. Moreover, it is believed that this antenna will have potential applications in such wireless communication and point to point communication due to its cost effective, light weight, and low profile.

Figures 13-14 (a) and (b) show two calculated orthogonal plane radiation patterns of y-z plane (E-plane) and x-z plane (H-plane), respectively. These results demonstrate that the gain of the antenna is significantly improved due to the introduction of the periodic grooves, enhancing the gain from 5.27 to 16.3 dB. It can still be seen that this antenna produces an extraordinary beaming effect in the E-plane since the angle width of the half-power beam is 5.4° only. The angle width of the half-power beam in the H-plane is about 28° and is significantly reduced in comparison to the conventional waveguide slit-based antenna. The far field radiation patterns of the slit array-based antenna with periodic grooves were simulated and compared to the conventional slit array-based antenna. From the simulation results shown in Fig. 13-14, it follows that the gain of a slit array antenna on a non-corrugated flat metallic plane can be dramatically enhanced by 10.8 dB and 11 dB for two and four slits, respectively, by using a corrugated flat metallic plane.

The comparison of far field patterns between the single slit (see Fig. 13-14 (b)) and four slits (see Fig. 13-14 (d)) on a metallic plane surrounded with periodic grooves shows that the angular width of the half-power beam in the H-plane is 13.1° only, which is considerably reduced by the constructive superposition of fields radiated by the four slit element. The counterpart for the E-plane is slightly reduced to be 5°, since the angular width in the E-plane mainly depends on the six corrugated grooves. It is expected that a higher gain and a narrower angular width can be achieved by the application of more slit elements.

13.4 Summary

Metamaterials-based microwave antennas were introduced from theory to design and testing. There are two analyzing approaches for the antennas: resonant approach and transmission line approach. Classified the antenna types in physical, there are three types antennas: RHM, LHM, and composite RHM-LHM. If classify it according to structures, there are four types of metamaterials-based antennas: (1) dominant leaky-wave antenna, (2) small metamaterial antenna, (3) dual-band hybrid coupler, and (4) negative refractive index flat lens.

Considering length of this chapter, the low refractive index flat lens was highlighted. Theoretical background for the matematerials-based structures was described in detail. EBG structure-based antenna and waveguide slit array-based antenna were cited as two examples to illustrate their design issues and performance. Strictly speaking, they are positive near zero refractive index ($n \approx 0$) only, not negative refractive index. Testing results were given for the EBG structure antennas. Compared to the theoretical calculation results, the experimental results are in good agreement to that of the theoretical calculations.

Reference

[1] V.G. Veselago, "The electrodynamics of substances with simultaneously negative values of ε and μ," Soviet Physics Usp., 10(4), 509-514 (1968).

[2] R. A. Shelby, D. R. Smith, S. Schultz, "Experimental verification of a negative index of refraction," Science 292, 77-79 (2001).

[3] A.K. Iyer, G. V. Eleftheriades, "Negative refractive index metamaterials supporting 2-D waves," IEEE MTT-S International Microwave Symposium Digest, Vol. 2, pp. 1067-1070, June 2-7, 2002, Seattle, WA.

[4] G. V. Eleftheriades, A. K. Iyer, P. C. Kremer, "Planar negative refractive index media using periodically L-C loaded transmission lines," IEEE Trans. on Microwave Theory and Tech., Vol. 50, No. 12, pp. 2702-2712, Dec. 2002.

[5] D. Sievenpiper, L. Zhang, R.F.J. Broas, N.G. Alexopolous, and E. Yablonovitch, "High-impedance surface electromagnetic surfaces with a forbidden frequency band," IEEE Trans. Microwave Theory Tech., Vol. 47, No. 11, pp. 2059-2074, Nov. 1999.

[6] R.A. Shelby, D.R. Smith, and S. Schultz, "Experimental verification of a negative index of refraction," Science, Vol. 292, pp. 77-79, Apr. 2001.

[7] A. Lai, C. Caloz, and T. Itoh, "Composite right/left-handed transmission line metamaterials," IEEE Microwave Magazine, Vol.

5, No.3, pp. 34-50, Sep. 2004.

[8] C. Caloz and T. Itoh, Electromagnetic Metamaterials: Transmission Line Theory and Microwave Applications, Wiley and IEEE Press, Hoboken, NJ, 2005.

[9] L. Liu, C. Caloz, and T. Itoh, "Dominant mode (DM) leaky-wave antenna with backfire-to-endfire scanning capability," Electron. Lett., Vol. 38, No. 23. pp. 1414-1416, Nov. 2002.

[10] A. Lai, K.M.K.H. Leong, and T. Itoh, "Infinite wavelength resonant antennas with monopolar radiation patterns based on periodic structures," IEEE Trans. Antennas Propag., vol. 55, no. 3, pp. 868-876, Mar. 2007.

[11] I. Lin, C. Caloz, and T. Itoh, "A branch-line coupler with two arbitrary operating frequencies using left-handed transmission lines," IEEE-MTT Int. Symp. Dig., Philadelphia, PA, Jun. 2003, vol. 1, pp. 325–327.

[12] D. R. Smith, S. Schultz, P. Markos, and C. M. Soukoulis, "Determination of effective permittivity and permeability of metamaterials from reflection and transmission coefficients". Phys. Rev. B, 2002, 65: 195104.

[13] X. D. Chen, T. M. Grzegorczyk, B. I. Wu, "Robust method to retrieve the constitutive effective parameters of metamterials", Phys. Rev. E, 2004, 70: 016608.

[14] D. R. Smith, D. C. Vier, Th. Koschny, and C. M. Soukoulis, "Electromagnetic parameter retrieval from inhomogeneous metamaterials". Phys. Rev. E, 2005, 71: 036617.

[15] J B Pendry, A J Holdenz, D J Robbinsz and W J Stewartz, "Low frequency plasmons in thin-wire structures", J. Phys.: Condens. Matter 10, 4785–4809 (1998).

[16] Liyuan Liu, Master thesis, "Design and Application of Index-Near-Zero metamaterials in Microwave Band", Chinese Academy of Sciences, June 2009.

[17] D. Smith, S. Schultz, P. Marko, C. Soukoulis, and S. Bratislava, "Determination of effective permittivity and permeability of metamaterials from reflection and transmission coefficients," Physical Review B 65, 195104 (2002).

[18] D. R. Smith, D. C. Vier, Th. Koschny, and C. M. "Soukoulis, Electromagnetic parameter retrieval from inhomogeneous metamaterials". Phys. Rev. E, 2005, 71: 036617.

[19] Huiliang Xu, Ph.D dissertation, "EBG structure and low refractive index materials for application of antenna design." Chinese Academy of Sciences, June 2008.

[20] Ziolkowski R.W.; Engheta N., "Metamaterial special issue introduction", IEEE Trans. Antennas. Propagat. 51, 2546- 2549 (2003).

[21] Kildal, P.-S., Kishk, A.A.; Maci, S., Special Issue on Artificial Magnetic Conductors, Soft/Hard Surfaces, and Other Complex Surfaces IEEE Trans. Antennas. Propagat. 53, 2-7 (2005).

[22] G. Cakir and L. Sevgi, "A new ultra-wideband antenna for UWB applications", Microwave Opt. Technol. Lett. 46, 399–401 (2005).

[23] D. Sievenpiper, L. Zhang, F.J. Broas, N.G. Alexopolous, and E. Yablonovitch, "High-impedance electromagnetic surfaces with a forbidden frequency band", IEEE Trans. Microwave Theory Technol. 47, 2059–2074 (1999).

[24] P. de Maagt, R. Gonzalo, Y.C. Vardaxoglou, and J.M. Baracco, "Electromagnetic bandgap antennas and components for microwave and (Sub) millimeter wave applications", IEEE Trans. Antennas. Propagat. 51, 2667–2677 (2003).

[25] R. Coccioli, F.R. Yang, K.P. Ma, and T. Itoh, "Aperture-coupled patch antenna on UC-PBG substrate", IEEE Trans Microwave. Theory. Technol. 47, 2123–2130 (1999).

[26] F. Yang and Y.Rahmat-Samii, "Microstrip antennas integrated with electromagnetic band-gap (EBG) structures: a low mutual coupling design for array applications", IEEE Trans. Antennas. Propagat. 51, 2936–2946 (2003).

[27] F. Yang and Y. Rahmat-Samii, "A low profile circularly polarized curl antenna over electromagnetic band-gap (EBG) surface", Microwave Opt. Technol. Lett. 31, 478–481 (2001).

[28] J. M. Bell and M.F. Iskander, "A low-profile Archimedean spiral antenna using an EBG ground plane", IEEE Antennas Wireless Propagat. Lett. 3, 223–226 (2004).

[29] A.P. Feresidis, G. Goussetis, S.H. Wang, and J.C. Vardaxoglou, "Artificial magnetic conductor surfaces and their application to low-profile high-gain planar antennas", IEEE Trans. Antennas. Propagat. 53, 209–215 (2005).

[30] Huiliang Xu, Yueguang Lv, Xiangang Luo, and Chunlei Du, "Method for Identifying The Surface Wave Frequency and gap of EBG Structures", Microw. Opt. Technol. Lett. 49, 2668-2672 (2007).

[31] Huiliang Xu, Yueguang Lv, Xiangang Luo, Chunlei Du, "Metamaterial Superstrate and Electromagnetic Band-Gap Substrate for High Directive Antenna", International Journal of Infrared and Millimeter Waves, Vol. 29, No.5, pp. 493-498, May 2008.

[32] Fan Yang, "EBG Structure and Reconfigurable Technique in Antenna Designs: Applications to Wireless Communications", Ph.D thesis, University of California, Los Angeles, 2002.

[33] Cheng Huang, Chunlei Du, and Xiangang Luo, "A waveguide slit array antenna fabricated with subwavelength periodic grooves", Appl. Phys. Lett. 91, 143512 (2007).

14 PLASMONIC STRUCTURES FOR DATA STORAGE

Abstract: Plasmonic structures for another novel application: data storage was presented in this chapter. Total three type structures: thermal assisted magnetic head, plasmonic lens array, metallic nanoparticles-based recording, and atomic force microscope configuration-based system were discussed. A shortcut for readers getting a quick view of the plasmonic structure-related data storage was provided through this chapter.

14.1 Introduction

Data storage as a major production in industry has been developed near half century. With its omnipresent computers, all connected via the Internet, the Information Age has led to an explosion of information available to users. The decreasing cost of storing data, and the increasing storage capacities of the same small device footprint, has been key enablers of this revolution. However, both magnetic and conventional optical data storage technologies, where individual bits are stored as distinct magnetic or optical changes on the surface of a recording medium, are approaching physical limits beyond which individual bits may be too small or too difficult to be stored. Storing information throughout the volume of a medium, not just on its surface, offers an intriguing high-capacity alternative. Holographic data storage is a volumetric approach which, although conceived decades ago, has made recent progress toward practicality with the appearance of lower-cost enabling technologies, significant results from longstanding research efforts, and progress in holographic recording materials. [1]

Holographic recording technology records data on discs in the form of laser interference fringes, and enables existing discs the same size as today's DVDs to store as much as one terabyte of data (200 times the capacity of a single layer DVD), with a transfer speed of one gigabyte per second (40 times the speed of DVD). This approach is rapidly gaining attention as a high-capacity, high-speed data storage technology for the age of broadband. Optware has announced that they have successfully achieved the world's first recording and play back of digital movies on a holographic recording disc. Holographic data storage currently suffers from the relatively high component and integration costs faced by any emerging technology. In contrast, magnetic hard drives, also known as direct access storage devices (DASD), are well established, with a broad knowledge base, infrastructure, and market acceptance.

Plasmonic devices-based optical head for reading and writing is a new approach recently. In this chapter, optical domain methods are focused, especially for the near-field and plasmonic structure-based approaches.

14.2 Plasmonics in data storage

14.2.1 Thermal assisted magnetic recording

A thermal assisted heating method for magnetic recording was reported. [2] It used an extremely small beam spot (as small as 50 nm) and the beam spot determines the write width. Near field optics is required to obtain the small beam spot because ordinary optics cannot produce an optical beam smaller than the diffraction limit.

The laser spot size must be extremely focused to achieve optical dominant recording. Therefore, it is necessary to develop a heating element that emits near-field light. The requirements for the heating element are as follows: a) the beam spot size must be smaller than 50 nm; b) the optical efficiency must be as high as about 2% to heat the media to the required temperature; c) because of the magnetic spacing (*i.e.*, the relatively large distance between the head and magnetic film), the attenuation length of the near-field light must be greater than 10 nm; and d) integration of the heating element with the magnetic head must be possible. Concerning requirement d), the fabrication process must be compatible with the process for the magnetic head, and precise positioning between the beam spot and write pole must also be assured. For the near-field-light heating element, ridge waveguide,[3,4] bow-tie antenna,[5,6] zone plate grating,[7] and SMASH head[8] versions were proposed before. However, the writing fields of these proposals will be limited to about 100 Oe because they were combined with a coil without a magnetic core.[2] Hasegawa *et. al.* proposed the butted grating structure for the heating element,[9] which was designed by the software of Poynting,[10] which analyzes electromagnetic waves using the Finite Difference Time Domain (FDTD) method. This structure is suitable for thermally assisted magnetic recording because the process used to fabricate it is compatible with the process for the current magnetic head. This means that the head, which must integrate the heating element and read/write elements, and can be fabricated on an AlTiC substrate using a planer process. Consequently, a strong magnetic field is available. The

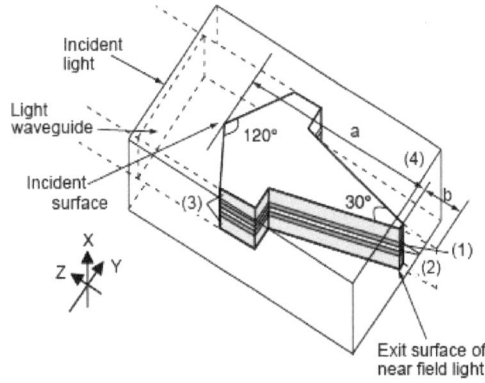

Fig. 14-1 Heating element with multi-layer butted grating of SiO_2 (1), Al (2), and diamond (3), Al (4). Reprinted with permission from "Koji Matsumoto, Akihiro Inomata, Shin-ya Hasegawa, Thermally assisted magnetic recording, FUJITSU Sci. Tech. J. 42(1), p.158-167(January 2006)" with copyright ©2006 of FUJITSU Inc.

heating element with the butted grating is shown in Figure 14-1. In the figure, the x-y plane is parallel to the surface of the media, the X-axis corresponds to the circumferential direction, and the Y-axis corresponds to the radial direction. The arrow-shaped polyhedron is a multi-layer grating of Al/diamond/Al/SiO2/Al/diamond/Al. The 400 nm light is incident from the upper left, and the near-field light is emitted from the lower right. The basic idea of the butted grating is to butt a one-period high-transmission efficiency grating (Al/SiO_2) in the central part of the structure with very low-transmission-efficiency Al/diamond gratings that have a small number of periods at either side for increasing the optical transmission efficiency. As a result, a high optical transmission of the nano-beam is achieved through the SiO_2, which is 30 nm in the X-direction. Furthermore, as the light, which is polarized in the X-direction, propagates in the minus Z direction, it becomes narrow in the Y-direction due to the interference of multiple reflections from the sidewall. The calculated beam spot size is 45 nm in the X-direction and 60 nm in the Y-direction. The Z-direction tolerance is 15±5 nm, indicating that the attenuation length is large enough compared to the magnetic spacing. The optical efficiency is 1.6%, which is lower than that of required, but this will be improved by optimizing the structure's design. The butted grating heating element can be formed by means of sputtering and dry etching, and it has good affinity with

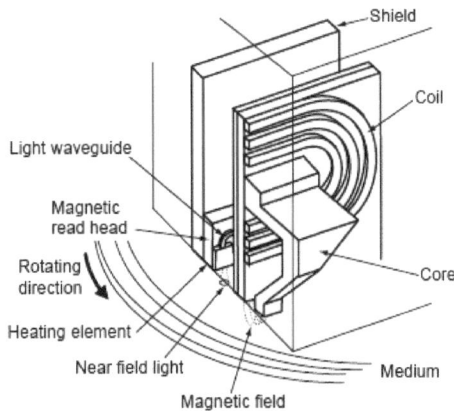

Fig. 14-2 Conventional integrated head with butted grating heating element. Reprinted with permission from "Koji Matsumoto, Akihiro Inomata, Shin-ya Hasegawa, Thermally assisted magnetic recording, FUJITSU Sci. Tech. J. 42(1), p.158-167(January 2006)" with copyright ©2006 of FUJITSU Inc.

the conventional magnetic head. Figure 14-2 shows a conventional magnetic head that integrates a heating element and butted grating.

Besides the heating element, there are many other challenges to be overcome, for example, how to integrate the heating element with the magnetic head, thermal issues about the integrated head and the media, and the characterization of media with the very large anisotropy constant of FePt. The growth rate of areal density for both longitudinal and perpendicular recording has slowed down since 2002, and a big breakthrough is now strongly needed. We believe that thermally assisted magnetic recording is the only way to achieve 1 Tbit/in^2.

14.2.2 AFM-based data storage

A new atomic force microscope (AFM)-based concept for data storage called "Millipede" was reported. [11] It has a potentially ultrahigh density, terabit capacity, small form factor, and high data rate. Its potential for ultrahigh storage density has been demonstrated by a new technique of thermo mechanical local-probe to store and read back data in very thin polymer films. With this new technique, 30–40 nm-sized bit indentations of similar pitch size were made by a single cantilever/tip in a thin (50-nm) polymethylmethacrylate (PMMA) layer, resulting in a data storage density of 400–500 Gb/in.^{2} High data rates were achieved by parallel operation of large two-dimensional (2D) AFM arrays that had been batch fabricated by silicon surface-micromachining techniques. The very large scale integration (VLSI) of micro/nanomechanical devices (cantilevers/tips) on a single chip leads to the largest and densest 2D array of 32×32 (1024) AFM cantilevers with integrated write/read storage functionality ever built. Time multiplexed electronics control the write/read storage cycles for parallel operation of the Millipede array chip. Initial areal densities of 100–200 Gb/in.^{2} were achieved with the 32×32 array chip, which has potential for further improvements. In addition to data storage in polymers or other media, and not excluding magnetics, we envision areas in nanoscale science and technology such as lithography, high-speed/large-scale imaging, molecular and atomic manipulation, and many others in which Millipede may open up new perspectives and opportunities.

14.2.3 Plasmonic lens-based recording system

A novel plasmonic lens array-based optical system was reported by X. Zhang *et. al.* recently. [12] They used an array of plasmonic lenses that 'flies' above the surface to be patterned, concentrating short-wavelength surface plasmons into sub-100 nm spots. However, these nanoscale spots were only formed in the near field, which makes it very difficult to scan the array above the surface at high speed. To overcome this problem, they have designed a self-spacing air bearing that can fly the array just 20 nm above a disk that is spinning at speeds of between 4 and 12 m/s, and experimentally demonstrated patterning with a linewidth of 80 nm. Recording material is photoresist.

Rotation of the substrate creates an air flow along the bottom surface of the plasmonic flying head, known as the air bearing surface (ABS). The ABS generates an aerodynamic lift force and it is balanced with the force supplied by the suspension arm to precisely regulate a nanoscale gap between the plasmonic lens arrays and the rotating substrate, which is covered with photoresist. With the high bearing stiffness and small actuation mass, this self-adaptive method can provide an effective bandwidth up to 120 kHz. The use of an ABS eliminates the need for a feedback control loop and therefore overcomes the major technical barrier for high-speed scanning. It can be seen in Fig. 14-3 that high throughput nanolithography is accomplished using the plasmonic flying head at a relatively high speed. The plasmonic flying head is made of a specially designed transparent air-bearing slider with arrays of plasmonic lenses fabricated on its bottom surface. Using large arrays of plasmonic lenses enables parallel writing for high throughput. The flying plasmonic lens array in the optical near field is inspired by the magnetic recording head in hard disk drives (HDD). Unlike a conventional HDD ABS, which uses only the trailing-edge-mounted transducer to serially read and write the magnetic bits, we designed the plasmonic head to contain a relatively large area filled by plasmonic lenses that enable parallel writing and high throughput. Owing to the rapid decay of light intensity in a plasmonic lens, all plasmonic lenses need to maintain the distance to the rotating substrate within 30 nm, which requires the bottom surface to be parallel to the substrate to within 100 mrad tilt. This stringent parallelism requirement made the design of the plasmonic

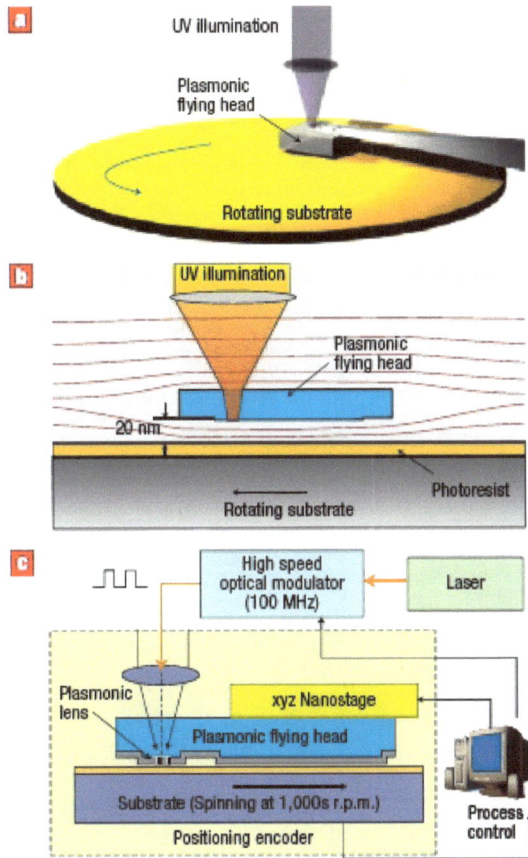

Fig. 14-3 High-throughput maskless nanolithography using plasmonic lens arrays. (a) Schematic showing the lens array focusing ultraviolet (365 nm) laser pulses onto the rotating substrate to concentrate surface plasmons into sub-100 nm spots. However, sub-100 nm spots are only produced in the near field of the lens, so a process control system is needed to maintain the gap between the lens and the substrate at 20 nm. (b) Cross-section schematic of the plasmonic head flying 20 nm above the rotating substrate which is covered with photoresist. (c) Schematic of process control system. The laser pulses are controlled by a high-speed optical modulator according to the signals from a pattern generator. The writing position is referred to the angular position of the disk from the spindle encoder and the position of a nano-stage along the radial direction. Reprinted with permission from "Werayut Srituravanich, Liang Pan, Yuan Wang, Cheng Sun, David B. Bogy, and Xiang Zhang, Nature Nanotechnology 3(12), 733-737(2008)" with Copyright © 2008 of Macmillan Publishers Limited.

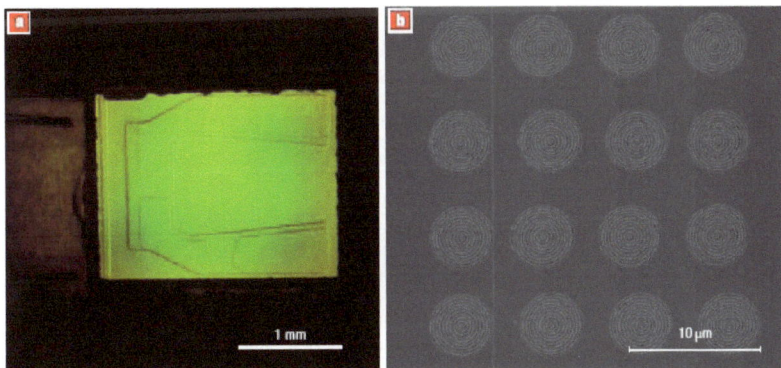

Fig. 14-4 Fabricated plasmonic flying head and flying height measurement. a, Optical micrograph of a plasmonic flying head assembled with suspension. b, Scanning electron microscopy (SEM) image of an array of plasmonic lenses fabricated on an ABS. Reprinted with permission from "Werayut Srituravanich, Liang Pan, Yuan Wang, Cheng Sun, David B. Bogy, and Xiang Zhang, Nature Nanotechnology 3(12), 733-737(2008)" with Copyright © 2008 of Macmillan Publishers Limited.

head both very challenging and also different from magnetic head sliders. For example, to fly 1,000 lenses within the 30 nm gap tolerance over the usable area of 800 × 20 mm on the rear pads with each plasmonic lens of size 4 mm in diameter, the ABS has to be designed with pitch angle <100 mrad and 2 mrad roll angle. Also, the ABS needs a larger air-bearing stiffness, higher damping ratio, and better contamination insensitivity than conventional ABS (see Supplementary Information). In addition, the plasmonic head must be transparent to light.

Figure 14-4 (a) shows an optical microscope image of the fabricated plasmonic flying head where the sapphire ABS coated with a metal film was assembled with the suspension, and a scanning electron microscopy (SEM) image of a two-dimensional array of plasmonic lenses (4 × 4) fabricated on the ABS in a square lattice (see Fig. 14-4 (b)).A spindle was used to rotate the disk at speed of 2,000 rpm., which is equivalent to a linear speed of 10 m s^{-1} at the outer radius. After pattern writing and development in diluted KOH solution, the patterns were examined using an atomic force microscope (AFM). The result demonstrated that we can achieve high-speed patterning with 80 nm linewidths at

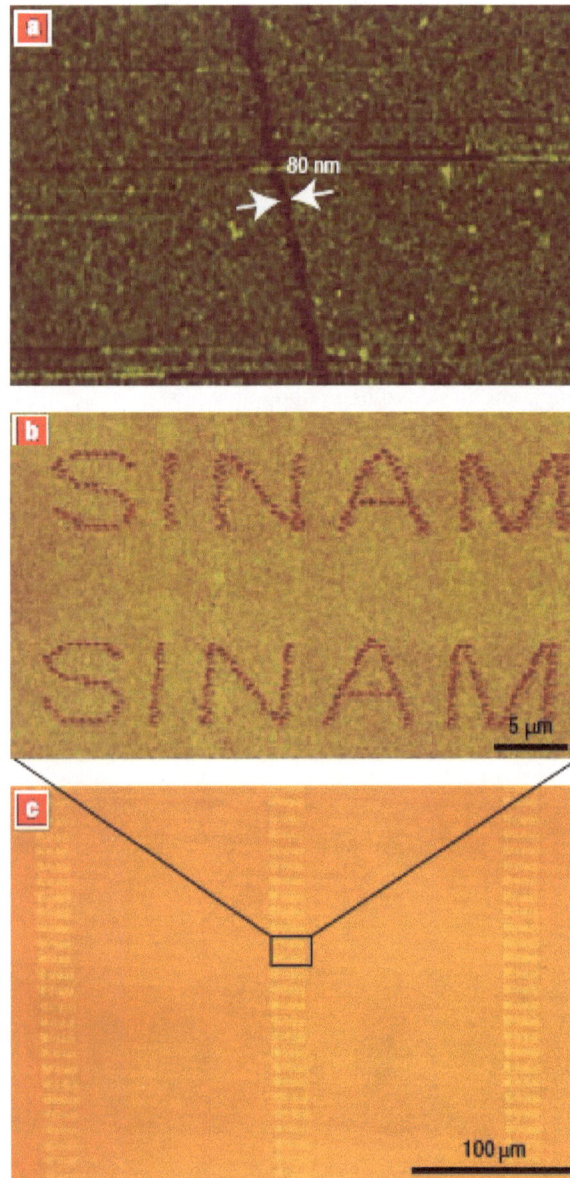

Fig. 14-5 Maskless lithography by flying plasmonic lenses at the near field. (a) AFM image of a pattern with 80 nm linewidth on the TeOx-based thermal photoresist. (b) AFM image of arbitrary writing of 'SINAM' with 145 nm linewidth. (c) Optical micrograph of patterning of the large arrays of 'SINAM'. Reprinted with permission from "Werayut Srituravanich, Liang Pan, Yuan Wang, Cheng Sun, David B. Bogy, and Xiang Zhang, Nature Nanotechnology 3(12), 733-737(2008)" with Copyright of © 2008, Macmillan Publishers Limited.

10 m s^{-1} (see Fig. 14-5 (a)). Figure 14-5 (b, c) demonstrates the successful patterning of arrays of the acronym 'SINAM' with a feature size of 145 nm. The pattern writing in the radial direction involved a coordinate transformation with Cartesian coordinates. The resolution can be improved by the careful design using shorter plasmon wavelength and guiding mechanisms, and theoretical simulation shows it is possible to reach a beam spot size down to 5–10 nm [13].

14.2.4 Metallic nanorods-based recording

The ever-increasing data storage of the digital age requires reliable high-density, and high-speed storage solutions. The demand for increased storage capacity was fueled by the improvement in high-definition digital television and was set to quickly exceed the capacities of current optical-storage platforms such as the digital versatile disc (DVD) and blue-ray DVDs. Currently, there is several optical storage technologies proposed to tackle the issue. Amongst, a spectral encoding technique [14,15–17] is particularly promising for its potential to increase storage capacity by an order of magnitude by introducing a new dimension into the recording. This technique is not limited by the spatial resolution of data bits and can be incorporated into various, currently available optical-storage technologies.

For successful implementation of the spectral encoding method, the recording medium needs to be recordable at multiple wavelengths, and there should be no interference or cross talk with any features recorded at other wavelengths

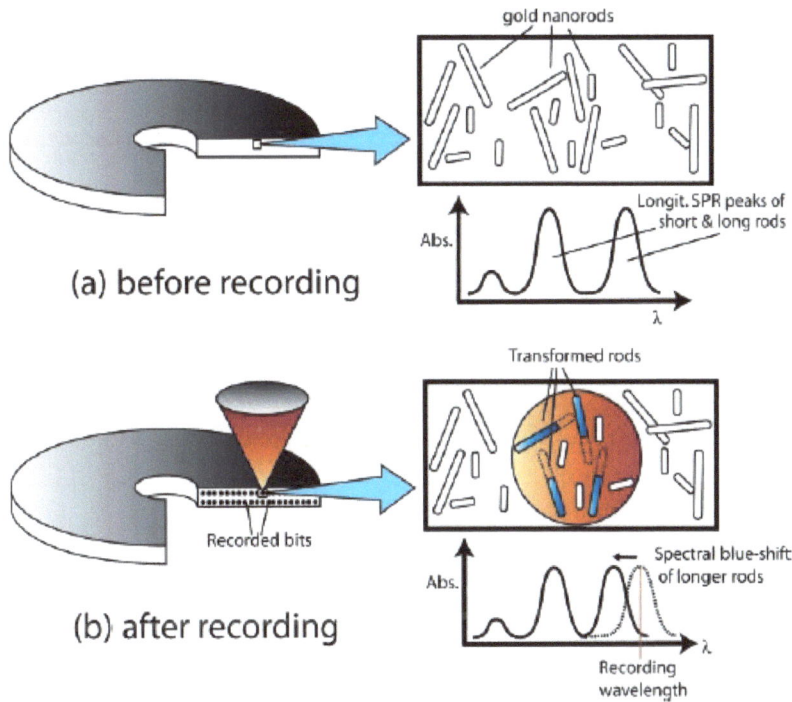

Fig. 14-6 Concept of spectral encoding on two aspect ratios of metallic nanorods. a) Before recording: the nanorods are randomly but homogeneously distributed in a host matrix with an absorbance spectrumthat shows one transverse SPR peak at a shorter wavelength and two longitudinal SPR peaks at longer wavelengths, arising from the two aspect ratios of nanorods. b) After recording by pulsed laser irradiation tuned at one of the longitudinal SPR bands: the transformation of the rod shape is induced, which results in a spectral shift of the SPR peak. If the readout is at the same wavelength at lower power density, the absorbance change could be detected. However, no change could be detected at the other SPR wavelength; hence, another spectral layer of information could be recorded at the other SPR wavelength on the same physical layer. This provides another dimension (on three spatial dimensions) to increase the data density by a factor of 2^2. Reprinted with permission from James W. M. Chon, Craig Bullen, Peter Zijlstra, and Min Gu, Advanced Functional Materials 17, 875-'880 (2007) with copyright © 2007 of Wiley-VCH Verlag GmbH & Co. KGaA, Weinheim.

during recording/ reading. To date, organic dye molecules [15,16] and metal nanoparticles [17] have been used as recording media for spectral encoding. More recently, spectral encoding using semiconductor nanocrystals was explored.[14] While these media have successfully demonstrated the proof-of-concept spectral encoding experiments, challenges of drive design and cost must still be overcome for practical device applications. Here, some researcher proposed to use metallic nanorods doped in a silica sol–gel matrix (NR-SG) as a new recording media for spectral encoding, which complemented and improved on the existing methods. Metallic nanorods exhibit fascinating optical properties due to surface plasmons: the oscillation of the electron cloud within a particle. They exhibit two principal absorption bands that correspond to surface plasmon resonance (SPR) along the long and short axes of the nanorods. The dominant longitudinal band can be tuned with the aspect ratio of the rods, making it a spectrally tunable optical material.[18–26] Furthermore, the nanorods irradiated with pulsed lasers at an SPR wavelength are known to transform their shape into spherical, ϕ-shaped, or shorter rods,[19, 22–24] which results in a significant reduction or shift in the SPR peak position. By selectively transforming a subpopulation of nanorods with plasmon bands in resonance with the recording laser wavelength, it is possible to achieve spectral encoding. Figure 14-6 illustrates the concept of the current proposed method which was presented by M. Gu *et. al.*.[27]

Nanorod reshaping due to melting is a more likely cause of the observed reduction in nanorod aspect ratio. In order to quantitatively assess the temperature reached by the nanorods during the absorption of a pulse, we calculate the total number of photons absorbed (N_{abs}) within a focal volume per pulse, using the following equation:

$$N_{abs} \sim CV\sigma I \tag{1}$$

where C is the concentration of nanorods in the matrix, V is the focal volume, r is the absorption cross section of Au nanorods, and I is the energy fluence of a single pulse [J cm^{-2}].

The temperature [K] increases due to the absorbed energy per nanorod is then calculated using the following equation:

$$T = \frac{Q - \Delta H_{melt}}{c_{p,rod}} + 293 \tag{2}$$

where Q is the energy absorbed per nanorod, ΔH_{melt} is the latent heat of melting of a single nanorod, and $c_{p,rod}$ is the specific heat capacity of a single rod.

Heat dissipation into the surrounding matrix via phonon–phonon relaxation is also a competing process. A simple spherically symmetric heat equation is used to estimate the characteristic time of heat diffusion into the surroundings: [27]

$$t = c_{p,m} \rho_m R^2 / \kappa_m \tag{3}$$

where $c_{p,m}$, ρ_m, κ_m is the specific heat capacity, density, and thermal conductivity of the matrix material, respectively, and R is the heat diffusion length.

Previously, metallic nanostructures were doped into dielectric materials, such as silica or titania, by an in situ ion-exchange method.[28] The dielectric host matrix is advantageous over a flexible polymer matrix because of its rigidity and stability under high-power laser exposure. However, because the embedded nanoparticles were produced in situ, this method did not provide control over the shape of these nanoparticles nor the freedom of incorporating the prefabricated metallic nanorods. The silica sol–gel host matrix used in this work combines the advantages of both techniques: it provided flexibility in doping and rigidity under laser exposure. The nanocomposite NR-SG material provided the spectral functionality of the nanorods and the stability of the host matrix, which is an ideal platform for the optical storage application. If the proposed material is applied for three color spectral encoding, the data density can be increased up to 2^3 times that of the current limit. Furthermore, the current method can also be applied to volumetric recording techniques such as three-dimensional bit-by-bit recording and holographic storage methods. In the case of holographic storage, the NR-SG material will provide stability, high index contrast, and no shrinkage.

Most recently, M. Gu *et. al.* [29] reported their improved recording approach: a multiplexed optical recording method. It provided an unparalleled approach to increasing the information density beyond 10^{12} bits per cm^3 (1 Tbit cm^{-3}) by storing multiple, individually addressable patterns within the same recording volume. Although wavelength, polarization and spatial dimensions have all been exploited for multiplexing, these approaches have never been integrated into a single technique that could ultimately increase the information capacity by orders of magnitude. The major hurdle is the lack of a suitable recording medium that is extremely selective in the domains of wavelength and polarization as well as the three spatial domains, so as to provide orthogonality in all five dimensions. They showed their true five-dimensional optical recording by means of exploiting the unique properties of the longitudinal surface plasmon resonance (SPR) of gold nanorods. The longitudinal SPR exhibits an excellent wavelength and polarization sensitivity, whereas the distinct energy threshold required for the photo-thermal recording mechanism provides the axial selectivity. The recordings were detected using longitudinal SPR-mediated two-photon luminescence, which they demonstrated to possess an enhanced wavelength and angular selectivity compared to the conventional linear detection mechanisms. Combined with the high cross-section of two-photon luminescence, this enabled non-destructive, and crosstalk-free readout. This technique can be immediately applied to optical patterning, encryption and data storage, whereas higher data densities are pursued.

The concept of five-dimensional patterning is illustrated in Fig. 14-7. The sample consists of a multilayered stack in which thin recording layers (~1 μm) are separated by a transparent spacer (~10 μm). In both the wavelength and polarization domains, three-states multiplexing were illustrated to provide a total of nine multiplexed states in one recording layer. The key to successfully realizing such five dimensional encoding is a recording material that (1) is orthogonal in all dimensions, in both recording and readout, (2) is able to provide multiple recording channels in each dimension, and (3) is stable in ambient conditions and can be read out non-destructively. Existing multiplexing techniques [1–8] are only orthogonal in one dimension (either wavelength or polarization), and often work in ambient conditions with readout degrading the recorded patterns through the unwanted isomerization or photobleaching.

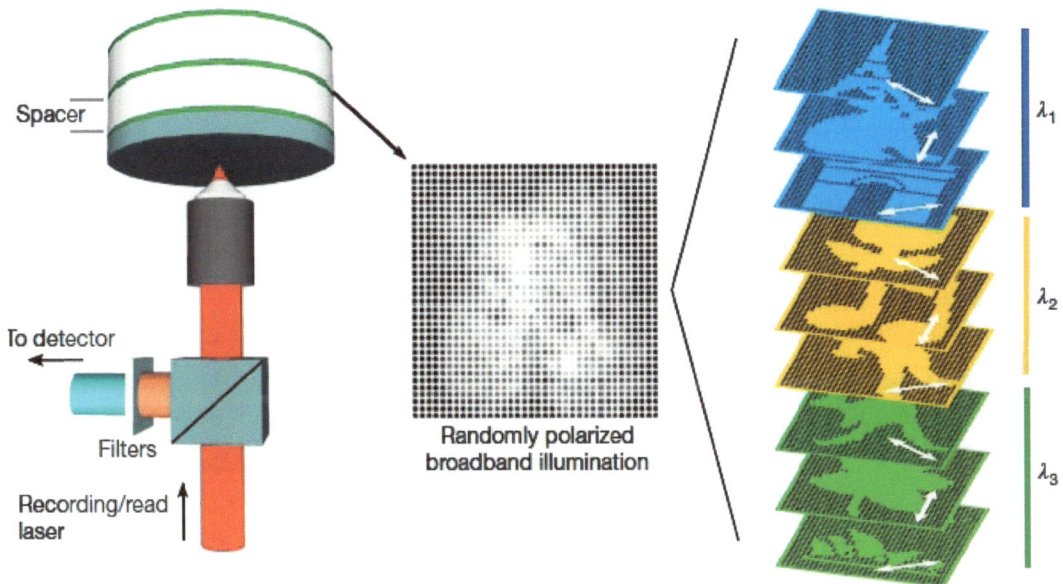

Fig. 14-7 Sample structure and patterning. Left, the sample consists of thin recording layers of spin-coated polyvinyl alcohol doped with gold nanorods, on a glass substrate. These recording layers were spaced by a transparent pressure-sensitive adhesive with a thickness of 10 μm. In the recording layers, we patterned multiple images using different wavelengths (λ_{1-3}) and polarizations of the recording laser. Middle, when illuminated with unpolarized broadband illumination, a convolution of all patterns will be observed on the detector (filters attenuate the reflected readout laser light). Right, when the right polarization and wavelength is chosen, the patterns can be read out individually without crosstalk. Reprinted with permission from Peter Zijlstra, James W. M. Chon & Min Gu, Nature 459, 410-413 (2009) with copyright © 2009 of Macmillan Publishers Limited.

A recording material based on plasmonic gold nanorods meets all the above criteria. Gold nanorods have been extensively used in a wide range of applications because of their unique optical and photo-thermal properties. The narrow longitudinal SPR linewidth of a gold nanorod (100–150 meV, or ~45–65 nm in the near-infrared [30, 31]), combined with the dipolar optical response, allows us to optically address only a small subpopulation of nanorods in the laser irradiated region. They use this selectivity to achieve longitudinal SPR mediated recording and readout governed by photo-thermal reshaping and two-photon luminescence (TPL) detection, respectively.

Using this technique, improved security imprinting and encryption can be realized, where the added dimensions can act as an extended and counterfeit-proof encryption key. In such applications, it would be highly beneficial to have immediate access to the patterns without the need for raster scanning. In order to test this, they recorded polarization multiplexed images in three layers spaced by 40 mm, which were then read out using a charge coupled device (CCD) and a white light source. This technique allows for one-shot readout of patterns when multiple CCDs were used. It can provide instant and simultaneous readout of all recorded patterns. Additionally, the modulation of the transmission introduced by the patterns can be used as a polarization and wavelength-dependent signal modulator for optical devices.

14.3 Summary

Four state-of-arts technologies for data storage: thermal assisted magnetic recording, AFM-based data storage, plasmonic lens array-based recording system, and metallic nanoparticles-based recording, were described in this chapter.

Optical recording was for a long time, and is still, considered a future replacement for magnetic recording. Optical recording systems potentially have much greater reliability than magnetic recording systems since there is a much larger distance between the read/write element and the moving media. Therefore, there is no wear associated with repeated use of the optical systems. However, there are other possible souses of trouble: the lifetime and stability of the laser, mechanical damage to the relatively soft and exposed-to-the-environment media, mechanical damage due to shock and vibration etc. Another advantage of the optical recording systems over the best performing magnetic recording systems (hard drives) is their removability.

The main disadvantage of optical storage in comparison to magnetic systems is slower random data access. This partially comes from the design of the relatively large (and heavy) optical heads. Moving 100 grams over the disks at the high acceleration and speed needed to match the 5 ms to 10 ms average access time for magnetic hard disk drives is a real challenge at present, since the effective weight of the moving parts of a hard drive's head-stack assembly (actuator arm with suspensions and sliders) is in the order of only a few grams. In addition, unlike the hard disk, an optical disk is usually removable, which limits rotational velocities and, correspondingly, limits access time. Increasing rpm causes the relatively loosely fixed CD-ROM disk to vibrate significantly in comparison to a stiff, fixed, and balanced hard disk. Optical drives of all kinds operate on the same principle of detecting variations in the optical properties of the media surface. CD and DVD drives detect the changes in light intensity, and MO drives - the changes in light polarization. All optical storage systems work in reflected light. It is reasonable to believe that with rapid development of subwavelength optics, plasmonic devices and near-field nanofocusing-based recording/reading systems for practical applications will appear soon.

Reference

[1]　J. Ashley, M.-P. Bernal, G. W. Burr, H. Coufal, H. Guenther, J. A. Hoffnagle, C. M. efferson, B. Marcus, R. M. Macfarlane, R. M. Shelby, G. T. Sincerbox, "Holographic data storage," IBM J. Res. Develop. Vol. 44, No. 3, 341-368 (May 2000).

[2]　Koji Matsumoto, Akihiro Inomata, Shin-ya Hasegawa, "Thermally assisted magnetic recording," FUJITSU Sci. Tech. J., 42,1,p.158-167 (January 2006).

[3]　T. E. Schlesinger, T. Rausch, A. Itagi, J. Zhu, A. Bain, and D. D. Stancil, "An Integrated Read/Write Head for Hybrid Recording". Jpn. J. Appl. Phys., 41, Pt. 1, 3B, p.1821-1824 (2002).

[4]　F. Chen, A. Itagi, J. A. Bain, D. D. Stancil, and T. E. Schlesinger, "Imaging of optical field confinement in ridge waveguides

fabricated on very-small-aperture laser". Appl. Phys. Lett., 83, 16, p.3245-3247 (2003).

[5] R. D. Grober, R. J. Schoelkopf, and D. E. Prober, "Optical antenna: Towards a unity efficiency near field optical probe". Appl. Phys. Lett., 70, 11, p.1354-1356 (1997).

[6] T. Matsumoto, T. Shimano, and S. Hosaka, "An efficient probe with a planar metallic pattern for high-density near-field optical memory". Technical Digest of 6th International Conference on Near Field Optics and Related Technics, p.55 (2000).

[7] J. Fujikata, T. Ishii, H. Yokota, K. Kato, M. Yanagisawa, K. Ohashi, T. Thio, and R. Linke: Digest of Magneto Optical Recording International Symposium 2004, Mo-S-6, p.84.

[8] S. Miyanishi, N. Iketani, K. Takayama, K. Inami, I. Suzuki, T. Kitazawa, Y. Ogimoto, Y. Murakami, K. Kojima, and A. Takahashi, "Near field assisted magnetic recording". Digest of INTERMAG ASIS 2005, AB-03, p.13.

[9] S. Hasegawa and F. Tawa, "Generation of nanosized optical beams by use of butted gratings with small numbers of periods". Appl. Optics, 43, 15, p.3085-3096 (2004).

[10] S. Hasegawa, W. Odajima, and T. Namiki, "Optics Simulator for Use in Nano-Optics Analysis". (in Japanese), FUJITSU, 56, 4, p.299-306 (2005).

[11] P. Vettiger, M. Despont, U. Drechsler, U. Dü¨rig, W. Ha¨berle, M. I. Lutwyche, H. E. Rothuizen, R. Stutz, R. Widmer, G. K. Binnig, "The "Millipede"—More than one thousand tips for future AFM data storage", IBM J. RES. DEVELOP. VOL. 44, NO. 3, 323-340 (MAY 2000).

[12] Werayut Srituravanich, Liang Pan, Yuan Wang, Cheng Sun, David B. Bogy, and Xiang Zhang, "Flying Plasmonic Lens In The Near Field for High-Speed Nanolithography," Nature Nanotechnology 3(12), 733-737 (2008).

[13] Stockman, M. I. "Nanofocusing of optical energy in tapered plasmonic waveguides". Phys. Rev. Lett. 93, 137404 (2004).

[14] J. W. M. Chon, J. Moser, M. Gu, presented at the *Int. Symp. Opt. Mem. Opt. Data Storage* (ISOM/ODS 2005), Honolulu, HI, July 2005.

[15] I. Gourevich, H. Pham, J. E. N. Jonkman, E. Kumacheva, "Multidye Nanostructured Material for Optical Data Storage and Security Labeling", Chem. Mater. 16, 1472 (2004).

[16] M. Saito, T. Miyata, A. Murakami, S. Nakamura, M. Irie, K. Uchida, "", Chem. Lett. 33, 786 (2004).

[17] H. Ditlbacher, J. R. Krenn, B. Lamprecht, A. Leitner, F. R. Aussenegg, "Spectrally coded optical data storage by metal nanoparticles", Opt. Lett. 25, 563 (2000).

[18] Y.-Y. Yu, S.-S. Chang, C.-L. Lee, C. R. C. Wang, "Gold Nanorods: Electrochemical Synthesis and Optical Properties", J. Phys. Chem. B 101, 6661 (1997).

[19] S. S. Chang, C. W. Shih, C. D. Chen, W. C. Lai, C. R. C. Wang, "The Shape Transition of Gold Nanorods", Langmuir 15, 701 (1999).

[20] N. R. Jana, L. Gearheart, C. J. Murphy, "Wet Chemical Synthesis of High Aspect Ratio Cylindrical Gold Nanorods", J. Phys. Chem. B 105, 4065 (2001).

[21] N. R. Jana, L. Gearheart, C. J. Murphy, "Wet chemical synthesis of silver nanorods and nanowires of controllable aspect ratio", Chem. Commun.7, 617 (2001).

[22] S. Link, M. B. Mohamed, M. A. El-Sayed, "Laser Photothermal Melting and Fragmentation of Gold Nanorods: Energy and Laser Pulse-Width Dependence", J. Phys. Chem. A 103, 1165 (1999).

[23] S. Link, Z. L. Wang, M. A. El-Sayed, "How Does a Gold Nanorod Melt?", J. Phys. Chem. B 104, 7867 (2000).

[24] S. Link, C. Burda, B. Nikoobakht, M. A. El-Sayed, "Laser-Induced Shape Changes of Colloidal Gold Nanorods Using Femtosecond and Nanosecond Laser Pulses", J. Phys. Chem. B 104, 6152 (2000).

[25] O. Wilson, G. J. Wilson, P. Mulvaney, "Laser Writing in Polarized Silver Nanorod Films", Adv. Mater. 14, 1000 (2002).

[26] S. Link, M. B. Mohamed, M. A. El-Sayed, "Simulation of the Optical Absorption Spectra of Gold Nanorods as a Function of Their Aspect Ratio and the Effect of the Medium Dielectric Constant", J. Phys. Chem. B 103, 3073 (1999).

[27] James W. M. Chon, Craig Bullen, Peter Zijlstra, and Min Gu, "Spectral encoding on gold nanorods doped in a silica sol–gel matrix and Its application to high-density optical data storage". Advanced Functional Materials 17, 875-880 (2007).

[28] U. Kreibig, M. Vollmer, *Optical Properties of Metal Clusters*, Springer Series in Materials Science, Vol. 25, Springer, Berlin 1995, Ch. 3.

[29] Peter Zijlstra, James W. M. Chon & Min Gu, "Five-dimensional optical recording mediated by surface plasmons in gold nanorods" Nature 459, 410-413 (2009).

[30] Carolina Novo, Daniel Gomez, Jorge Perez-Juste, Zhenyuan Zhang, Hristina Petrova, Maximilian Reismann, Paul Mulvaney, and Gregory V. Hartland, "Contributions from radiation damping and surface scattering to the linewidth of the longitudinal plasmon band of gold nanorods: a single particle study". Phys. Chem. Chem. Phys. 8, 3540–3546 (2006).

[31] C. Sönnichsen, T. Franzl, T. Wilk, G. von Plessen, J. Feldmann, O. Wilson, and P. Mulvaney, "Drastic reduction of plasmon damping in gold nanorods". Phys. Rev. Lett. 88, 077402 (2002).

15 FUTURE TENDENCIES AND CHALLENGES

Abstract: This chapter briefly discusses future tendencies and challenging in the fields of metamaterials-based imaging, nanoholes array for biosensing and detection, metamaterials-based antenna for communications, nanophotonic devices-based optical nanocircuits for quantum computer, LSPR-based immunoassay for nano-biosensing, and nanometrology for real-time dynamic inspection/testing/measurement.

Future tendencies and challenging may appear in the following areas:

(1) Metamaterials-based imaging

- How to effectively convert near field evanescent wave to far field propagation wave?
- How to collect the objective information with high spatial frequency in near/far fields?
- How to deliver high frequency messages involved in near-field into far-field?
- How to realize amplification of the SPP wave-induced extraordinary transmission in far-field?
- Theoretical model for nanoparticles with complex shapes and dimensions.

A bio-friendly nano-sized light source capable of emitting coherent light across the visible spectrum has been invented by a team of the researchers with U.S. Department of Energy's Lawrence Berkeley National Laboratory, and the University of California at Berkeley. Among many of the potential applications of this nano-sized light source, once the technology is refined, are single cell endoscopy and other forms of subwavelength bio-imaging, integrated circuitry for nanophotonic technology, and new advanced methods for cyber cryptography.

Although the field of negative index materials has shown enormous progress over the past few years, some critical issues remain unresolved. Particularly important issue is the problem of high losses that originates from the design required to achieve a negative magnetic response. Since the demonstration of the first optical NIM, the corresponding figure of merit (the ratio of the real to the imaginary part of the refractive index) was improved by more than an order of magnitude. However, the losses are still too high for the negative index-based superlens to become practical. Yet another issue is the development of truly three-dimensional (3D) optical NIMs. Although the task of placing the nanoscale meta-atoms into a uniform 3D arrangement of the desired topology can be thought of as a purely technological challenge, its complexity is mind-boggling. Perhaps self organization will be needed for such fabrication to a certain extent. It is not clear how and when these issues will be resolved, but the quest for the optical NIMs and superlens has already initiated the whole new field of metamaterials. These artificial structures, with electromagnetic response tailored to a particular objective (such as the 'magnetic mirror' or an electromagnetic 'cloaking device' and their applications), may well eclipse the area of negative index materials that stimulated their development.

(2) Nanoholes array for biosensing and detection

Nanoholes array has been widely applied in optofluidic waveguides and nanostructures for cancer cell detection and biosensing. The optofluidic waveguides and nanostructures have the following significant functions:

- Demonstrate liquid core waveguides with nanoparticles.
- Demonstrate SPP sensing with nanohole array and optofluidic delivery exploiting ultrashort pulse SPP confined in space and time.
- Enable efficient interfaces between optofluidic and dielectric waveguides.
- Induce near field interactions in resonant structures.
- Enable cost effective compact flow cytometer and biochemical sensors.

In addition, the nanoholes arrays can be used as optofluidic components/devices in adaptive optics with the following applications and functions:

- Develop composite membrane technology for adaptive optics.
- Demonstrate adaptive cladding in resonant structures.
- Enable fluidically tunable compact optical devices robust to fabrication tolerances.

(3) Metamaterials-based antennas

EBG-based antenna introduced in Chapter 13 of this book regarding the metamaterials-based antenna is highlighted. Further study is still required for the following two issues:

- Multifunctioning and miniaturization of EBG-based antenna.
- NRI metamaterials-based antenna for the antennas working in large power.

Future applications and research in this area are listed below:

- High-gain leaky-wave antennas (embedded amplifiers in unit-cell) [1]
- Distributed amplifiers [2]
- Tunable Phase Shifters [3]
- Metamaterial multiple-input-multiple-output (MIMO) arrays for 802.11 nm wavelength-based applications [4]
- Active composite right -left-hand Metamaterials

It is anticipated that homogenization of current metamaterial structures by a factor of one order of magnitude will lead to spectacular improvements in their performances, which will pave the way for a new generation of quasi all-optical devices, beam-formers and antennas. This contribution has revealed a key challenge in this homogenization procedure: larger inductances and capacitances will have to be confined within smaller volumes, which will require novel technologies in the next generation of metamaterials.

(4) Nanophotonic devices-based optical nanocircuits

Currently, data rates also have exponential growth. Surface plasmons with wavelengths less than 100 nm still can work at optical frequencies. In addition, nanophotonics have the following promising applications:

- Nanophotonic circuit elements
- Nanophotonic transistors
- Nanophotonic logic gates

(5) LSPR-based immunoassay

Experimental proof-of-concept was completed with great success. Future works will focus on the following issues:

- Portable and one-time use biochips with metallic nanoparticles arrays.
- Development of mass production and cost effective biochips.
- Practicability for cancer cell detections.

Apart from immunoassay, LSPR-based sensors have promising applications in food safety, environment monitoring, and medical diagnosis.

(6) Nanometrology

Although "nanometrology" is a term which has been used since the late of 1980s [5], its definition still remains unclear. In 1992, E. C. Teague of NIST defined "nanometrology" as the science of measuring the dimensions of objects or object features to uncertainties of one nanometer or less [6]. In that paper a Japanese group from the National Research Laboratory of Metrology used precision measurements to determine fundamental constants and test physical theory which relates to silicon lattice spacing and magnetic flux quantum. Various other papers have followed from that original paper using the word "nanometrology" and a number of different techniques and approaches have been used. In the original paper, a scanning tunnelling microscopy (STM) was used to provide a physical measurement of distance. Various other techniques, such as near field scanning optical microscopy (NSOM), AFM, and SEM/TEM, have been used to provide the measurement in nano-scale. In 2005, the Institute of Physics published a dedicated edition in Journal of Measurement Science and Technology on the topic of nanometrology. As an alternative approach to electron microscopy, a fluorescence resonance energy transfer (FRET) has been used to provide measurement of the distances between fluorophores between 1 nm and 10 nm. In addition, in material science, electron back-scattering diffraction (EBSD) and x-ray diffraction (XRD) are extensively used for the nanometrology of molecular structures of the materials. Texture has a large influence on many properties of thin films. In the future, plasmonic devices-based superlenses and nanoprobes may play an important role in nanometrology and dynamic online nano-inspection, especially the negative refraction index (NRI)-based imaging and nanofocusing/nanolensing technique.

Reference

[1]   F. P. Casares-Miranda, C. Camacho Peñalosa, and C. Caloz, "High-gain active composite right/left-handed leaky-wave antenna," IEEE Trans. Antennas Propag. 54(8), 2292-2300 (2006).

[2]   J. Mata-Conteras, T. M. Martìn-Guerrero, and C. Camacho-Peñalosa, "Distributed amplifiers with composite right/left-handed transmission lines," Microwave Opt. Technol. Lett. 48(3), 609-613 (2006).

[3]   E.S. Ash, "Continuous phase shifter using ferroelectric varactors and composite right-left handed transmission lines," Master Thesis, Dept. E.E., UCLA, Los Angeles, CA 2006.

[4]   Rayspan Corporation, http://www.rayspan.com

[5]   Verdee, M. S., "Nanometrology of optical flats by laser autocollimation," Surface Topography 1, 415-425 (1988).

[6]   Teague, E. C., "Nanometrology," Proceedings of AIP Conference, pp. 371-407 (1992).

Index